Environmental Impact Assessment

A Comparative Review

Environmental Impact Assessment

A Comparative Review

Christopher Wood

Prentice
Hall

An imprint of **Pearson Education**

Harlow, England · London · New York · Reading, Massachusetts · San Francisco · Toronto · Don Mills, Ontario · Sydney
Tokyo · Singapore · Hong Kong · Seoul · Taipei · Cape Town · Madrid · Mexico City · Amsterdam · Munich · Paris · Milan

Pearson Education Limited
Edinburgh Gate
Harlow
Essex CM20 2JE
England

and Associated Companies throughout the world

Visit us on the World Wide Web at:
http://www.pearsoneduc.com

First published 1995

British Library Cataloguing in Publication Data
A catalogue entry for this title is available from the British Library.

ISBN 0-582-23696-7

Library of Congress Cataloging-in-Publication data
A catalog entry for this title is available from the Library of Congress

10   9   8   7   6
04   03   02   01   00

Set in 10/11 Palatino by 3
Printed in Malaysia, PP

Contents

List of figures

List of tables

List of boxes

Preface

Environmental impact assessment (EIA): 25 years old in 1995. Born in the United States, it was initially ignored then (in turn) caused great disturbance and antagonism, began to change people's lives for the better, settled down and learned from experience, became respectable and, eventually, was extensively imitated all over the world. As concern about the environment has grown, EIA has been widely seen as a panacea to environmental problems. It is not. EIA is an anticipatory, participatory, integrative environmental management tool which has the ultimate objective of providing decision-makers with an indication of the likely consequences of their decisions relating to new projects or to new programmes, plans or policies. Effective EIA alters the nature of decisions or of the actions implemented to reduce their environmental disbenefits and render them more sustainable. If it fails to do this, EIA is a waste of time and money.

Interest in EIA has burgeoned and there are now numerous EIA systems in existence worldwide. While the various EIA systems all differ in detail, their basic principles are very similar and demonstrate many common problems. Different jurisdictions have used different means to try to solve these problems and to improve the effectiveness of their EIA systems. There is growing interest in learning from the experience of others, and not just from that of the United States, where their EIA systems have elements worthy of emulation. While there is now a very considerable literature on EIA, there is no book which presents a comparative step-by-step review of international EIA procedures and practice. This book, by reviewing seven EIA systems in detail (the United States, California, the United Kingdom, The Netherlands, Canada, the Commonwealth of Australia and New Zealand) and EIA in developing countries in general, is intended to fill the gap.

Each EIA system is evaluated against a set of criteria which enables comparisons to be made easily. The EIA process is analysed step-by-step in a succession of chapters. Each chapter contains a discussion of appropriate evaluation criteria and methods for the relevant step in the EIA process and then goes on to an analysis of its treatment in each of the seven EIA systems. Numerous tables, boxes and figures are employed to summarise the comparative findings. The evaluation criteria are intended to be generally applicable and could be used to analyse other EIA systems.

EIA is interdisciplinary and involves large numbers of different practitioners but is most closely associated with the professionals concerned with siting new development. This book should therefore be of interest to practitioners involved in government, development and land use planning, landscape design, the environment, law and engineering. The book is intended also to be read by those studying EIA on undergraduate and postgraduate courses and on short courses and by academics and researchers. Students of land use planning, law, geography, environmental studies, development studies and engineering should also find the book helpful.

Abbreviations and national EIA terms

	Assessment report (Australia)
CEAA	Canadian Environmental Assessment Act 1992
CEA Agency	Canadian Environmental Assessment Agency
CEARC	Canadian Environmental Assessment Research Council
CEC	Commission of the European Communities
CEPA	Commonwealth Environment Protection Agency (Australia)
CEQ	Council on Environmental Quality (USA)
CEQA	California Environmental Quality Act 1970
	categorical exclusion (USA)
	categorical exemption (California)
	class screening (Canada)
	comprehensive study (Canada)
DOE	Department of the Environment (UK)
	direction (UK)
EA	environmental assessment (document: USA) (process: Canada, UK)
EARP	Environmental Assessment and Review Process (Canada)
ECW	Evaluation Committee on EIA (The Netherlands)
EIA	environmental impact assessment
EIA Commission	Environmental Impact Assessment Commission (The Netherlands)
EIR	environmental impact report (California, New Zealand)
EIS	environmental impact statement (Australia, Canada, The Netherlands, USA)
EPA	Environmental Protection Agency (Australia)
	Environmental Protection Agency (USA)
EPEP	Environmental Protection and Enhancement Procedures (New Zealand)
ES	environmental statement (UK)
	environmental information (UK)
FEARO	Federal Environmental Assessment Review Office (Canada)

FONSI	finding of no significant impact (USA)
	findings (California)
	follow-up program (Canada)
IEE	initial environmental evaluation (Canada)
	initial assessment (Canada)
	initial study (California)
LNV	Ministry of Agriculture, Nature Management and Fisheries (The Netherlands)
LPA	local planning authority (UK)
ME	Minister of the Environment (Canada)
MER	EIA (milieu-effectrapportage) (The Netherlands)
MfE	Ministry for the Environment (New Zealand)
	mediation (Canada)
NEPA	National Environmental Policy Act 1969 (USA)
NOI	notice of intent (USA)
	negative declaration (California)
	notice of determination (California)
	notice of intention (Australia)
	notice of preparation (California)
	notification of intent (The Netherlands)
OECD	Organisation for Economic Cooperation and Development
	opinion (UK)
PCE	Parliamentary Commissioner for the Environment (New Zealand)
PER	public environment report (Australia)
	panel review (Canada)
	public registry (Canada)
	public review (Canada)
RA	responsible authority (Canada)
RMA	Resource Management Act 1991 (New Zealand)
ROD	record of decision (USA)
	recommendations (Australia)
SEA	strategic environmental assessment
	screening (Canada)
	self-directed assessment (Canada)
	specified information (UK)
	statement of overriding considerations (California)
UNEP	United Nations Environment Programme
VROM	Ministry of Housing, Physical Planning and the Environment (The Netherlands)
Wm	Environmental Management Act 1994 (The Netherlands)

Acknowledgements

Much of this book was written while I was on sabbatical leave from the Department of Planning and Landscape, University of Manchester. I wish to thank my colleagues at Manchester, and especially Norman Lee, Carys Jones, Fiona Walsh and Jo Hughes of the EIA Centre, for their support. I owe a great deal to Abigail Shaw who patiently typed innumerable drafts of the book and also to Margaret Barrow, Mary Howcroft, Sue Massey and Anita Tomlinson. Elsewhere in the book, I have acknowledged all those who provided me with information or opinions and who commented on draft sections but I wish to reiterate my gratitude to these many practitioners and academics here.

My sabbatical leave in Australia and my visits to New Zealand, The Netherlands, Canada and the United States were partially funded by the Sir Herbert Manzoni Scholarship Trust, the University of Manchester and the British Council, to whom I am most grateful. I wish to record my thanks to Professor Brian McLoughlin (sadly since deceased) and others at the School of Environmental Planning, University of Melbourne for making me welcome there.

I am grateful for permission to use material from the following sources in the book: Ron Bass and Al Herson, Jones and Stokes Associates, Sacramento (Fig. 2.1, Boxes 8.2 and 15.1); Blackwell Publishers and the authors, William Sheate (Imperial College, London) and Richard Macrory (University of Oxford) (Table 3.1); Department of the Environment and the Controller of Her Majesty's Stationery Office, London (Box 11.1 [1994b] and Fig. 19.3 [1993]); Environmental Impact Assessment Commission, Utrecht (Fig. 5.2 and Box 12.4); Federal Environmental Assessment Review Office, Hull, Quebec (Boxes 8.3, 10.3, 11.3, 14.2 and 19.2 also Tables 11.1 and 13.1 and Figs 19.2, 19.4 and 20.1); Richard Morgan and Ali Memon, University of Otago (Box 12.5); United Nations Environment Programme, Bangkok (Fig. A.1).

Last, but by no means least, I wish to thank Josephine Wood for her patience and support. I dedicate this book to her and to Jason, Tessa and Norah Wood.

CHAPTER 1

Introduction

Nature of environmental impact assessment

Environmental impact assessment (EIA) refers to the evaluation of the effects likely to arise from a major project (or other action) significantly affecting the natural and man-made environment. Consultation and participation are integral to this evaluation. EIA is a systematic and integrative process, first developed in the United States as a result of the National Environmental Policy Act of 1969 (NEPA), for considering possible impacts prior to a decision being taken on whether or not a proposal should be given approval to proceed. NEPA requires, *inter alia*, the publication of an environmental impact statement (EIS) describing in detail the environmental impacts likely to arise from an action.

The EIA process should supply decision-makers with an indication of the likely consequences of their actions. Properly used, EIA should lead to informed decisions about potentially significant actions, and to positive benefits to both proponents and to the population at large. As the UK Department of the Environment (1988b, para. 7) put it, formal EIA:

is essentially a technique for drawing together, in a systematic way, expert qualitative assessment of a project's environmental effects, and presenting the results in a way which enables the importance of the predicted effects, and the scope for modifying or mitigating them, to be properly evaluated by the relevant decision-making body before a decision is given. Environmental assessment techniques can help both developers and public authorities with environmental responsibilities to identify likely effects at an early stage, and thus to improve the quality of both project planning and decision-making.

In principle, EIA should lead to the abandonment of environmentally unacceptable actions and to the mitigation to the point of acceptability of the environmental effects of proposals which are approved. EIA is thus an anticipatory, participatory environmental management tool, of which the EIA report is only one part. The objectives of the Californian EIA system make this very clear (Bass and Herson, 1993b, p. 1):

1. To disclose to decision makers and the public the significant environmental effects of proposed activities.
2. To identify ways to avoid or reduce environmental damage.
3. To prevent environmental damage by requiring implementation of feasible alternatives or mitigation measures.
4. To disclose to the public reasons for agency approvals of projects with significant environmental effects.
5. To foster interagency coordination.
6. To enhance public participation.

Appropriately employed, EIA is a key integrative element in environmental protection policy, but only one element in that policy (Lawrence, 1994).

EIA is not just a procedure, or for that matter just a science. Its nature is dichotomous, rather like the duality of matter. As Kennedy (1988b, p. 257) has put it, EIA is both science and art, hard and soft:

> EIA as 'science' or a planning tool has to do with the methodologies and techniques for identifying, predicting, and evaluating the environmental impacts associated with particular development actions.
>
> EIA as 'art' or procedure for decision-making has to do with those mechanisms for ensuring an environmental analysis of such actions and influencing the decision-making process.

Caldwell (1989c, p. 9) has summarised the significance of EIA as follows:

1. Beyond preparation of technical reports, EIA is a means to a larger end – the protection and improvement of the environmental quality of life.
2. It is a procedure to discover and evaluate the effects of activities (chiefly human) on the environment – natural and social. It is not a single specific analytic method or technique, but uses many approaches as appropriate to a problem.
3. It is not a science, but uses many sciences (and engineering) in an integrated inter-disciplinary manner, evaluating relationships as they occur in the real world.
4. It should not be treated as an appendage, or add-on, to a project, but regarded as an integral part of project planning. Its costs should be calculated as a part of adequate planning and not regarded as something extra.
5. EIA does not 'make' decisions, but its findings should be considered in policy- and decision-making and should be reflected in final choices. Thus it should be part of decision-making processes.
6. The findings of EIA should focus on the important or critical issues, explaining why they are important and estimating probabilities in language that affords a basis for policy decisions.

It should be emphasised that EIA is not a procedure for preventing actions with significant environmental impacts from being implemented. Rather the intention is that actions are authorised in the full knowledge of their environmental consequences. EIA takes place in a political context:

it is therefore inevitable that economic, social or political factors will outweigh environmental factors in many instances. This is why the mitigation of environmental impacts is so central to EIA: decisions on proposals in which the environmental effects have palpably been ameliorated are much easier to make and justify than those in which mitigation has not been achieved.

This chapter briefly describes the diffusion and evolution of EIA from its origins in the United States National Environmental Policy Act of 1969. It goes on to discuss the elements of the EIA process, the effectiveness of EIA systems and to suggest a number of criteria against which EIA systems can be evaluated. The purpose of the comparative review of the selected EIA systems presented in this book is then explained. Finally, an overview and explanation of the structure of the book is presented.

Diffusion and evolution of EIA

California was the first of the American states to introduce an effective 'little NEPA', in 1970 (Bass and Herson, 1993b). (The majority of the US states have still not done so.) International attention was soon being directed to EIA as a result of several celebrated legal cases in the United States, which clarified NEPA's significance. The ramifications of NEPA were beginning to be accepted at a time of unprecedented interest in the environment occasioned by the United Nations' conference on the environment in Stockholm in 1972. The problems of burgeoning development, pollution and destruction of the natural environment which NEPA was intended to address were perceived as universal. The rigorous project-by-project evaluation of significant impacts inherent in EIA was seized upon as a solution to many of these environmental problems by many other jurisdictions and elements of the US EIA process were adopted by them. Most were, however, very cautious about importing NEPA-style litigation with EIA and made strenuous efforts to avoid doing so.

The methods of adoption varied: cabinet resolutions, advisory procedures, regulations and laws were employed. Probably the first overseas jurisdiction to declare an 'extremely rudimentary environmental impact policy' (Fowler, 1982, p. 8) was the Australian state of New South Wales in January 1972. The Commonwealth of Australia announced an EIA policy in May 1972 and passed the Environment Protection (Impact of Proposals) Act in December 1974. Canada preceded Australia, approving a federal cabinet directive on EIA in 1973. New Zealand instituted EIA procedures by cabinet minute in 1974. Columbia (Verocai Moreira, 1988) and Thailand (Nay Htun, 1988) established EIA systems through specific legislation in 1974 and 1975 respectively, followed by France in 1976. Ireland passed legislation which permitted, but did not require, EIA in 1976 and the cabinet of the West German government approved an EIA procedure by minute in the same year. The Netherlands governmental standpoint on EIA followed in 1979. There was also considerable EIA activity in numerous Third World countries (Biswas and Agarwala, 1992). The diffusion of EIA was gathering pace and has continued unabated.[1]

Several international agencies have involved themselves with EIA. In 1974 the Organisation for Economic Cooperation and Development (OECD) recommended that member governments adopt EIA procedures and methods (Council on Environmental Quality (CEQ), 1975) and more recently, that they use EIA in the process of granting aid to developing countries (OECD, 1992). In addition, in 1985, the Council of the European Communities adopted a directive which required member states to implement formal EIA procedures by 1988. The United Nations Environment Programme (UNEP) has also made recommendations to member states regarding the establishment of EIA procedures and has established goals and principles for EIA (Fookes, 1987a). It subsequently issued guidance on EIA in developing countries (UNEP, 1988). Similarly (and somewhat belatedly) the World Bank ruled in 1989 that EIA should normally be undertaken by the borrower country under the Bank's supervision (CEQ, 1990, p. 45) and has prepared a sourcebook on EIA (World Bank, 1991).

This diffusion of EIA has resulted in a diverse vocabulary. In The Netherlands, EIA is known as MER (milieu-effectrapportage) and in Canada and the United Kingdom as environmental assessment. The EIS (EIA report) becomes an environmental statement in Britain, and an environmental impact report in California and under the original New Zealand provisions. The Commonwealth of Australia has both an EIS and a public environment report.

As EIA has spread, so has its nature been elaborated and clarified. There have been, perhaps, six main themes as EIA has evolved over the years:

1. An early concern with the methodology of impact forecasting and decision making gave way first to an emphasis on administrative procedures for EIA and then, more recently, to a recognition of the crucial relationship of EIA to its broader decision-making and environmental management context.
2. A tendency to codification and away from discretion. This is evident in CEQ graduating from the use of guidelines to regulations in the United States and in the enactment of federal Canadian EIA legislation after almost two decades of experience with administrative EIA procedures.
3. The refinement of EIA systems by the adoption of additional elements as experience has been gained. These include procedures for determining the coverage of EIAs (scoping) (first in the United States and then in, for example, New Zealand) and for monitoring the effects of implemented actions (for example, in California).
4. A concern to increase the quality of EIA by, for example, improving EIA reports, providing more opportunities for consultation and participation and increasing the weight given to EIA in decision making.
5. A concern to increase the effectiveness of EIA in reducing environ-

mental impacts and to ensure efficiency in terms of its costs in time, money and manpower.

6. A recognition that many variables are already resolved by the time the EIA of projects takes place and thus that some form of EIA of policies, plans and programmes (strategic environmental assessment) is necessary.

Elements of the EIA process

While not all EIA systems contain every element, the EIA process emanating from NEPA and subsequently diffused around the world can be represented as a series of iterative steps:

* consideration of alternative means of achieving objectives
* designing the selected proposal
* determining whether an EIA is necessary in a particular case (screening)
* deciding on the topics to be covered in the EIA (scoping)
* preparing the EIA report (i.e., *inter alia*, describing the proposal and the environment affected by it and assessing the magnitude and significance of impacts)
* reviewing the EIA report to check its adequacy
* making a decision on the proposal, using the EIA report and opinions expressed about it
* monitoring the impacts of the proposal if it is implemented.

As indicated by Fig. 1.1, which summarises these steps, the EIA process is cyclical. Thus, the consideration of the environmental effects of alternative means of achieving the proponent's aims and the detailed design of the action are inextricably linked. Again, the results of consultation at the scoping stage or later may require the proponent to return to the design stage to increase the mitigation of impacts. Consultation and public participation should be important inputs at each stage in the EIA process, though the people and bodies invited to comment on the proposal may vary. Equally, the mitigation of environmental impacts should take place at each step in the process. Not every step in the EIA process shown in Fig. 1.1 takes place overtly (or indeed, at all) in every EIA system. As mentioned above, scoping and project monitoring were not part of the original conception of EIA in NEPA and are still not required in many EIA systems. Indeed, there is a very considerable diversity of views about the essential elements of an effective EIA system.

EIA system effectiveness

Much of the debate about the effectiveness of EIA systems emanates from North America. It centres not so much on whether or not EIA can be viewed as effective, but on the factors which can be advanced to explain

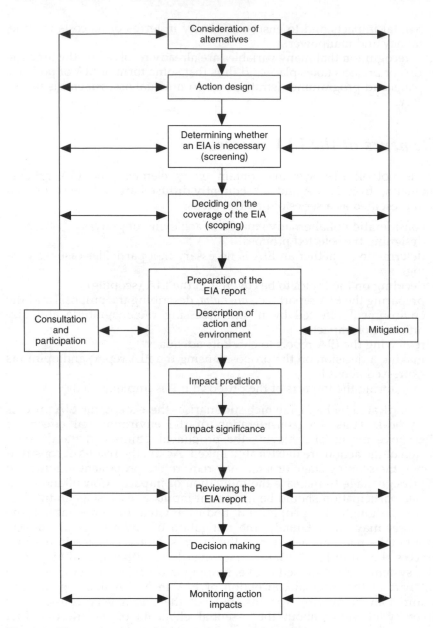

Fig. 1.1 The environmental impact assessment process.

why an EIA system is effective, on which evaluation criteria are appropriate in judging the effectiveness of an EIA system and on how EIA can be improved.

While the view is not unanimous (see, for example, Fairfax, 1978; Renwick, 1988; Rees, 1990, pp. 349–52), it is accepted very widely that:

At the US federal level, impact assessment works. We know how it works to influence project selection and design and to mitigate environmental impacts.

(Wandesforde-Smith and Kerbavaz, 1988, p. 162)

CEQ (1990, p. 16) concurred with this view of the effectiveness of EIA:

The act unquestionably has had a profound effect on attitudes within the federal government, and its influence outside the federal government is almost as impressive.

Taylor (1984) believed that the US EIA system worked because it was an administrative reform in tune with its times: supportive forces both inside and outside government worked together to ensure the effective implementation of EIA, and the changes in organisational behaviour associated with it. Caldwell (1989c, p. 10) has accepted this view:

To the question: has EIA reformed administration, my reply is yes – but as an instrument of a public opinion demanding administrative and policy reform.

Wandesforde-Smith and Kerbavaz (1988) and Wandesforde-Smith (1989) have gone further and emphasised the role of federal and state elections and of personnel ('public sector entrepreneurs') in implementing changes to both the US federal and the Californian EIA systems. They also emphasise the importance of financial and manpower resources and skill in bargaining. Perhaps more important is Caldwell's (1989c, p. 12) point about environmental values:

EIA will be most effective where environmental values (1) are implicit and consensual in the national culture and (2) are explicit in public law and policy.

It is apparent that the success of EIA depends upon a large number of factors in addition to the precise nature of the procedures in force. As Hollick (1986, p. 159) has stated:

outside the USA it is commonly assumed not only that introducing procedural changes will change decision making, but also that agency procedures will change in accordance with promulgated procedures without some measure of coercion.

This assumption is clearly over-optimistic.

This is not to say that EIA procedure is not important in determining effectiveness. It is clearly of crucial importance but, while it is a necessary condition for EIA success, it is not a sufficient condition. Every EIA procedure operates within a policy, political, legal and administrative context peculiar to the jurisdiction concerned. To be successful in achieving a real shift in the weight given to the environment in decisions, the EIA procedure needs to interact positively with its jurisdictional context. As in the United States, this may not happen immediately. Wandesforde-Smith (1989, p. 165) has summarised these points as follows:

EIA effectiveness is associated with changing political regimes and with the changing level of support for the EIA process among courts,

7

chief executives, and senior agency managers that this implies. The way an EIA process is formally structured and the way structure taps informal incentives for administrative behaviour are, equally clearly, important variables.

Ortolano *et al.* (1987) felt that, to be effective, an EIA system needed to have a number of characteristics, including:

Utilization of proper methods in assessing impacts;

Influence of environmental information on various aspects of planning and decision making, including formulation of alternative plans, selection of a proposed plan, and mitigation of adverse impacts;

Placement of appropriate weight on environmental impacts relative to economic and technical factors.

Ortolano (1993) has subsequently emphasised the need to include procedural compliance and the completeness of EIA documents among the dimensions of EIA effectiveness.

Ortolano *et al.* (1987) believed that the effectiveness of EIA systems could be explained by reference to 'control mechanisms': intraorganisational and interorganisational processes and structures to ensure that the procedures actually worked. They advanced six types of control as causative: 'judicial; procedural; evaluative; instrumental; professional; and direct public and outside agency'. They suggested that two or more of these mechanisms usually operated simultaneously and that opportunities for public involvement played a key role in the exercise of each.

This emphasis on the role of public involvement in the success of NEPA is widespread. Fairfax and Ingram (1981, p. 43) felt that:

considerable public discussion and support has come less because of its uniquely cogent approach to the problems of fragmented decision-making, than because of the vocal and powerful environmental constituency which came to support the legislation after a spate of expansive judicial readings of its requirements.

CEQ (1990, p. 42) has also emphasised the crucial role of external review (which it takes to mean external agency review, public participation and judicial review) in the success of EIA.

Kennedy (1988b, p. 262) reached the following conclusion to the question of which EIA procedures work:

Generally speaking, however, it would appear that EIA works best when it is instituted in a formal–explicit way. That is to say, it works when there is a specific legal requirement for its application, where an environmental impact statement is prepared, and where authorities are accountable for taking its results into consideration in decision-making.

In addition, for EIA to be successfully integrated in the project planning process it would appear that procedures for screening, scoping, external review and public participation need to be a part of it.

Evaluation of EIA system effectiveness

There has been, as yet, no reliable quantification of the effectiveness of EIA. It may be that this is not possible. As CEQ (1990, p. 15) has stated:

> Because NEPA was not designed to control specific kinds or sources of pollution, its benefit to society is difficult to quantify. The act was designed primarily to institutionalize in the federal government an anticipatory concern for the quality of the human environment, that is, an attitude, a heightened state of environmental awareness that, unlike pollution abatement, is measurable only subjectively and qualitatively.

Bartlett and Baber (1989, pp. 148, 149) endorsed the difficulty of empirically examining the effects of EIA on decision making within organisations:

> For that reason, it may be more desirable to judge the impact of impact assessment on bureaucratic decision making by examining the attitudes and opinions of those immediately responsible.

While the difficulties of reaching an objective overall judgement about any EIA system are apparent, there is a need for an evaluative framework for comparing the formal legal procedures, the arrangements for their application, and practice in their implementation in EIA systems. This evaluative framework could be provided by analysing the extent to which various principles are met by EIA systems. Such principles might, for example, be based upon NEPA provisions, or upon the requirements of the European EIA Directive, or upon the more detailed EIA principles for assessing authorities, for proponents, for the public and for government put forward by the Australian and New Zealand Environment and Conservation Council (1991). Thus, Fookes (1987a) used the United Nations Environment Programme goals and principles for EIA as a basis for an evaluation of the South Australian EIA system. Perhaps the most rigorous example of the use of this type of evaluative framework is Gibson's (1993) analysis of the Canadian federal and Ontario EIA systems on the basis of eight 'interdependent principles for the design of effective environmental assessment processes' (Box 1.1).

The Canadian Environmental Assessment Research Council (1988a, pp. 1, 2) advanced the following criteria for evaluating EIA:

> An EIA may be considered *effective* if, for example:
>
> - information generated in the EIA contributed to decision making
> - predictions of the effectiveness of impact management measures were accurate, and
> - proposed mitigatory and compensatory measures achieved approved management objectives.
>
> *Efficiency* criteria are satisfied if, for example:
>
> - EIA decisions are timely relative to economic and other factors that determine project decisions, and

Box 1.1 Eight basic principles for evaluating EIA processes

---

1. An effective environmental assessment process must encourage an integrated approach to the broad range of environmental considerations and be dedicated to achieving and maintaining local, national and global sustainability.

2. Assessment requirements must apply clearly and automatically to planning and decision making on all undertakings that may have environmentally significant effects and implications for sustainability within or outside the legislating jurisdiction.

3. Environmental assessment decision making must be aimed at identifying best options, rather than merely acceptable proposals. It must therefore require critical examination of purposes and comparative evaluation of alternatives.

4. Assessment requirements must be established in law and must be specific, mandatory and enforceable.

5. Assessment work and decision making must be open, participative and fair.

6. Terms and conditions of approvals must be enforceable, and approvals must be followed by monitoring of effects and enforcement of compliance in implementation.

7. The environmental assessment process must be designed to facilitate efficient implementation.

8. The process must include provisions for linking assessment work into a larger regime including the setting of overall biophysical and socio-economic objectives and the management and regulation of existing as well as proposed new activities.

Source: Gibson (1993).

- costs of conducting EIA and managing inputs during project implementation can be determined and are reasonable.

Fairness criteria are satisfied if, for example:

- all interested parties (stakeholders) have equal opportunity to influence the decision before it is made, and
- people directly affected by projects have equal access to compensation.

While many of these criteria relate to an individual EIA rather than to EIA systems, they are nevertheless helpful in deriving a set of evaluation criteria for a comparative review.

Various alternative approaches for evaluating EIA systems have been advanced (for example, by Hollick (1986) and the North Atlantic Treaty Organization (1993)). Evaluation criteria are, in effect, shorthand versions of principles for EIA and, carefully articulated, have considerable advantages in terms of brevity and clarity.

Box 1.2 presents a set of evaluation criteria which are based upon the representation of the stages in the EIA process shown in Fig. 1.1, the aims of EIA, and the various evaluation frameworks discussed above. The focus of the criteria is on the requirements and operation of the EIA process. Only the penultimate criterion involves an overall evaluation of the EIA system. For the reasons outlined above, this relies mainly on the opinions of those involved in the EIA process. These criteria can be employed to judge the effectiveness of any EIA system and to enable an international comparison to be made between EIA systems. Such a comparative review provides the basis for suggesting how the effectiveness of EIA can be improved, a goal which is attracting considerable interest (Sadler, 1994a).

Comparative review of EIA systems

Because every EIA system is unique and each is the product of a particular set of legal, administrative and political circumstances, the examination of several EIA systems comparatively by analysing each element in the EIA process should achieve three objectives. The first is explanatory. By placing the EIA process and the stages in EIA procedures in their international context, it should be possible to explain their nature much more clearly than by studying the system in a single jurisdiction. Second, analysis across EIA systems provides a means of better understanding practice in any particular jurisdiction. It is known that some EIA systems work better than others and step-by-step comparative analysis may help to throw more light on the factors which are essential to the success of EIA processes. The third objective stems from the first two. As Lundquist (1978) has stated:

> comparative studies of national approaches to solving environmental problems have often led to valuable and practical suggestions to improve the effectiveness of the national processes examined.

If this comparative review leads to one such suggestion it will have been successful.

Seven different EIA systems are compared in this book: those in the United States, the state of California, the United Kingdom, The Netherlands, Canada, the Commonwealth of Australia and New Zealand. The US and UK systems chose themselves. The United States possesses the original EIA system and, as with so much else in the environmental policy field, examination of American experience is often a pointer to the future elsewhere. Many of the problems currently facing, for example, the United Kingdom in improving the quality of EIA have been apparent in the United States over the years since 1970, and attempts have been made to resolve them which are relevant to experience elsewhere.

The Californian state EIA system is, after the NEPA process, the oldest in the world. It is much modified from its original form. It applies not only to state permits and developments, but to local permits. Hundreds of environmental impact reports are prepared each year, frequently by consultants retained by local authorities at the developer's expense. In many ways the Californian EIA system is more directly comparable with that in, for example, the United Kingdom than is the federal EIA process, since it is largely locally administered and is closely integrated with the land use planning system. It differs from the NEPA process, upon which it is modelled, in several important respects.

The United Kingdom is the last of the seven jurisdictions to have introduced a formal EIA system, and the only one to have done so with initial reluctance. Comparison of the UK's system with longer-established, more mature, EIA systems should provide a valuable insight into the remedies for problems which are already apparent, many of which have been experienced elsewhere. However, such comparisons need not only indicate improvements in the UK system: there are some respects in which the British system may provide pointers to others.

It is important to see EIA in the United States and in the United Kingdom in their international context. The Netherlands is generally

Box 1.2 EIA system evaluation criteria

1. Is the EIA system based on clear and specific legal provisions?

2. Must the relevant environmental impacts of all significant actions be assessed?

3. Must evidence of the consideration, by the proponent, of the environmental impacts of reasonable alternative actions be demonstrated in the EIA process?

4. Must screening of actions for environmental significance take place?

5. Must scoping of the environmental impacts of actions take place and specific guidelines be produced?

6. Must EIA reports meet prescribed content requirements and do checks to prevent the release of inadequate EIA reports exist?

7. Must EIA reports be publicly reviewed and the proponent respond to the points raised?

8. Must the findings of the EIA report and the review be a central determinant of the decision on the action?

9. Must monitoring of action impacts be undertaken and is it linked to the earlier stages of the EIA process?

10. Must the mitigation of action impacts be considered at the various stages of the EIA process?

11. Must consultation and participation take place prior to, and following, EIA report publication?

12. Must the EIA system be monitored and, if necessary, be amended to incorporate feedback from experience?

13. Are the financial costs and time requirements of the EIA system acceptable to those involved and are they believed to be outweighed by discernible environmental benefits?

14. Does the EIA system apply to significant programmes, plans and policies, as well as to projects?

acknowledged in Europe and throughout the world as having a sophisticated system of environmental controls, including an EIA system regarded by many observers as the most effective in Europe (CEQ, 1990). The Netherlands, following numerous studies, had almost put its EIA system in place when the European Directive on EIA was adopted. Its system is, therefore, in marked contrast to that in the United Kingdom, which was instituted as a direct response to the Directive.

The Canadian federal environmental assessment (EA) system was established in 1973 and has been refined substantially over the years. It provided the model for The Netherlands EIA system and, in particular, for the Dutch use of panels to review EIA reports. It also had an influence on the design of the EIA system in New Zealand. The introduction of formal Environmental Assessment Review Process (EARP) guidelines in 1984 was followed in 1992 by the Canadian Environmental Assessment Act. The provisions of this Act establish a formal and tightly prescribed second generation EIA system and are supported by considerable financial and manpower resources and a substantial research programme. It is likely that the Canadian EA system will continue to provide a model for other jurisdictions.

Formal provisions for EIA in the Commonwealth of Australia date from 1974, 14 years before the UK system was introduced. The requirements are derived, with amendment to avoid frequent recourse to the courts, from NEPA. The Commonwealth procedure has evolved over the years but, like NEPA, remains largely unchanged with the significant exception (again like the US system) of the addition of scoping. The six states and two main territories of Australia have each subsequently put their own legislative and/or administrative EIA procedures in place to extend the scope of EIA to their own activities. As a federal EIA system, the Commonwealth of Australia's provides a valuable contrast to those in the United States and Canada.

New Zealand first introduced EIA procedures by means of a cabinet minute in 1974, the same year as Australia. From the outset it employed formal 'audit' (review) procedures to provide an independent check on the EIA reports prepared. After very considerable debate, environmental management in New Zealand generally, and EIA in particular, were revolutionised in 1991. One of the aims of this far-reaching reform was 'sustainable management'. EIA is now inextricably interwoven into regional and local authority procedures for determining various types of applications. In principle at least, EIA is much more comprehensive than it was in that it applies, at the appropriate level of detail (as determined by the regional and local authorities), to all projects. EIA in New Zealand is thus largely locally administered (as in the United Kingdom and in California) and has become an almost infinitely flexible approach: reason enough for its inclusion in a comparative study. New Zealand, Australia and Canada have held several biennial tripartite meetings on EIA.

Apart from reviewing the literature (including many unpublished documents), the interview was the main research method employed in this comparative study. Interviews were conducted with government and agency officials at various levels, with researchers in universities and research establishments, with representatives of industry, with lawyers, with consultants and with pressure group campaigners in each of the jurisdictions analysed. A structured approach was employed, interviews being conducted on the basis of a set of questions derived from the criteria set down in Box 1.2.

In compiling the reports on EIA in particular jurisdictions, an attempt was made to overcome inaccuracies by cross-checking participants' accounts with those of other participants in the EIA process and with documentary evidence wherever possible. Drafts of parts of earlier versions of much of the material in this book were reviewed by many of those interviewed.

Structure of the book

The first chapters of the book provide the background to, and an overview of, the seven EIA systems. Chapter 2 describes the first of the EIA systems: it explains the evolution of the NEPA provisions, recounts the main features of the US EIA system and discusses its implementation. The chapter also provides a brief description of the Californian EIA system.

Chapter 3 deals with the European Directive on EIA, using the same format as the American chapter. This provides the context for analysis of the UK and The Netherlands EIA systems. The following chapter covers EIA in the United Kingdom, again using the same format, but commencing with an account of pre-European Directive British activity in EIA. Chapter 5 presents an overview of the EIA systems in The Netherlands, Canada, the Commonwealth of Australia and New Zealand.

The next 14 chapters each follow the same pattern. Chapter 6 deals with the legal basis of EIA systems and Chapter 7 with their coverage of proposals and impacts. The subsequent nine chapters review, in turn, the various steps in the EIA process shown in Fig. 1.1. They cover: (Chapter 8) the consideration of the environmental impacts of alternative actions in the design process; (Chapter 9) screening; (Chapter 10) scoping; (Chapter 11) the preparation and content of the EIA report; (Chapter 12) reviewing the EIA report; (Chapter 13) the consideration of EIA in decision making; (Chapter 14) monitoring the impacts of projects; (Chapter 15) the mitigation of environmental impacts; (Chapter 16) consultation and participation. There follow chapters on EIA system monitoring (Chapter 17), the costs and benefits of EIA systems (Chapter 18) and, in Chapter 19, the assessment of the environmental impacts of programmes, plans and policies. In each case a discussion of the relevant aspect of the EIA process is presented. This is followed by a description of how this aspect of the EIA process is dealt with in the United States, California, the United Kingdom, The Netherlands, Canada, the Commonwealth of Australia and New Zealand, and the extent to which the appropriate evaluation criterion is met. Finally, a comparative summary table is presented.

Chapter 20 draws the main threads of the earlier chapters together by summarising the performance of each of the seven EIA systems against the evaluation criteria, and discussing their shortcomings. Finally, the chapter puts forward a number of suggestions, based upon the comparative review, for improving the various EIA systems and EIA generally.

This review would be incomplete without a discussion of EIA in developing countries. A brief synopsis of the relevant generalised literature has therefore been undertaken. Since the analysis was not based on detailed investigations of particular developing country EIA systems, it is of a different nature to the accounts of the seven EIA systems presented in the book. For this reason it is presented as an appendix rather than in the main text. However, the same structure as in the remainder of the book is followed, starting with the legal basis of developing country EIA systems and progressing through each stage or aspect of the EIA process to a conclusion.

Note

1. See O'Riordan and Sewell (1981), Wathern (1988a) and Commission of the European Communities (1993b) for valuable reviews of EIA in a range of jurisdictions.

CHAPTER 2

EIA in the United States of America

Introduction

The United States has a total area of 9.4 million square kilometres and a population of over 255 million people. With a current population density of only 27 people per square kilometre (almost a ninth of the United Kingdom's) it is hardly surprising that a frontier ethic developed in which land was seen as a disposable asset and in which controls over land use were regarded as a curtailment of the individual liberty which was one of the principal goals of the original settlers. Partly as a result of this frontier ethic there is a historic distrust of government institutions in the United States and a consequent desire for decision making which is open to inspection and intervention by the public.

The history of environmental control in the United States is remarkably brief, but typically vigorous. Prior to 1970 there was no effective federal control over the environment. While federal control over land use remains very weak (and while much state and local control over the use of land is not much stronger) the United States now has an imposing array of detailed and complex controls over air pollution, water pollution, hazardous wastes, etc. (Wood, 1989b). The Environmental Protection Agency, formed in 1970, is now the largest federal regulatory agency with 19,000 employees and an annual budget of US$ 7 billion. There are now estimated to be some 20,000 lawyers specialising in environmental matters in the United States (a country with more lawyers per capita than any other) (Andreen, 1992). Mandelker (1993a) cautioned that the complexity, duplication, jurisdictional fragmentation and expense of US environmental regulation were not viable indefinitely.

In its fifth report, the Council on Environmental Quality (CEQ, 1974, p. 54), reflecting the difficulties of using zoning to control US land use change effectively, stated:

> There is an increasing recognition that development proposals must be examined on an individual basis under a system of review that has both clearly defined standards and the flexibility to take into account

changing community values and the special characteristics of each project.

Environmental impact assessment is perhaps the best known technique for individual project appraisal. The EIA system was introduced in the United States on 1 January 1970, under the provisions of broad enabling legislation, the National Environmental Policy Act of 1969 (NEPA). In retrospect, NEPA can now be seen as the first step in an environmental revolution in the United States. One of its authors said at the time that it was 'the most far reaching environmental and conservation measure ever enacted by the Congress' (Jackson, quoted in Fogleman, 1990, p. 1).

It is remarkable that what Fairfax and Ingram (1981, p. 43) described as a 'standard administrative reform measure' should have received so much attention and been so widely imitated around the world. This chapter describes the evolution of the NEPA provisions and their subsequent refinement through the use of regulations. It provides an overview of the US EIA system at the federal level and comments on the implementation of NEPA in practice.[1] Finally, a brief description of the EIA system in California is presented.

Evolution of the NEPA provisions

During the 1960s it became apparent to many in the US Congress that pollution and other environmental problems were both complex and interrelated.[2] It was clear to some that a comprehensive approach to the environment was needed, one that was capable of anticipating environmentally disruptive activities and avoiding them, rather than merely reacting to episodes of pollution by passing specific abatement laws. Because of activities like international airport and interstate highway construction, the federal government was perceived to be a major cause of environmental degradation. However, environmental responsibilities were divided and lacking in enforcement power. An advisory council to coordinate the prevention of environmental degradation was proposed in a bill introduced by Representative Dingell in 1969. This became the Council on Environmental Quality, one important outcome of NEPA.

A second important element was the national environmental policy introduced by Senator Jackson. This 'motherhood and apple-pie' policy (Box 2.1) can now be seen to have anticipated the world's concern about sustainable development, inter-generational equity, resource usage and the integration of environmental considerations into decision making generally (World Commission on Environment and Development, 1987). The inclusion of the policy was intended to provide guidance in making decisions where environmental values were in conflict with other values. However, the policy by itself, though laudable, was seen to be insufficient if environmental degradation was to be reduced.

An 'action-forcing' mechanism was needed to ensure implementation of the policy. The belatedly introduced and justifiably famous Section 102(2)(C) of NEPA required a detailed statement by federal agencies evaluating the effect of their proposals on the state of the environment (Box 2.1). Caldwell, a consultant to the responsible Senate Committee, has stated that:

> The impact statement was required to force the agencies to take the substantive provisions of the Act seriously, and to consider the

Box 2.1 The US National Environmental Policy Act of 1969: ends and means

---

Sec. 101.(a) The Congress, recognizing the profound impact of man's activity on the interrelations of all components of the natural environment, particularly the profound influences of population growth, high-density urbanization, industrial expansion, resource exploitation, and new and expanding technological advances and recognizing further the critical importance of restoring and maintaining environmental quality to the overall welfare and development of man, declares that it is the continuing policy of the Federal Government, in cooperation with State and local governments, and other concerned public and private organizations, to use all practicable means and measures, including financial and technical assistance, in a manner calculated to foster and promote the general welfare, to create and maintain conditions under which man and nature can exist in productive harmony, and fulfil the social, economic, and other requirements of present and future generations of Americans.

(b) In order to carry out the policy set forth in this Act, it is the continuing responsibility of the Federal Government to use all practicable means, . . . to the end that the Nation may:

(1) fulfil the responsibilities of each generation as trustee of the environment for succeeding generations; . . .

Sec. 102. The Congress authorizes and directs that, to the fullest extent possible:

(1) the policies, regulations, and public laws of the United States shall be interpreted and administered in accordance with the policies set forth in this Act, and

(2) all agencies of the Federal Government shall . . .

(c) include in every recommendation or report on proposals for legislation and other major Federal actions significantly affecting the quality of the human environment, a detailed statement by the responsible official on:

   (i)   The environmental impact of the proposed action,

   (ii) Any adverse environmental effects which cannot be avoided should the proposal be implemented,

   (iii) Alternatives to the proposed action,

   (iv) The relationship between local short-term uses of man's environment and the maintenance and enhancement of long-term productivity, and

   (v) Any irreversible and irretrievable commitments of resources which would be involved in the proposed action should it be implemented.

---

environmental policy directives of the Congress in the formulation of agency plans and procedures.

(Caldwell, 1976, quoted in CEQ, 1990, p. 21)

This detailed statement, the 'environmental impact statement' (EIS), was the third important element in the Act and became the central document in the environmental impact assessment process.

Interestingly, while most environmental legislation in the United States has become increasingly prescriptive, detailed and complex, NEPA was short, simple and comprehensive. Blumm (1988) has argued that it was, perhaps, the last of the 'New Deal' legislation. However, Senator Muskie, who guided the Clean Air Act through Congress, insisted on far more detailed directives and left less scope for agency discretion. He forced the newly created Environmental Protection Agency (EPA) to become the environmental evaluator of all other agencies' actions by requiring it to review and comment on the impact of those projects for which EISs were prepared (Clean Air Act 1970, Section 309). The Office of Federal Activities within EPA undertakes this function. It therefore provides important support to CEQ's lead role in overseeing the operation of NEPA (Yost and Rubin, 1989).

Operationalising the Act was no simple matter, however. While the Council on Environmental Quality, located in the Executive Office of the President, issued guidelines dealing with the preparation of EISs, federal agencies reacted to the new requirements in ways varying from avoidance to amateurism. Although Congress had not anticipated that the courts would have a major role in implementing NEPA, a celebrated series of cases clarified the substantive nature of the EIS requirement. Most importantly, it was determined that the 'action-forcing' procedural provisions were not ends in themselves but were designed to ensure that the environmental policy, from which the Act takes its name, was implemented. However, application of the essentially procedural requirements of NEPA was believed to be almost certain to affect the decisions made by agencies.

One effect of this litigation (apart from making foreign observers quake at the prospect of importing it) was to lead to the writing of voluminous documents designed to resist legal challenge rather than to meet the policy objectives of NEPA. The EIS for the Trans-Alaska Pipeline was reputed to be more than 2 metres thick. There were two separate sets of problems in relation to EISs:

1. The usefulness of EISs reviewed was impaired by several common failings – inadequate discussion of the identified environmental impacts, inadequate treatment of the reviewing agencies' comments on environmental impacts, and inadequate consideration of alternatives and their environmental impacts.

(CEQ, 1973, p. 245)

2. Too many statements have been deadly, voluminous, and obscure and lacked the necessary analysis and synthesis. They have often

19

> been inordinately long, with too much space devoted to unnecessary description rather than to analysis of impacts and alternatives.
>
> (CEQ, 1975, p. 632)

As a result of the confusion caused by the various court decisions, President Carter instructed CEQ to prepare regulations to make the EIA process more relevant and to reduce the length of EISs.

The CEQ legal team led by Yost consulted widely and issued several drafts during the process of framing the regulations which were eventually put forward in 1978. They were given effect by an executive order and have remained almost unchanged for the 15 years since they were drafted (CEQ, 1978). As Yost (1984, p. 416) has noted, there are two important differences between the Regulations and the previous guidelines. First, they are mandatory requirements binding on all agencies. Second, the Regulations cover the whole of the EIA process, whereas the guidelines dealt only with EISs.

The Regulations standardised basic NEPA compliance practice throughout the federal government. Agencies responsible for the preparation of EISs adapted and supplemented the CEQ Regulations to meet their own needs. CEQ was entitled to be self-congratulatory:

> through mechanisms such as scoping, classification of actions, incorporation by reference, tiering, and other procedures, the Council's regulations have been instrumental in keeping the NEPA process focused and useful to both decision-makers and the public alike, while providing agencies substantial discretion to adapt those procedures to their programs.
>
> (CEQ, 1990, p. 26)

CEQ was subjected to savage staff cuts by the Reagan Administration in the early 1980s (Vig and Kraft, 1984). However, despite the hostility of the Administration, the breadth of support (including that of the US Chamber of Commerce) for NEPA and the Regulations ensured that they remained substantially unchanged. CEQ (1992, p. 134) was then 'reinvigorated' (with a one person Council) under the Bush Administration but its future was again questioned (and reaffirmed) during the Clinton regime. Figure 2.1 shows the fluctuations in CEQ staff levels over the years. That NEPA and CEQ survived is testament to the widespread perception of NEPA as a standard-bearer of US environmental protection policy. As Train, a past chairman of CEQ who was instrumental in its implementation, has stated:

> I can think of no other initiative in our history that had such a broad outreach, that cut across so many functions of government, and that had such a fundamental impact on the way government does business ... I believe I had a unique familiarity with the whole EIS process from inception to implementation and am qualified to characterize that process as truly a revolution in government policy and decision-making.
>
> (Train, quoted in Bartlett, 1989, p. 2)

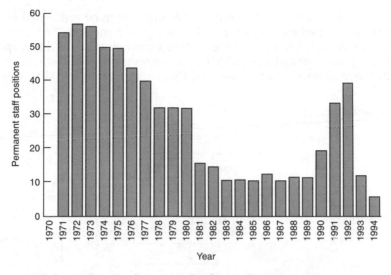

Year

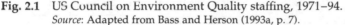

Fig. 2.1 US Council on Environment Quality staffing, 1971–94.
Source: Adapted from Bass and Herson (1993a, p. 7).

The US federal EIA system

The main steps in the US federal EIA system are shown in Fig. 2.2.[3] A
'lead' federal agency is designated to implement the various steps in the
EIA process. This agency is usually involved in actually constructing a
project or funding it, or granting a permit for it, or in proposing a
programme, plan or legislation. However, it relies heavily on the developer (if funding or permit-granting activities are involved) for
information and upon other agencies and the public for comment.

The first step in the EIA process is the identification of the proposal
leading to the action by the agency (i.e. construction of, or funding, or
permit granting for a project; or proposed programmes, plans or regulations). The agency will then undertake a preliminary environmental
analysis to determine whether there is a need for an environmental impact
statement (preparation of which can commence forthwith), whether the
environmental impacts are clearly so insignificant as to permit a categorical exclusion from the EIA process (for which documentation is optional)
or whether an 'environmental assessment' (EA) should be prepared so
that the significance of impacts can be more clearly identified.[4] This
environmental assessment, again prepared by the agency, is, in effect, an
abbreviated EIS as it covers many of the topics required in an EIS. However, it is not subject to the same rigorous consultation provisions.
Depending on the findings of the environmental assessment, an EIS may
be required or, as in the vast majority of cases (Chapter 9), the agency may
decide that none is necessary. In this case, a 'finding of no significant
impact' (FONSI) must be written, summarising the reasons for this
decision.

When an EIS is required, a 'notice of intent' has to be published by the agency in the Federal Register and scoping commences. Scoping is a procedure intended to bring those with different interests in the proposal (including members of the public) to an agreement about which of the environmental impacts associated with it are significant and thus require

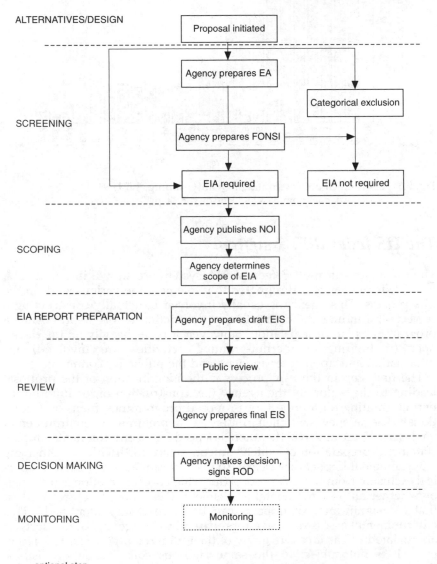

··· optional step
EA: environmental assessment
FONSI: finding of no significant impact
NOI: notice of intent
ROD: record of decision

Fig. 2.2 Main steps in the US federal EIA process.

investigation. Agency regulations may require analysis of some issues but other significant matters are agreed by consultation, and frequently at a meeting (or a series of meetings) at which the various interested parties are represented. These issues are then addressed in the draft EIS. This is written by or on behalf of the agency though the developer provides a great deal of the relevant information upon which it is based if funding or permitting is involved.

The draft EIS normally follows a set pattern dictated by the relevant agency's regulations or guidelines: it will describe the existing environment, explain what the proposed project is and analyse the effects of the project on the environment. It is these effects which constitute the substance of the draft EIS, which should not normally be more than 150 pages in length. They are generally discussed at some length and mitigation measures are usually proposed. In accordance with the requirements of NEPA (Box 2.1), it is usual to provide: a summary of probable adverse environmental effects which cannot be avoided; a discussion of alternatives to the action; a discussion of the relationship between local, short-term uses of the environment and maintenance and enhancement of long-term productivity; and a discussion of irreversible and irretrievable commitments of resources. Most agencies follow the tighter structure specified in the Regulations in organising these discussions in their EISs.

The draft EIS is sent to the Environmental Protection Agency for critical review and filing and is forwarded to all the relevant federal, state, tribal and local organisations likely to wish to comment. This review process involves reading the draft EIS and commenting both on the way the reviewing agency's interests are affected and on the content of the EIS generally (though this latter type of comment is less common than the former). There are arrangements for local groups and for the public to participate and there has to be a minimum period for deposit of the documents of 45 days to allow this participation to take place. Once the lead agency has received the comments of the various consulted agencies and bodies it is in a position to prepare the final EIS.

The final EIS describes the modified form of the proposed action, including any changes that have been made since the draft EIS was published, and responds to the comments received from the various bodies consulted. This document usually contains quite extensive proposals for mitigation of impacts. A 'record of decision' has also to be prepared, indicating the decision that has been made and the reasons for it. This is sometimes circulated for a period of time and agencies with an administrative appeals process can adopt a procedure whereby they release the final EIS and the record of decision simultaneously. Generally, however, the record of decision is issued after a 30 day waiting period following the filing of the final EIS with the Environmental Protection Agency.

There are somewhat inadequate provisions for monitoring the environmental impacts arising from an action and for ensuring that the various conditions or mitigation measures that have been included in the final EIS are implemented. This may be done in the form of conditions appended to permits that have to be obtained from the lead agency or in the form of conditions attached to grants that are made by the agency. If the agency itself is carrying through the measures there is usually a system of inspection to ensure that the project is actually constructed as described in the final EIS (unless, of course, there are overwhelming and unforeseen reasons for change, in which case a supplementary EIS may have to be prepared).

There are provisions for 'tiering' EISs, that is, the preparation of broad programme EISs followed by site-specific EISs cross-referenced to the overall document. The use of these is increasing. There are also provisions for mediation by the Council on Environmental Quality if EPA or other agencies such as the Department of the Interior are unable to agree that the impacts of the action are acceptable.

Implementation of NEPA

There is no doubt that the EIA process is biting.[5] NEPA nowhere provides for the termination of a major federal action because of the environmental consequences, but actions in the courts have stalled or stopped such projects if their consequences have not been properly documented (Callies, 1984, pp. 120–3). There has been substantial EIA litigation, initially and still predominantly by environmental or citizen groups, but more recently by the states and by industry. The volume of litigation has declined as issues have been clarified, but continues to be significant (Fig. 2.3). The most common causes of legal action are the absence or inadequacy of EISs. Of 94 cases in 1991, 14 resulted in injunctions (CEQ, 1993).

Several projects have been aborted as a result of the adverse impacts revealed in preparing an EIS and it appears that a majority of projects are modified as a result of the assessed impacts. This mitigation of impacts appears to be 'where the action is' and is widely cited as one of the main justifications of the process. To a large extent, EIA has been assimilated into federal decision-making processes and is meeting many (but not all) of the objectives of its proponents.

Renwick (1988) has argued that the role of NEPA has declined and become largely symbolic as more specific environmental legislation has been enacted, shifting the focus of project impact evaluation to, for example, the air pollution control process. It seems more likely that NEPA's current lack of notoriety may well be a measure of its success in internalising the consideration of environmental quality in federal agencies. However, this process is not complete, as CEQ (1992, p. 133) has admitted:

> While virtually all federal agencies have adopted regulations to ensure that NEPA procedural requirements are met, patterns of compliance vary widely. For example, some agencies have difficulties integrating environmental values in pursuit of their more immediate, mission-oriented goals.

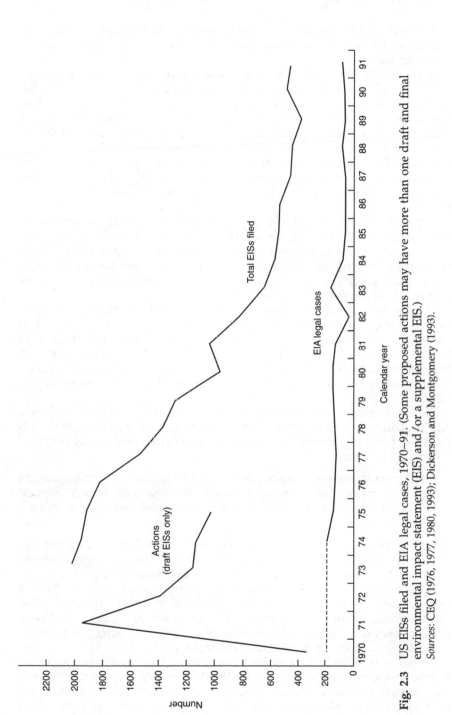

Fig. 2.3 US EISs filed and EIA legal cases, 1970–91. (Some proposed actions may have more than one draft and final environmental impact statement (EIS) and/or a supplemental EIS.)

Sources: CEQ (1976, 1977, 1980, 1993); Dickerson and Montgomery (1993).

CEQ (1992, p. 134) has stated that it is working with federal agencies to 'improve the usefulness of the law as an efficient, integrative policy planning tool'.

It is normal for the EIS to address the procedural requirements of NEPA, as refined in the agency guidelines, and to rely on scoping for the identification of issues, rather than to use any 'comprehensive EIA methodology'. Widespread use is, however, made of specialised technical methods for assessing particular impacts (e.g. air pollution modelling). The trend is to make greater use of the information generated for other purposes (for example, the granting of an air pollution permit) in preparing the EIS and to combine the granting of permits to reduce the number of hurdles applicants must negotiate in realising their proposals.

The number of EISs has fallen (Fig. 2.3). Well over 1000 EISs per annum were produced in the first decade (Environmental Law Institute, 1981) but the number has dropped steadily since. The emphasis in EIA is moving toward mitigated FONSIs in which negotiation takes place very early in the process, often on the basis of the environmental assessment and no EIS is produced (see Blaug, 1993).

Despite the generally accepted improvement in the quality of EISs over recent years, there is scope for further amelioration in their analytical content and for closer adherence to the spirit of NEPA rather than its letter (which has tended to be over-emphasised as a result of litigation (Bear, 1989; Caldwell 1989a,b,c)). There have been several attempts (so far unsuccessful) by Congress to strengthen NEPA's monitoring provisions.[6]

Environmental impact assessment has been mainly confined to projects and probably owes much of its success to the general weakness of the US land use planning system, especially at the federal level. Reilly (1974, p. 350) has stated that:

> the impact statement process reflects a more realistic understanding of the way major development is sited. No one any longer expects comprehensive plans to detail precisely the nature and location of new development.

Another factor in NEPA's success, which has been empirically demonstrated by Caldwell et al. (1983), is that it was directed at government agencies, particularly those responsible for the undertaking of development activities of potential environmental significance, rather than at private developers (Fowler, 1982; von Moltke, 1984). To implement NEPA, these agencies have recruited and developed interdisciplinary environmental review staffs which have grown increasingly influential over the years (Environmental Law Institute, 1981).

Rather more than half the states have enacted their own environmental impact assessment legislation (CEQ, 1992). Some states require EISs only for projects proposed within specific areas, others demand them for actions undertaken by state agencies or using state funds, and yet others require them for these categories together with actions which need state permits. The requirements of other states apply to all these types of actions plus a number of actions taken by local agencies, and three states, including California, have a comprehensive system covering local government and private activities as well as those of the state itself (Hart and Enk, 1980). The various state legal requirements differ from the federal system and most have proved weak and ineffective, the comprehensive systems being among the exceptions. Some counties and cities have also introduced their own EIA requirements.

The Californian EIA system

California is by far the most populous state in the Union with a 1991 population of over 30 million (12.2 per cent of that of the United States) and an area of 424,000 km^2 (about the size of Great Britain). The 'golden' state's population is still growing at over 2 per cent per annum. California's environmental regulation is rigorous, with stronger land use planning controls than most other states and the most stringent air pollution emission standards in the United States.

The Californian Environmental Quality Act (CEQA) was passed in 1970[7] as a 'little NEPA' which required state decisions to consider effects upon the environment. In fact, it was big (i.e. more detailed than NEPA) but, like its progenitor, its import was not clarified until a number of major court cases had been fought. One of these (the Friends of Mammoth case) resulted in a ruling that the term 'project' in CEQA, which governs the scope of the procedural requirements, included private activities which were subject to discretionary state or local government approvals. Thus, instead of applying only to major state actions, CEQA suddenly applied to thousands of local projects. CEQA is therefore administered by both state officials and by local government land use planners. It has come to govern the review and approval process of all large developments in California (Bass and Herson, 1993b).

It is unsurprising that CEQA has so many parallels with NEPA, the Act upon which it was closely modelled, though CEQA is now much larger (having incorporated various elements of the procedures set down in the federal regulations on EIA and the outcome of numerous court cases).[8] The main steps in the California EIA process are set down in Fig. 2.4.[9] Following a review for exemptions, screening takes the form of an 'initial study' (similar to the NEPA environmental assessment) which can lead to a 'negative declaration' (which, like the NEPA finding of no significant impact, can be 'mitigated'). If an environmental impact report (EIR) is undertaken which reveals significant impacts the project must be modified or the agency must include in its 'findings' (equivalent to the NEPA record of decision) a 'statement of overriding considerations'. The main substantive difference between the state and federal Acts is the requirement that Californian agencies mitigate impacts where feasible. There are also some procedural differences (Bass and Herson, 1993b, c).

There are two state agencies responsible for CEQA administration and oversight, as at the federal level. The Governor's Office of Planning and Research is responsible for drafting the state CEQA Guidelines (which

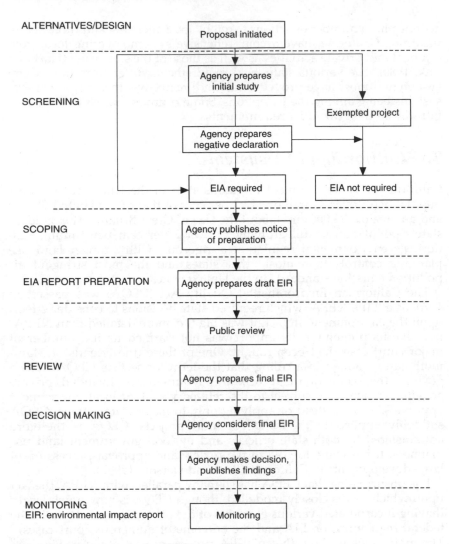

Fig. 2.4 Main steps in the Californian EIA process.

first appeared two years after CEQA itself) and the Resources Agency is responsible for rule making, verification and adoption of the Guidelines. The intention of the Guidelines (Office of Planning and Research, 1992) is to give detailed guidance on exemptions, the contents of initial studies and environmental impact reports, and on various other requirements of CEQA. A set of appendices gives various notes on checklists and on other detailed procedural elements of CEQA. Despite their name and uncertain legal status they are treated, in effect, as regulations. In addition, the Office of Planning and Research (1983) has produced a report on preparing an environmental impact report for general plans, together with a number of other items of guidance. Various jurisdictions within

California have produced guidelines on the operation of CEQA as it applies to them, and most have adopted appropriate procedures to implement the Act.

While the Guidelines have not been updated since 1986, the Act itself has been regularly amended by the California legislature in the light of practice and of court cases. CEQA was amended in 1989 to attempt to close a monitoring loophole which had proved to be the Act's Achilles' heel. Public agencies are now required to adopt monitoring and reporting programmes at the time they certify an EIR if the proposed project involves mitigation measures to reduce or avoid significant environmental impacts. Further reforms were being actively discussed in 1994, as a result of pressure from developers and industrialists to simplify, speed up and reduce the cost of CEQA procedures as the recession in the Californian economy deepened.

Most environmental impact reports are produced by consultants hired either by the local counties or cities or by the developer. Counties and cities frequently do not have the extensive environmental staff expertise of the kind available to most federal agencies, and therefore must rely on their own consultants (generally funded by the developer) to produce the environmental impact report. The EIR is then subject to consultation by interested agencies outside the local government concerned. There is a growing tendency to combine the federal EIS requirements with the state EIR requirements in a single document in the few cases where both apply.

More environmental impact reports are produced annually in California than EISs in the United States as a whole. Of the several hundred EIRs produced, more than half are deemed to be of 'state-wide interest' and hence are sent to the State Clearinghouse in the Office of Planning and Research for distribution to interested agencies. In 1992, over 850 EIRs were sent to the Clearinghouse, together with about 30 EISs. Olshansky (1993) estimated that about 1600–1800 draft EIRs were completed by local government in 1990. About 2300 negative declarations were also forwarded to the Clearinghouse (Office of Planning and Research, 1993), a number believed to be less than 10 per cent of the total (Kaplan-Wildman and McBride, 1992; Olshansky, 1992). Olshansky (1993) suggested that there were approximately 28,000–32,000 negative declarations by local government in 1990. As Wandesforde-Smith and Kerbavaz (1988, p. 181) have stated, the mitigated negative declaration is:

a licence for local authorities to bargain with developers within the framework of the CEQA process in order to reach the best deal they can very early in the review process.

The number of staff devoted to EIA at the Office of Planning and Research is less than one full-time, though about five people are employed in the Clearinghouse. Several thousand EIA professionals are employed in local government, state agencies and consultancies. In addition, because there are often over 100 CEQA related court cases each year, a number of lawyers earn their living from EIA in California. (The

Environmental Law section of the Californian Bar Association has about 2500 paid-up members.)

Notes

1. This account, and the subsequent treatment of the US EIA system in this book, is partially based on interviews. Apart from those I have acknowledged elsewhere, I am very grateful to Ron Bass, Dinah Bear, Ray Clark, Shannon Cunniff, Anne Miller, Ken Mittelholtz, Joseph Montgomery, David Powers, Russell Train and Nicholas Yost for their assistance.
2. This description owes much to the account of 20 years of NEPA (CEQ, 1990, pp. 15–51; see also Blumm, 1988, 1990; Caldwell, 1989a,b).
3. This description is derived from Wood, 1985. These steps are taken from the text of the relevant Regulations first promulgated in 1978 and now codified in the Code of Federal Regulations at *40 CFR 1500–1508* 'Regulations for Implementing the Procedural Provisions of the National Environmental Policy Act'. The Regulations are reproduced in CEQ (1992) together with the text of NEPA. For recent thorough descriptions of federal EIA procedures (which also reproduce both NEPA and the Regulations) see Fogleman (1990), Bass and Herson (1993a) and Mandelker (1993b). See also Canter (1977), Ortolano (1984), Baldwin (1985), Murthy (1988), Bear (1989), Environmental Law Institute (1989) and, for a British perspective, Glasson *et al.* (1994) and Sheate (1994).
4. Most agencies (which are frequently divided into regions or divisions) have promulgated rules or guidelines to help identify which projects can be categorically excluded or should move straight to an EIS.
5. This account is partially derived from Wood, 1989b, pp. 63–6.
6. Various major gatherings have been held to mark NEPA's maturity and to suggest improvements. See *Environmental Law* **20**: 447–810 (1990) (summarised by Blumm, 1988), *The Environmental Professional* **15** (1993) and Hildebrand and Cannon (1993).
7. This description is derived from Wood (1985).
8. This account, and the subsequent treatment of the Californian EIA system in this book, is partially based on interviews. Apart from those I have acknowledged elsewhere, I am very grateful to Ron Bass, Robert Cervantes, Don Collin, Al Herson, Jennifer Jennings, Bob Johnston, Richard Lyon, Robert Olshansky, Tom Pace, Terry Rivasplata, Michael Remy, Paul Sabatier, Tim Taylor and Geoffrey Wandesforde-Smith for their assistance.
9. Succinct descriptions of the Californian EIA system can be found in Bendix (1979), in Fogleman (1990) and in Mandelker (1993b). Bass and Herson (1989), (1993b), Remy *et al.* (1993) and Hernandez *et al.* (1993) provide detailed guides to CEQA in which CEQA and the Guidelines are produced. Roberts (1991) also provides a less detailed overview. See also Wandesforde-Smith (1981) and Wandesforde-Smith and Kerbavaz (1988).

CHAPTER 3

The European Directive on EIA

Introduction

The Treaty of Rome, dating from 1957, contained no reference to environmental policy but this oversight was addressed in the Paris Declaration on the Environment in 1972 (the year the United Kingdom, Ireland and Denmark joined the original six member states of the European Communities (EC)). The Commission of the European Communities (CEC) published its first 'Action Programme on the Environment' in 1973, justifying this on harmonisation and competition grounds.[1] From that time, a trickle of environmental legislation has become a stream, covering water and air pollution, waste disposal, control of chemicals, noise, wildlife protection and environmental impact assessment. EC legislation is directly applicable in national courts (regulations) or binding on government as to the ends to be achieved (directives) without the need for ratification. Furthermore, the Commission has a duty to enforce EC legislation, eventually bringing matters to the Court of Justice if necessary (Haigh, 1991).

Partly as a result of proportional representation, the environment has been higher on the political agenda in many influential European countries than in the United Kingdom. This concern is driving EC policy, so that the stream of environmental legislation is maturing to become a river. This legislation is impinging more and more on British practice, as is its enforcement in the Court of Justice.

Much of the pressure for tighter environmental controls stems from Germany where concern about the environment has generally been even greater than in the United Kingdom. Germany, of course, is largely surrounded by other countries and is only too aware that pollution originating elsewhere can affect it severely (witness the effects of acid rain on German forests and of accidents in Switzerland and France on the Rhine). Because of its powerful position in the European Communities, Germany has been able to translate its concern about the environment into action by the CEC. The range of people affected by this action is growing: it is no longer confined to a few specialists in industry and in the environmental control agencies.

One of the central tenets of European environmental policy is anticipatory action:

> The best environmental policy consists in preventing the creation of pollution nuisances at source, rather than subsequently trying to counteract their effects.
>
> (CEC, 1977)

The directive on environmental impact assessment exemplifies the way in which European influence has led to an increase in planning (and other) controls over the environment, affecting large numbers of people, by applying this principle.

The EIA Directive represents the first European Union intrusion into the planning domain, and has major repercussions on member state decision making and practice. This is undoubtedly the reason it took so long to move from the Commission's original proposal to adoption. However, this environmental directive is, though very far-reaching, but one of the many which can be expected with the changing nature of the European Union. Some of these directives are likely to prove just as significant in altering member state regulation as that on EIA. The implementation of the Single European Act, which introduced specific reference to environmental protection and provided that 'environmental protection requirements shall be a component of the Community's other policies' (Article 130r)(2), is unlikely to reduce the flow of the river of environmental regulation. Rather, it is likely to raise the importance of the environment in many fields.

This chapter describes the evolution of the European Directive and, in particular, the changes which took place between the published draft directive and the version eventually adopted. It then presents an overview of the provisions of the Directive and briefly mentions its implementation by the 12 member states.

Evolution of the Directive

The CEC has stated that 'too much economic activity has taken place in the wrong place, using environmentally unsuitable technologies' (CEC, 1979, p. 49) and that 'effects on the environment should be taken into account at the earliest possible stage in all the technical planning and decision-making processes' (CEC, 1977).[2] Because of this concern to anticipate environmental problems, and hence to prevent or mitigate them, the CEC became interested in environmental impact assessment in the early 1970s, like many other bodies.

The CEC commissioned research investigations on EIA in 1975. Lee and Wood (1976, 1978b) reported that many aspects of EIA procedure already existed within member states. As a consequence, it was suggested that the requirements of a European EIA system could be integrated into member state decision-making processes without the disruption or the litigation which characterised early American experience. Lee and Wood felt that project EIA should be the first stage of a European EIA system which would eventually encompass policies and plans, once more than rudimentary experience of this type of assessment had been gained. They suggested a set of procedures to form the basis of a European EIA system, including broad criteria for the selection of projects to be assessed.

Following this research programme, the Commission decided that an EIA system should meet two objectives:

1. To ensure that distortion of competition and misallocation of resources within the European Economic Community (EEC) was avoided by harmonising controls.
2. To ensure that a common environmental policy was applied throughout the EEC.

The Commission issued its first preliminary draft directive in 1977. After 20 such drafts, not all of which were released, and substantial consultation (this is reliably reported to have been the most discussed European draft directive to date (Wathern, 1988b, 1989; Sheate and Macrory, 1989; Sheate, 1994)), the Commission put forward a draft to the Council of Ministers in June 1980 (CEC, 1980).

The draft directive specified that projects likely to have a significant effect on the environment were to be subject to EIA. Such an assessment was obligatory for nearly all projects, other than modifications to existing installations, in certain specified categories which were listed in Annex 1. There were some 35 of these types of projects, grouped under the headings: extractive industry; energy industry; production and preliminary processing of metals; manufacture of non-metallic mineral products; chemical industry; metal manufacture; food industry; processing of rubber; and building and civil engineering.

EIA was also required for certain projects in other specified categories listed in Annex 2 of the Directive and for substantial modifications to Annex 1 projects, subject to criteria and thresholds to be established by member states. Annex 2 included agricultural and forestry practices as well as many industries not encompassed by Annex 1. In addition, EIA was to be required for any other projects outside the above categories where a significant environmental impact was likely to occur. There were provisions for Commission coordination of criteria and thresholds and for a simplified form of assessment in certain cases. These proposals left considerable discretion with the member states in deciding the precise coverage of the EIA system to be adopted. Member states had to obtain the prior agreement of the CEC to exempt projects from the provisions of the directive, which included the requirement that the impacts upon the environment within another member state affected by the proposal be assessed.

The developer was to bear the primary responsibility for supplying all of the relevant basic information required in the EIA report. At the same time, it was envisaged that the 'competent authority' would often need to assist the developer in the preparation of the study. The authority also had the responsibility for checking the information supplied, which was to include:

- a description of the proposed project and reasonable alternatives to it
- a description of the environment likely to be significantly affected by the project
- an assessment of the project's likely significant effects on the environment
- a description of any environmentally mitigating measures that are proposed
- an indication of the likely compliance with existing environmental and land-use plans and standards for the area
- a justification of the rejection of reasonable alternatives to the proposed project where these are expected to have less significant adverse effects on the environment
- a non-technical summary.

Annex 3 specified the required content of the assessment in more detail. The kinds of impact to be considered included those arising from the physical presence of the project, the resources it used, the wastes it created and its likely accident record.

There were provisions for consultation. The competent authority had to publish the fact that the application had been made, make all the environmental documentation available to members of the public and make arrangements for concerned parties to present their views. The competent authority had then to make its own final assessment and publish it (unless permission was refused on other than environmental grounds). This publication was to contain the assessment itself, a summary of the main comments received, the reasons for granting or refusing planning permission and the conditions, if any, to be attached to the granting of the permission. In the event of the project being authorised, the competent authority was expected to check periodically whether any conditions attached to the approval were being satisfied, and whether the project was having any unexpected environmental effects that might necessitate further measures to protect the environment. It was intended that assessment should eventually be extended from projects to plans and programmes. The system outlined bore a strong resemblance to many comprehensive EIA systems, such as the US and (particularly) the Australian federal EIA processes.

The Council of Ministers did not approve the draft directive in June 1980. The British Government, for reasons elaborated in Chapter 4, was reluctant to accept the imposition of a mandatory system of EIA, at least in the form set out in the published draft directive. This draft was sent for formal deliberation by interested parties both within member states and within European institutions. This consultation resulted in numerous representations to the CEC. As a result of a sitting of the European Parliament in 1982, the Commission accepted a number of alterations to the draft directive. These were mostly minor in nature and, on balance, strengthened rather than weakened the EIA provisions (CEC, 1982).

During subsequent negotiations between the Commission and the

British Government several of the more controversial aspects of the draft directive were deleted to meet the British position. In the course of the amendments, many types of industry were shifted from Annex 1, where their assessment would be compulsory, to Annex 2, where their assessment would be much more discretionary (Sheate and Macrory, 1989; Wathern, 1989; Sheate, 1994). The British Government withdrew its objections to the amended draft directive late in 1983, only for the Danish Government to continue to express serious reservations about the undermining of the sovereign power of the Danish Parliament to approve development projects. A provision exempting projects approved by specific Acts of national legislation (Wathern, 1988b) paved the way to adoption of a much modified version of the draft directive in June 1985. Table 3.1 shows how one aspect of the Directive, the selection of projects for assessment, evolved between 1976 and 1985.

The adopted version of the Directive (CEC, 1985) limited the Annex I projects to oil refineries, large coal gasification and liquefaction plants, large power stations, radioactive waste disposal sites, integrated steel works, asbestos plants, integrated chemical plants, motorways, railways and large airports, ports, canals and toxic waste disposal facilities. The list of Annex II projects has grown substantially but the requirement for Commission coordination of criteria and thresholds has been dropped.

The main text of the Directive no longer mentions the need to discuss alternatives and deletes the provision involving compliance with land-use plans, standards, etc. The requirements for publication of the authority's own assessment and of its synthesis of public comments, and for the provision of competent authority assistance have also been substantially weakened. Those for assessment of unlisted projects, for consideration of impacts upon neighbouring member states in the mandatory developer's documentation and for monitoring were deleted. Further, the requirement relating to Commission approval of exemptions has been replaced by duties upon member states to provide information. There is no mention of the EIA, of plans or programmes in the adopted Directive. Annex III specifying the desirable (rather than the required) content of the information supplied by the developer reflects the reduced scope of the requirements.

The net effect of these changes has been the emasculation of the provisions in the earlier drafts of the Directive. These early versions were themselves criticised as being over-cautious, and for not containing provisions for the Commission to monitor and oversee the EIA system effectively, let alone use it to make substantive and constructive inputs to problem solving (Wandesford-Smith, 1979). As Brouwer (quoted in Wathern, 1988b, p. 201) stated, from a Dutch perspective:

> This EC-directive, like so many others, is a very weak compromise. It is more the result of the cumulative resistance from the development promoters and bureaucracies in the member countries than a synthesis of the best ideas for the protection of the environment.

The Directive in its adopted form provides a flexible framework of basic EIA principles to be implemented in each member state through national

Table 3.1 Selection of projects for assessment in the European Directive and its progenitors

1976 Lee and Wood ENV/197/76	1977 EIE/OU/10	1979 EIE/OU/14	EIE/OU/18
No list system. Projects subject to EIS determined by 'applicability' guidelines.	List of projects subject to mandatory assessment. Criteria for selection of other projects.	List of projects subject to mandatory assessment. Criteria for selection of other projects.	List of projects subject to mandatory assessment (Annex 1). List of projects (and modifications to Annex I projects) subject to assessment when so required according to criteria set by competent authority (Annex 2). Provision for simplified form of assessment. Screening criteria for selection of other projects.

1980 COM (80) 313	1982 COM (82) 158	1985 85/337/EEC
List of projects subject to mandatory assessment (Annex 1). Provision for exemption and simplified assessment where appropriate. List of projects (and modifications to Annex 1 projects) subject to assessment when so required (Annex 2). Competent authority(ies) establish criteria and thresholds. Provision for determining other projects.	As previous draft except more detailed provision for exemption.	As before except detailed exemption clause (paragraph). Commission to report annually to Council on the application of the paragraph. Provisions for simplified assessment and determining other projects deleted. Transfer of projects from mandatory to more discretionary annex

Source: Sheate and Macrory (1989) with last two paragraphs added.

legislation. While many of the original provisions have been amended, there is, of course, nothing to prevent member states from instituting EIA systems which are more comprehensive and rigorous than the provisions put forward by the Commission. Several countries, including The Netherlands, have considerably exceeded the requirements of some of the articles in the Directive in their national EIA systems (Coenen, 1993). As the Council on Environmental Quality (1990, p. 46) has stated, the European Directive represents a first step 'towards establishing effective, efficient EIA processes to help reconcile economic growth and development with maintenance and enhancement of environmental quality ...'.

The European Directive EIA system

The *legal basis* of the EIA system, a European directive, is clear. It is left to member states to implement the requirements of the EIA Directive in whatever legislation they consider to be appropriate. The Directive, as mentioned above, provides a skeletal framework and leaves a great deal of detail to be determined by member states (Coenen, 1993).

The Directive consists of 14 articles and three annexes.[3] The main steps in the EIA process are shown in Fig. 3.1. The Directive places a general obligation on each member state to ensure that, before consent is given, projects likely to have significant effects on the environment by virtue, *inter alia*, of their nature, size or location, are made subject to an assessment (Article 2(1)). This assessment may be integrated into existing project consent procedures or into other procedures (Article 2(2)). Acts of national legislation (Article 1(5)) are excluded and specific projects may be exempted in exceptional cases, after making relevant information available to the public and to the Commission (Article 2(3)).

The *coverage* of the Directive is confined to projects. While the lists of projects to which the Directive applies in Annex I and Annex II is lengthy, it is not comprehensive and certain environmentally sensitive projects (e.g. water treatment plants, types of land reclamation from the sea) are omitted.

The word 'environment' is used to mean the physical environment. The social and economic environments are not overtly included in this definition, as they are in many other jurisdictions, e.g. in the United States (Mandelker, 1993b). Article 3 of the Directive requires that:

> the environmental impact assessment shall identify, describe and assess, in an appropriate manner ... the direct and indirect effects of a project on:
> - human beings, fauna and flora,
> - soil, water, air, climate and the landscape,
> - the inter-action between the[se] factors ...,
> - material assets and the cultural heritage.

Other types of effect (below) are consigned to Annex III where their use is largely discretionary.

There is no mention of *alternatives* in the main text of the Directive but, subject to member state requirements, the information specified in Annex III should be provided: 'Where appropriate, an outline of the main alternatives studied by the developer and an indication of the main reasons for his choice, taking into account the environmental effects'.

The European Directive requirements regarding *screening* (i.e. choosing which projects should be subject to EIA) by virtue 'of their nature, size or location' (Article 2(1)) are summarised in Fig. 3.1. All projects listed in Annex I are subject to assessment (Article 4(1)). Projects listed in Annex II are also subject to EIA where member states 'consider their circumstances so require' (Article 4(2)). These project types are grouped under 12 broad headings: agriculture; extractive industry; energy industry; processing of metals; glass manufacture; chemical industry; food industry; textile, leather, wood and paper industries; rubber industry; infrastructure

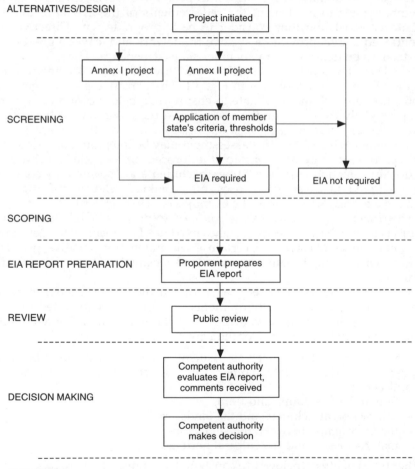

Fig. 3.1   Main steps in the European Directive EIA process.

projects; other projects; and modifications to Annex I projects. Member states may specify certain types of projects or establish screening criteria and/or thresholds to determine which projects should be subject to this requirement (Article 4(3)). There is, however, no requirement in the Directive for the provision of preliminary information equivalent to the US environmental assessment to assist in screening. Nor is there any provision for different levels of EIA in the Directive, though proposals for a simplified level of environmental assessment were advanced at one stage in the discussions leading to the adoption of the Directive (above).

There is no provision for *scoping* (i.e. for determining the content of EIA information for particular projects) in the European Directive. Nor is there any provision in the Directive that the commencement of work on an EIA be announced.

The general nature of the content of the EIA information to be utilised by the proponent in the *preparation of the EIA report* is, however, specified. Article 5(2) of the Directive sets down the minimum information which must be provided by the developer:

- a description of the project comprising information on the site, design and size of the project,
- a description of the measures envisaged in order to avoid, reduce and, if possible, remedy significant adverse effects,
- the data required to identify and assess the main effects which the project is likely to have on the environment,
- a non-technical summary ...

Article 5 also indicates that the developer should furnish all the information listed in Annex III where member states consider that it is relevant and reasonable to do so. This includes a description of the project, of alternatives to the project (where appropriate), of baseline environmental conditions, of the likely significant environmental effects of the project, and of mitigating measures, a non-technical summary and an indication of difficulties encountered in compiling the information. This information corresponds closely to that specified in most other EIA systems. Annex III makes it clear that the description of the likely significant effects of the project should cover the direct effects and any indirect, secondary, cumulative, short, medium and long term, permanent and temporary, positive and negative effects, of the project.

In order to facilitate the assessment, Article 5(3) provides that 'where they consider it necessary' member states should ensure that the authorities holding information relevant to the assessment make this available to the developer. However, there is no requirement that liaison between the developer and relevant authorities takes place while the assessment is being undertaken. Having carried out the assessment, the developer is obliged to supply the competent authority responsible for the authorisation of the project with the resulting information. The form in which this information is submitted is not specified in the Directive. (It is referred to hereafter as the 'EIA report'.)

The Directive does not provide for a formal *review* of the EIA report by the competent authority (or any other body) or for the preparation of draft and final EIA reports. However, Article 6 of the Directive provides for the EIA report to be made widely available as a basis for consultation and public participation. There is no provision for the developer to respond to the points raised by the public or consultees on the content of the EIA report, or for these comments to be made public.

Article 8 requires that the results of this exercise, together with the developer's EIA report, must be taken into consideration in taking the *decision* on the project. The Directive requires, when the competent authority has reached a decision on the consent application, that the public (and any member state that was consulted under Article 7) be informed and that any conditions attached to that decision be made public. Somewhat ambiguously, Article 9 of the Directive also requires that the reasons upon which the decision has been based should be provided 'where the Member States' legislation so provides'.

The European Directive is silent on the question of the *monitoring* of project impacts.

The *mitigation* of project impacts is one of the main aims of the European Directive. As mentioned above, it is a requirement of Article 5(2) that mitigation measures be specified in the proponent's EIA report. These mitigation measure requirements are also listed (almost word for word) in Annex III. This is, however, the only point in the EIA process where mitigation measures must be considered, although the precautionary principle underlies the whole Directive.

Consultation and participation is limited, under the provisions of the Directive, to commenting upon the EIA report. Member states are required to designate the environmental authorities which should receive copies of the environmental information and who must be consulted for their opinion on the consent application (Article 6(1)). Similarly, member states must ensure that both the consent application and the environmental information are made available to the public and that the public concerned is given an opportunity to comment before the project is initiated (Article 6(2)). In addition, member states are required to provide the above information, as a basis for consultation, to another member state where the project is likely to have significant effects on its environment (Article 7). As in other EIA systems, there are provisions for the protection of industrial and commercial secrecy (Article 10). There is no provision for third party appeals against decisions involving EIA.

There is provision in the Directive for *EIA system monitoring*. Article 11 of the Directive provides for the exchange of information between member states and the Commission on experience in applying the Directive and for member states to inform the Commission on the criteria and thresholds they have used in the selection of Annex II projects, and for the preparation of a five year review of the Directive's application and effectiveness. Article 11(4) commits the Commission, should it be necessary, to submit additional proposals, to ensure that the Directive is 'applied in a sufficiently co-ordinated manner'.

Article 12 requires member states to take the necessary measures to comply with the Directive within three years of its notification (i.e. by 3 July 1988) and to send copies of the national laws by which this has been done to the Commission. Article 13 empowers member states (to whom the Directive is addressed (Article 14)) to lay down stricter legal rules than those required by the Directive. The European Commission has not issued any published guidance as to how the Directive is to be implemented (as is its usual practice) but has arranged expert meetings and information exchange and funded a substantial EIA training programme, together with a five year review.

The *costs and benefits* of the EIA system are discussed below.

The European Directive contains no provisions relating to *strategic environmental assessment*. Further consideration was being given to a form of policy, plan and programme assessment by the European Commission in 1994.

The compromises made in the gestation of the Directive are very evident in its final 'minimax' form. At its minimum, it requires that a limited list of projects be subjected to a limited form of EIA. At its maximum, it recommends that a much longer list of projects be subjected to a more universally recognised form of EIA. The Commission no doubt hoped that practice in member states would prove to be well above the minimum required. More realistically, it may also have hoped that, once the benefits of the flexible EIA system had become as apparent as they have in the United States (Council on Environmental Quality, 1990) it would become possible to strengthen it.

Implementation of the Directive

A directive is binding in that it specifies ends which must be achieved, while leaving member states the choice of means. Member states must not only introduce the necessary legal provisions but ensure that they work, i.e. that the ends specified in the Directive are achieved in practice (Haigh, 1991). Several countries (e.g. Belgium, France, Ireland, Luxemburg, The Netherlands) had already introduced some legal requirements for EIA by 1985 but these were insufficient to implement the Directive fully.

There is now some indication of how the European Directive is being implemented in member states (CEC, 1993b). By July 1991, all member states had incorporated some EIA provisions within their own legislation. However, three years after it was required, transposition of the Directive into national legislation was still not complete. (This was still true in 1994.) It was possible to observe considerable achievements in carrying into practice both the letter and the spirit of the Directive, but these varied considerably between member states.

Member states were submitting numerous projects to EIA each year (Table 3.2). However, the total number of projects subjected to EIA, and their composition, differed greatly between member states, even after differences in size and population had been taken into account. Although the Directive is silent on scoping, a number of member states (including Germany) had made provision for scoping in their legislation. Other member states (for example, Ireland) either had some non-mandatory

Table 3.2 Approximate numbers of EIA reports submitted in member states annually, 1988–91

Member state	Number of EIA reports p.a.
Belgium	40
Denmark	5
France	5500
Germany	1000 (Future estimate)
Greece	— (No data)
Ireland	50
Italy	30
Luxemburg	15
Netherlands	70
Portugal	10
Spain	150
United Kingdom	230
TOTAL	7000

Source: Derived from Commission of the European Communities, 1993b, p. 38.

arrangements for scoping or encouraged developers to use this practice. Generally, access was provided to the developer to environmental data in member states, but instances of lack of cooperation and, more seriously, lack of data existed.

While some member states had transposed the requirements of Article 5 and Annex III regarding the content of EIA reports into law, others had confined EIA report content to the minimum requirements specified in Article 5(2) and made inclusion of Annex III requirements fully discretionary. Some good quality EIA reports were being prepared but a substantial proportion of EISs, in many member states, were of an unsatisfactory quality. Access by the public to copies of EIA reports varied from the provision of complimentary copies, through purchase of copies and rights of reference, to instances where considerable persistence was necessary to consult even the non-technical summary. Generally, practice relating to participation and consultation varied from satisfactory to unsatisfactory.

It is clear that, for some projects, both the decisions themselves and the form in which the projects have been approved have been influenced by the EIA process. However, there are also significant numbers of cases where the EIA is considered, as yet, to have had little or no effect. Although the Directive contains no requirements for the monitoring of project impacts, some member states have made specific legal provision for this. While other member states have relied on existing monitoring procedures and practice, there appear to be serious deficiencies in these arrangements in many member states.

A number of member states, like the Commission, have been active in the provision of training to support the implementation of the Directive.

Many have also issued procedural guidance and some have published operational EIA guidance. However, there is a widely perceived need, in many member states, for further guidance on different aspects of EIA.

While it was hard to generalise from the varied experience of EIA both within and between member states, CEC (1993b, p. 59 emphasis added) found that:

> The financial *costs* of carrying out an assessment for an EIS are typically a small fraction of one percent of the capital cost of the project. . . . The overall timescale of implementing projects does not appear to be significantly affected by EIA.

These costs, together with the administrative costs of involvement in the EIA process, appeared to be broadly acceptable to the principal participants.

While mitigation practice varied, it was apparent that amelioration was not yet always taking place, and certainly not at all the various stages of the EIA process. Despite the briefness of the implementation period examined, CEC (1993b, p. 61) reported that:

> there is clear evidence that project modifications have and are taking place, due to the influence of the EIA process. However, there is also evidence that, as yet, its impact is not as widespread as intended and that modifications are mainly confined to those of a minor or non-radical nature. . . .

CEC (1993b, p. 63, emphasis added) cite this as evidence that the 'planning, design and authorisation of projects are beginning to be influenced by the EIA process and that environmental *benefits* are resulting'.

However, it was apparent that, three years after the Directive finally came into effect, EIA practice frequently left much to be desired. A number of measures were necessary before the full realisation of the benefits obtainable from the implementation of the Directive could be achieved. These included better coverage of projects, commencing EIA early in the project design process, improving the quality of EIA reports and reviews, improving consultation and participation, increasing the importance of EIA in decision making and strengthening project monitoring (CEC, 1993b, pp. 61–2). Commission proposals to initiate improvements in the treatment of alternatives, stronger screening, scoping and better international consultation were put forward in 1994. Ironically, if they are enacted, these proposals would restore many of the provisions originally contained within the draft directive published in 1980.

Notes

1. This account is drawn largely from Wood (1990).
2. This section is derived, in large part, from Wood (1988b).
3. This account is drawn, in part, from Wood and McDonic (1989). See also CEC (1993b).

CHAPTER 4

EIA in the United Kingdom

Introduction

The United Kingdom has possessed a land use planning system since 1948 which allows considerable discretion in the consideration of the environmental implications of new development. It is possible for local planning authorities (LPAs) to prepare plans in which environmental policies are emphasised and to refuse development, or to impose conditions to planning permissions, for environmental reasons. LPAs can thus make a powerful contribution to environmental protection by determining the nature and location of new development and redevelopment (Miller and Wood, 1983; Wood, 1989b). They are now being encouraged to use their long-standing powers actively to assist in environmental protection (Her Majesty's Government, 1990) and in achieving the aims of sustainable development (Her Majesty's Government, 1994).

The flexibility of the town and country planning system and the ability of LPAs, and central government following public inquiry, to take environmental factors into account in decision making is the principal reason for the belated acceptance that environmental impact assessment (EIA) could provide a valuable additional safeguard in the United Kingdom. The UK's initial reluctance to accept the imposition of EIA by Brussels and its later implementation of the Directive almost to the letter by integration into existing decision-making procedures can be seen as a reflection of the larger British relationship with Europe, and especially of its attitude to environmental regulation (Wathern, 1989). The United Kingdom has traditionally tended to take an insular view of environmental policy (accentuated under the Thatcher government) assuming that nature (wind and an encircling sea) would provide solutions to environmental problems without the need for rigid regulation. Once European directives have been adopted, however, British governments have prided themselves on their proper implementation.

This chapter describes the early interest in EIA in the United Kingdom, including the Government's response to draft versions of the European Directive on EIA. It then explains the evolution of the various UK regulations made to implement the Directive. The chapter goes on to provide an overview of the UK EIA system and describes briefly how the UK regulations have been implemented in practice.

Early EIA interest in the United Kingdom

There has been a largely bipartisan attitude to EIA in the United Kingdom since the early 1970s.[1] The development of onshore oil facilities in Scotland, where planning authorities were not accustomed to handling complex projects, was one of the factors leading the Scottish Development Department, with the support of the Department of the Environment (DOE), to commission Aberdeen University in 1973 to develop a systematic procedure for planning authorities to make a balanced appraisal of the environmental, economic and social impacts of major industrial developments within the existing statutory planning framework. The authors recommended a procedure relying on a discretionary, flexible and cooperative approach by developers and local planning authorities (Clark *et al.*, 1976). Their report was distributed to every local planning authority in the country, free of charge, and 'commended for use by planning authorities, government agencies and developers'.

Concurrently, a further study was commissioned by DOE to consider the administrative, rather than the methodological, aspects of EIA. Catlow and Thirwall (1976) recommended legislative changes to ensure the formal integration of EIA within the existing UK planning system. DOE delayed the publication of this report by a year, eventually distributed it to a much smaller audience than the previous study, and expressed grave reservations: 'any new procedures involving additional calls on central or local government finance and manpower are unacceptable ...' (DOE, 1977).

Central Government (a Labour administration until 1979) was very careful not to commit itself to EIA, despite its positive reception of the Aberdeen report. In its response to the Dobry (1975) report on the development control system (in which it was recommended that, in the case of specially significant development proposals, the LPA should be able to insist that an environmental impact study be submitted) DOE was neutral. The Government stated that environmental impact analyses might have a part to play in assisting in the consideration of major applications and that the Secretary of State would study Catlow and Thirwall's proposals (DOE, 1975). It made no response to the Royal Commission on Environmental Pollution's (1976) endorsement of Dobry's proposal that 'developers provide an assessment of the effects of air, water, wastes and noise pollution of certain major developments'.

Eventually, after much deliberation, the Government guardedly announced that it favoured the limited use of EIA:

We should therefore wish to encourage use of this [Catlow/Thirwall] approach where its use is worthwhile in the circumstances ... [and where it will] improve the practice in handling these relatively few large and significant proposals.

(Department of the Environment, 1978)

There was no hint, however, of implementing any legislative changes to encourage the use of EIA. The incoming Conservative administration's attitude to EIA did not differ markedly from its predecessor's. It did not actively discourage the growing acceptance of the value of EIA by local planning authorities, but it was most reluctant to introduce new procedures or statutory requirements into the existing planning system it had criticised for causing unnecessary delay.

As the various drafts of the European Commission's directive on EIA (Chapter 3) began to appear, the Government continued to endorse the principle that EIA was a useful element in the planning process for considering large and significant proposals while rejecting the idea of any mandatory EIA system. The Government's reception of the updated Aberdeen assessment manual which it had commissioned contrasted with that of its predecessor to the original document. While the manual did not require legislative change and was deemed 'of value', ... 'it is important that the approach suggested in the report should be used selectively to fit the circumstances of the proposed development and with due economy' (Clark *et al.*, 1981, p. v). Local authorities were left to purchase the document themselves, though some had been given drafts of the manual earlier.

The European Commission's draft directive was deliberated by a Select Committee of the House of Lords which heard a great deal of evidence from both supporters and antagonists of the proposed directive. The Committee expressed some reservations on points of detail but, in an important report which summarised the various views on EIA, came down firmly in its favour:

The present draft directive strikes the right kind of balance: it provides a framework of common administrative practices which will allow Member States with effective planning controls to continue with their system, possibly with some modifications of detail, while containing enough detail to ensure that the intention of the draft cannot be evaded ...

(House of Lords, 1981a, pp. xxv–xxvi)

The Royal Town Planning Institute, among other bodies, responded positively to the House of Lords Select Committee's report in its submission to members of that House:

The Royal Town Planning Institute welcomes the Report of the Select Committee and is appreciative of its thorough investigation of the subject. The Institute is delighted that many of the points it made in evidence to the committee have received support in the report and urges the House in its debate to commend the report to the Government and to press the Government actively to encourage the use of environmental assessment as soon as possible.

(Royal Town Planning Institute, 1981)

This report was debated in the House of Lords and received widespread support from those peers who spoke.

The lone voice of dissent was that of the Minister for the Environment. He felt that assessments of the potential environmental effects of major development projects should be carried out but that this was inherent in the existing planning system. He did not rule out the notion of a future directive on environmental assessment proving satisfactory but that now proposed was certainly not acceptable. 'The Government do not believe that the present draft directive yet gets it right. As a first step in [a] new field we consider it to be over-ambitious and likely to fail in its intention' (House of Lords, 1981b). It is unusual for the Government to reject the recommendations of a well-informed Select Committee but it took the position that the European requirements for environmental impact assessment might duplicate or complicate the existing town and country planning procedures which it was striving to simplify. Meanwhile, considerable experience of the use of non-mandatory EIA was being gained in the United Kingdom (Petts and Hills, 1982).

As a result of the period of intensive negotiation in Brussels which ensued, the major concessions limiting both the coverage of an EIA report and the range of projects to be subject to mandatory EIA were felt to reduce the ramifications of EIA to the point where the British minister could accept the Directive at a Council of Ministers' meeting. It seems probable that the House of Lords committee report was one factor influencing this decision. Another factor seems to have been the view that the British Government need not extend the coverage of EIA to Annex II projects (Wood, 1988b; Macrory, 1989; Sheate and Macrory, 1989; Wathern 1989).

Evolution of the UK regulations

The Department of the Environment set up a working party in 1984 with members drawn from industry, local government, the planning profession, environmental groups and other government departments to consider the implementation of the forthcoming Directive, preferably within existing procedures, now that Britain had withdrawn its objections.[2] The working party made a number of recommendations to implement the directive largely within the existing planning system with minimum procedural complexity. Suggestions by the town planners on the working party for extension of EIA to Annex II projects and for formal pre-submission consultations (scoping), to take place over a prescribed period, were not accepted but are reflected in the advice contained in the DOE circular on EIA (below). A consultation paper reflecting these recommendations was circulated in 1986, accompanied by a draft advisory booklet which the working party had drawn up.

The consultation paper excluded Annex II projects, except on the direction of the Minister, from mandatory EIA and stated that LPAs should not demand an EIA for such developments (DOE, 1986a). However, the paper sanctioned, and even encouraged, the use of voluntary EIA by authorities and developers. It thus reflected government policy since the early 1970s, though it was expected that there would now be perhaps half-a-dozen mandatory Annex I project EIAs per annum.

The draft advisory booklet indicated, perhaps for the first time, a positive attitude by Government towards EIA. The EIA report was not to be a minimalist document:

> There is no prescribed form of assessment report, provided that the requirements of Article 3 and Annex III of the Directive are met. The aim should be to provide as objective a statement of the environmental effects of the project as possible.
>
> (Department of the Environment, 1986b)

The working party suggested a checklist of factors, some or all of which might need to be included in an EIA. This consisted of three parts: information describing the project; information describing the site and its environment; and an assessment of effects. This was to cover effects on land, on water, on air and climate, on humans and artefacts and on flora and fauna, together with indirect and secondary effects, risks of accident, mitigatory measures and the effects of the interrelationship of these factors. Particular emphasis was placed on the need for consultation between all concerned at an early stage.

Following the publication of the consultation paper, legal advisers at DOE and at the Commission of the European Communities (which responded in a well-publicised letter) independently indicated that the exclusion of Annex II projects could not be justified in law. The British acceptance of the Directive may therefore have been based on a misconception and DOE officials had to return to the drafting board. They did not recall the members of the working party but took informal advice in producing a further consultation paper in early 1988, which represented a much less reluctant approach to implementing the Directive. This draft extended to Annex II projects 'likely to have significant effects on the environment' and advanced indicative criteria to help determine which Annex II projects were to be subject to EIA (DOE, 1988a). The expectation was that the number of these would be no more than a few dozen a year. The emphasis on voluntary EIA had disappeared.

The Department of the Environment officials publicised the proposals energetically and more than 100 responses were received. Many of these focused on the indicative criteria but others pointed out problems associated with proposals relating to conformity with land use plans, to the proposed screening appeal to the Minister, to resource implications, to the availability of environmental statements and ministerial decisions and to the lack of a check on adequacy. Other problems connected with procedures in enterprise zones and simplified planning zones and with arrangements for 'voluntary' environmental statements were identified.

Many of these problems were addressed in the regulations and circular published on 15 July 1988 (Wood and McDonic, 1989).

The UK EIA system

For projects requiring planning permission, the Directive was given legal effect in England and Wales through the Town and Country Planning (Assessment of Environmental Effects) Regulations 1988, in Scotland through the Environmental Assessment (Scotland) Regulations 1988 and in Northern Ireland through the Planning (Assessment of Environmental Effects) Regulations (Northern Ireland) 1989.[3] The environmental assessment (EA) regulations apply to two separate lists of projects, based on Annexes I and II of the Directive. It is significant that DOE adopted the term 'environmental assessment', rather than the US 'environmental impact assessment', given its earlier opposition to a formal EIA system. Whether it took the term from the name of the US preliminary (screening) document (Chapter 2) or from the Canadian name for EIA (Chapter 5) is a matter of conjecture.

Advice on procedures and the implementation of the EA Planning Regulations in England and Wales is presented in DOE Circular 15/88 (Welsh Office 23/88) (DOE, 1988b) and in a *Guide to the Procedures* (DOE, 1989). An equivalent circular applies in Scotland (Scottish Development Department 13/88). These circulars set out indicative criteria and thresholds to help determine whether certain projects (Annex II projects) should be subject to EA.

The Planning Regulations contain provisions for LPAs to give a formal 'opinion' that EA is required where they are requested to do so by developers. They may also notify developers that EA is required where a planning application is submitted without an environmental statement (ES). In either event, the Regulations permit the developer to ask the relevant Secretary of State for a 'direction' that EA is, or is not, required. Certain statutory consultees (including Her Majesty's Inspectorate of Pollution) are required to provide the developer with information should it be requested. The Regulations also set down the nature of prescribed consultation and publication arrangements and extend the amount of time available to LPAs to reach a decision on planning applications involving EA.

The Planning Regulations implement the provisions of the European Directive almost to the letter, though there is some disagreement about whether the content requirements for environmental statements specified by the Regulations accurately reflect the Directive. The three annexes to the Directive become Schedules 1, 2 and 3. Several changes to these are made. In particular, certain projects are excluded from Schedules 1 and 2 where they are covered by other regulations (notably nuclear power stations). Further, Schedule 3 contains a list of the mandatory information requirements (Paragraph 2 defines 'specified information' or the information required by Article 5 of the Directive), together with a list of the information set down in Annex III of the Directive which may be provided 'by way of explanation or amplification' (paras 3 and 4). An 'environmental statement' is defined by reference to Schedule 3 and 'environmental information' consists of this statement, together with the representations of consultees and members of the public about the impacts of the developments (Wood and McDonic, 1989).

An outline of the main steps in the EA process for planning decisions is shown in Fig. 4.1. There are special provisions relating to developments in enterprise zones and simplified planning zones (DOE, 1988c).

The Planning and Compensation Act 1991 contains a section which enables the Secretary of State for the Environment to require environ-

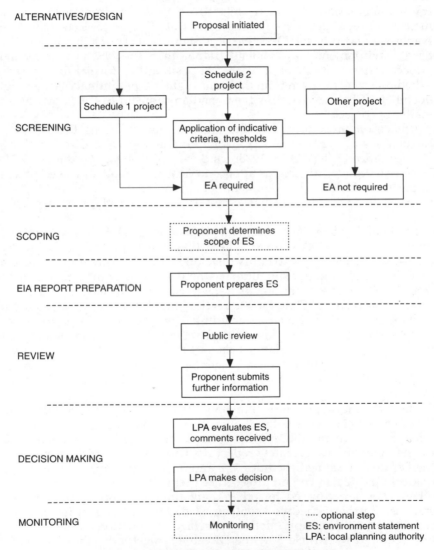

Fig. 4.1 Main steps in the EA process for UK planning decisions.

mental assessment of planning projects other than those listed in the European Directive. The Planning Regulations have now been extended to cover the EIA of private motorways, motorway service areas, wind generators and coast protection works (DOE, 1994a).

The Circular (DOE, 1988b) and the Guide (DOE, 1989) together provide detailed guidance on the operation of the procedures. The indicative criteria and thresholds to be used by LPAs in reaching a judgement about whether EA is to be required for Schedule 2 projects are contained in the advisory Circular, and not in the Regulations. The criteria (and the other advice contained in the Circular) can therefore be changed relatively easily and, in any event, do not have regulatory force. However, the Regulations provide a right of appeal against an LPA determination that EA is required.

Normal town planning appeal provisions against the non-determination of planning applications (and against planning decisions) apply and the Secretaries of State can call applications in for determination by central government. There is therefore relatively little discretion left to LPAs in determining applications to which the Regulations apply.

Far more projects require environmental assessment under the provisions of the Town and Country Planning Regulations than under those of all the other regulations combined (below). These other regulations were necessary because certain types of projects listed in Annex I and Annex II of the European Directive (Chapter 3) are authorised outside the British planning system. Further details about the various stages of the UK EA system for planning decisions are contained in Chapters 6–20.

The arrangements relating to other regulations are broadly similar (DOE, 1989; Jones *et al.*, 1991; Glasson *et al.*, 1994; Sheate, 1994). The Highways (Assessment of Environmental Effects) Regulations 1988 require the Secretary of State for Transport (in England) to publish an ES for the preferred route at the time when draft orders are published. The Electricity and Pipeline Works (Assessment of Environmental Effects) Regulations 1990 require an ES to be submitted to the Secretary of State for Energy (in England) for the construction of nuclear power stations, for some other power stations and for certain overhead power lines and for many pipelines.

The Environmental Assessment (Afforestation) Regulations 1988 require the EA of afforestation projects prior to the provision of a grant in any case where, in the opinion of the Forestry Commission, the project is likely to have significant environmental effects. If no grant is applied for, there is no provision for EA. Equally, there is no requirement for the EA of the Forestry Commission's own planting projects. The United Kingdom did not include the Directive's Annex II rural land holdings restructuring projects and projects for the agricultural use of uncultivated land in any regulations, because the Government believed they were unlikely to occur in the United Kingdom in a form that would have significant environmental effects and hence require EA (Sheate and Macrory, 1989; Commission of the European Communities, 1993b). However, proposals were underway in 1994 to extend EA to these types of projects, to meet more fully the requirements of the Directive.

Improvements to existing land drainage works undertaken by drainage bodies and the National Rivers Authority do not require an express grant of planning permission but fall under the Land Drainage Improvement Works (Assessment of Environmental Effects) Regulations 1988. These require the drainage body to consider whether or not the proposed works would be likely to have significant environmental effects and ought therefore to be the subject of an environmental statement.

There are two sets of regulations relating to ports and harbours, to reflect the existing authorisation procedures. Both the Harbour Works (Assessment of Environmental Effects) Regulations 1988 and the Harbour Works (Assessment of Environmental Effects) (No. 2) Regulations 1989 empower the Minister of Agriculture, Fisheries and Food or the Secretary of State for Transport (in England) to decide whether EA is needed.

Offshore salmon farming facilities within 2 km of the coast require a lease from the Crown Estate Commissioners. The Environmental Assessment (Salmon Farming in Marine Waters) Regulations 1988 require the Commissioners to consider an ES provided by the developer before granting a lease in circumstances where the development may have significant effects on the environment. Dredging for minerals offshore requires a dredging licence from the Crown Estate Commissioners. In cases where dredging is likely to have significant environmental effects, the applicant is required to provide an ES.

The arrangements for non-planning projects in Wales are very similar to those in England, save that the Secretary of State for Wales is the responsible minister. The arrangements for Scotland are broadly similar to those for England and Wales, though there are some differences due to the different legal and administrative arrangements which apply in Scotland. Separate provision for EA has been made in Northern Ireland, which has its own legal and administrative procedures. However, the general principles of the EA system covering the rest of the United Kingdom apply. Arrangements for the provision of ESs with private and hybrid Parliamentary bills have also been made in the United Kingdom (Wood and McDonic, 1989; DOE, 1989; Jones *et al.*, 1991; Glasson *et al.*, 1994; Sheate, 1994).

Not only has the Commission of the European Communities (1993b) pursued the compatibility of UK rule-making with the requirements of the Directive with some vigour but it has publicised a letter sent under the provisions of Article 169 of the Treaty of Rome querying whether seven projects, including the Channel Tunnel Rail Link, ought to proceed without EIA (Salter, 1992a, b). This European intervention caused a considerable political storm, since the United Kingdom prides itself on the quality of its implementation of European directives (above). This misunderstanding was the only occasion to date that the subject of EIA and its integration within British decision-making procedures has reached prominence in the UK media.

Implementation of the regulations

Of over 1300 ESs prepared between 1988 and the end of 1993, over 1000 were produced under the planning regulations for England and Wales, for Scotland or for Northern Ireland. Less than 10 per cent of ESs related to projects falling within Annex I of the European Directive (Chapter 3). Most ESs related to Annex II infrastructure, 'other' and extractive industry projects (Fig. 4.2).

It should be observed that, unlike in the United States (Mandelker, 1993b), there have been very few court cases relating to EA to date. On this criterion, the introduction of EA into the UK land use planning system has clearly been successfully implemented. There is evidence of very considerable variations in practice in implementing the Planning Regulations (Wood and Jones, 1991, 1992). These are described in Chapters 6–20. There is less evidence in relation to the implementation of the other types of regulation but this conclusion is believed to hold equally true (Coles, 1991).

In relation to the planning system, DOE has committed itself to publish more advice both on the preparation of ESs[4] and on reviewing and dealing with ESs[5] (Her Majesty's Government, 1990). Other guidance has been published by the Department of Transport (1983, 1993), Cheshire County Council (1989), Fortlage (1990), the Countryside Commission

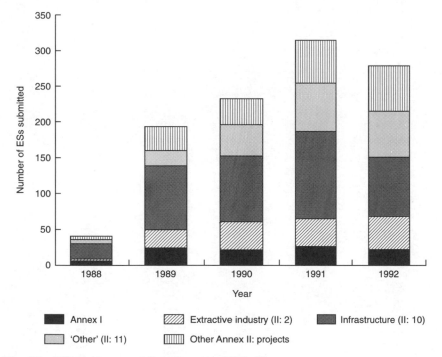

Fig. 4.2 UK environmental statements, 1988–92.
Source: University of Manchester EIA Centre statistics.

(Stiles *et al.*, 1991), Kent County Council (1991), the Passenger Transport Executive Group (1991), Essex County Council (1994) and Salford City Council (1994). Recent commentaries have been published by Glasson *et al.* (1994) and Sheate (1994). Various EA training activities have also been undertaken (Wood and Lee, 1991), some of which have been supported by speakers from DOE.

Notes

1. This account is derived largely from Wood (1982).
2. This account is derived largely from Wood and McDonic (1989). See also Macrory (1989) and Wathern (1989).
3. This account is derived partially from Jones *et al.* (1991).
4. A draft of this advice was released for consultation in 1994 (DOE, 1994b).
5. Addendum: Both the advice (Department of the Environment, 1994, *Evaluation of Environmental Information for Planning Projects – a Good Practice Guide*. HMSO, London) and the research findings (Department of the Environment, 1994, *Good Practice on the Evaluation of Environmental Information for Planning Projects*. Research Report. HMSO, London) were published late in 1994.

CHAPTER 5

EIA in The Netherlands, Canada, the Commonwealth of Australia and New Zealand

Introduction

This chapter provides an overview of the EIA systems in The Netherlands, Canada, the Commonwealth of Australia and New Zealand. In each case a brief account of the origins and development of the EIA system is presented. This is followed by a summary of the legislative provisions for EIA and an outline of the main procedural steps in the EIA process. The principal organisations responsible for EIA are also mentioned. The accounts of the EIA systems in this chapter are intended to provide an introduction to the discussion of the different stages or aspects of the EIA processes presented in Chapters 6–20.

The Netherlands EIA system

The Netherlands has a population of 15 million people, not much smaller than that of Australia, and an area of 41,000 km^2 (about 0.5 per cent of the size of Australia or the contiguous United States of America). The country is not only densely populated but low lying: a significant proportion of its land area has been reclaimed from the sea and is protected by dikes. It is not surprising, therefore, that there should be a deep national concern for the environment. This has manifested itself in a strong land use planning system and in a very high and long-standing level of public interest in environmental protection. The existence of a proportional representation system has led to the translation of this public concern into political interest and action.

The Dutch EIA (or MER – milieu-effectrapportage) system,[1] which owes its parentage more to Canada than to the United States, was very carefully considered before implementation. North American EIA specialists were employed by the Government to assist in this process (Jones, 1984). The Netherlands commissioned a series of research studies in the late 1970s, which resulted in proposals for the introduction of a formal EIA system. This work and the resulting proposals were largely independent of the European Commission's initiative on EIA, despite taking a similar time to come to fruition. Nine 'trial run' EIAs were undertaken to ensure that the procedures envisaged were practicable. A 'governmental standpoint' was submitted to the Dutch Parliament in 1979 and a bill was introduced in 1981. Twelve EIAs were initiated using the prospective legislation and an 'interim' EIA Commission was set up.

By the time the EIA provisions were incorporated into the amendments to the Environmental Protection (General Provisions) Act made in 1986, which came into effect in 1987, considerable experience of EIA had been gained (Ministry of Housing, Physical Planning and the Environment and Ministry of Agriculture, Nature Management and Fisheries, 1991). These provisions were consolidated, unamended, in the Environmental Management Act 1993 (abbreviated to Wm in Dutch) and this in turn has been replaced by the Environmental Management Act 1994. They are supported by the Environmental Impact Assessment Decree 1987 (No. 278, 20 May 1987)[2] which was amended slightly in 1992 and again in 1994, and by the Environmental Impact Assessment Notification of Intent Decree also dated 1987 (slightly amended in 1993). Provinces may pass their own EIA legislation but, by mid-1994, this had happened in only three cases.

The main features of the Dutch EIA system are shown in simplified form in Fig. 5.1 (see also Glasson *et al.*, 1994). Several features of what is generally acknowledged to be the strongest European EIA system should be noted:

- The EIA process is integrated into existing decision-making procedures.
- The EIA process is not confined to projects.
- There are statutory requirements relating to the treatment of alternatives, to scoping (including the preparation of project-specific guidelines), to the review of EIA reports and to the monitoring of the impacts of implemented projects.
- There are provisions for public participation at both the scoping and EIA report review stage and there is a third party right of appeal against decisions.
- The Dutch EIA Commission plays a central and very influential role in the EIA process generally and at the scoping and EIA report review stages in particular.

The EIA Commission (Commissie voor de milieu-effectrapportage) is an independent body able to call upon about 200 appointed members (experts) to provide guidance. The involvement of this body is required by law and the EIA Commission's opinions are made public (Verheem, 1991). The EIA Commission has a full-time secretariat of about 30 and furnishes part-time chairmen to guide small working parties of two to seven Commission members with expertise relevant to the activities assessed. Other experts may be co-opted where necessary. The working parties advise on the guideline recommendations prepared and assess the accuracy of the scientific content of the EIA report and its completeness in relation to the statutory requirements and the guidelines.

The Dutch Government is committed by law to submit a report to

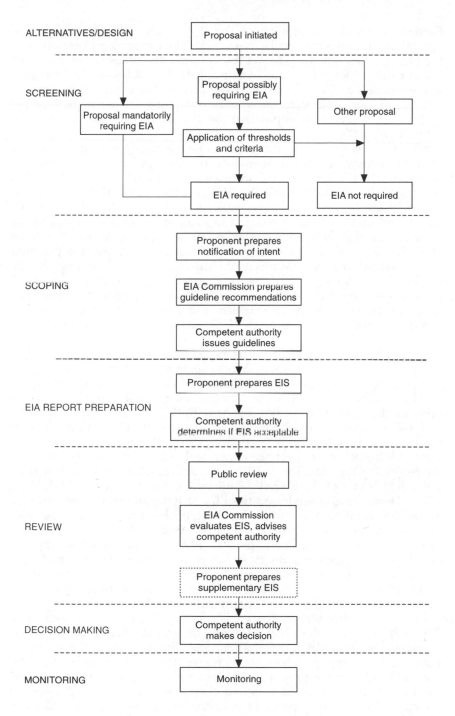

Fig. 5.1 Main steps in The Netherlands EIA process.

Parliament on the functioning of the EIA system after three years of operation (and thereafter every five years). The Evaluation Committee on the EIA provisions in the 1986 Act (ECW) consisted of four academics and a secretariat of four. Its report was published on schedule in 1990 and, while giving the EIA system a generally favourable review, it suggested a number of amendments to the system, including (at the request of the Government) several to implement the European Directive more faithfully (ECW, 1990).

Somewhat surprisingly, the original Dutch legal provisions did not fully implement the requirements of the European Directive on EIA (Chapter 3). The European Commission formally noted deficiencies relating to the failure to ensure the EIA of certain Annex I projects and to the exclusion of certain Annex II projects in 1990. Further, the Commission noted shortcomings in relation to exemption provisions and to the consultation of neighbouring countries. These deficiencies appear to have been due to misunderstanding and the Dutch Government committed itself to remedying them (Ministry of Housing, Physical Planning and the Environment, 1991) and has now done so. Changes relating to Annex I projects were made in the 1992 Decree and the remainder were due to be made in 1994. These latter amendments were designed to meet the comments both of the European Commission and of the Evaluation Committee on EIA.

A considerable body of literature on EIA has been established in The Netherlands. A comprehensive *Manual on Environmental Impact Assessment* has been produced by the Ministry of Housing, Physical Planning and Environment (VROM) and a new one was published in 1994. Various other guides, including one on scoping (VROM, 1981), one on alternatives and two on post-auditing have been released. In addition, a series of guides on the application of methods of impact forecasting, covering a general introduction, air, surface water, soil, plants/animals/ecosystems, landscape, noise, radiation, health and multi-factor methods has been published (some of these are in English: see for example VROM, 1984, 1985). There are also guides on the EIA of types of certain projects (e.g. mining). Several training courses on EIA have been run by the authorities involved in EIA and by universities (de Boer, 1991). There is now a growing community of perhaps 300 EIA professionals working for developers, competent authorities, ministries, consultancies and universities. While there remain some difficulties in the operation of the Dutch EIA system, experience, knowledge and skills are still increasing (Commission of the European Communities, 1993c).

The level of EIA activity has grown substantially over the years (see Fig. 5.2). In particular, the numbers of guideline recommendations prepared by the EIA Commission, and the number of reviews of environmental impact statements increased year by year until 1993 after which the number of projects for which EIA procedures were initiated levelled off. However, decisions had been made on only 116 projects by the end of 1993, with suspensions or cancellations taking place in 58 cases (EIA Commission, 1994). These cancellations usually result from non-environmental factors.

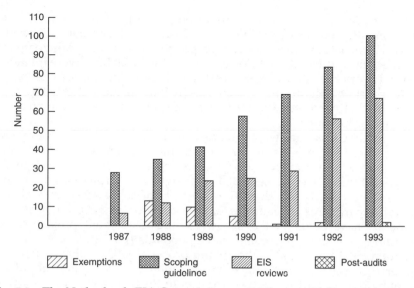

Fig. 5.2   The Netherlands EIA Commission exemptions, guidelines, reviews and audits 1987–93.

Source: Environmental Impact Assessment Commission (1994, p. 13).

The Canadian EIA system

Canada is a vast country, the second largest in the world, covering nearly 10 million km^2, with a population of 27 million, most of whom live in a band about 150 km wide immediately north of the US border. The population density of Canada, less than one-tenth of the United States, is thus somewhat misleading, since much of the mineral-rich north is very sparsely populated. Canada's fragile northern environment has sometimes fallen victim to the frontier ethic in the form of ill-considered logging, mining or so-called 'mega-projects' such as huge hydro-electric power stations. More recently, many of the multi-ethnic immigrant population of this outdoor country have joined the indigenous peoples (Inuit and Indian) in their traditional concern about the environment, resulting in environmental issues being placed consistently high on the political agenda (Rees, 1987). Inevitably, however, this marked environmental concern has been somewhat tempered by economic realities as the pendulum of recession has swung.

It was inevitable that interest in the National Environmental Policy Act should spill over the border from the United States to Canada.[3] A group of officials, individuals in the business community, and citizens' groups began to see the potential of EIA and the federal Environmental Assessment and Review Process (EARP) was established by a Cabinet decision on 20 December 1973. This was amended by a second decision in 1977 and the responsibility of the federal Minister of the Environment for the environmental assessment (EA, not EIA) of federal projects, programmes and activities was reaffirmed in the Government Organisation Act 1979.

In 1984 these provisions were formalised in the Environmental Assessment and Review Process Guidelines Order (Government of Canada, 1984) which clarified the roles and responsibilities of the participants in the EARP procedures (Couch, 1988). The EARP Guidelines were intended to be advisory but, following successful and highly publicised challenges in the courts by environmental groups in 1989 in the Rafferty/Almeda and Oldman River cases, the 1984 Order-in-Council was held to be a law of general application.

The EARP Guidelines were intended to ensure that the environmental consequences of proposals for which the federal government has decision-making authority, including projects to which it makes a financial contribution, were assessed. EARP, which is principally a project planning process, is based on the principle of self-assessment. All the federal departments or agencies having decision-making authority for proposals have developed their own initial assessment procedures to determine whether proposals have potentially significant adverse environmental impacts. If so, or if the proposal is the subject of public concern because of its potential environmental effects, the minister of the initiating department refers the proposal to the Minister of the Environment for a review by an environmental assessment panel (Couch, 1988; Glasson *et al.*, 1994).

EARP is administered by the Federal Environmental Assessment Review Office (FEARO), the Executive Chairman of which reports directly to the federal Minister of the Environment. FEARO, which now has a staff of over 90 and a budget of over $CD11 million, is responsible for providing policy direction and procedural information on EARP and for periodic reports on the implementation of EARP. It also provides the substantial secretariat for the reviews conducted by environmental assessment panels. These panels consist of three to seven independent members appointed by the Minister of the Environment for each referred proposal. Members are selected for their 'objectivity, public credibility, and special knowledge of factors associated with the proposed undertaking' (Couch, 1988, p. 13). The panels are chaired by a nominee of the Executive Chairman of FEARO and discourage legal representation. There were 40 completed panel reviews by mid-1993 (with 13 actively ongoing and another 6 being inactive), an average of just over 2 per year since 1974. The number of initial assessments was, however, much greater and the number actually reported to FEARO is widely viewed as an underestimate. Having started very slowly, the reported number of initial assessments was running at about 10,000 p.a. (with the number of initial environmental evaluations at perhaps 25–40 p.a. in 1992) (FEARO, 1992a).

FEARO is also responsible for research on EA in Canada. Between 1984 and 1992 research was administered through a grant to the Canadian Environmental Assessment Research Council (CEARC). The objective was to promote research to improve the scientific, technical and procedural basis for EA (Couch, 1988) and to advance the theory and practice of EA. This much admired independent body sponsored research and commissioned reviews, studies and workshops. It produced about 150 reports, many of which have been widely disseminated (see, for example, Jacobs and Sadler, 1990). It was disbanded in 1992 largely as a result of public expenditure cuts. FEARO continues to have a substantial research budget, however, much of which is directed to studies related to the operationalisation of policy initiatives such as the Canadian Environmental Assessment Act (below).

Most of the land north of 60° latitude (the northern boundary of most of the provinces) is under federal jurisdiction and some is subject to land claims settlements with aboriginal peoples. Special EA provisions apply within these settlement areas and within the James Bay and Northern Quebec Agreement area – the site of the Great Whale 10,000 MW hydroelectric mega project. In addition, each of the ten provinces and two territories has its own EA legislation. About a dozen municipalities, including the City of Ottawa, have also instituted their own formal or informal EA requirements or are putting them into place (Lawrence, 1994).

Smith (1990, 1993, pp. 40–1) conducted a comparative evaluation of the environmental assessment provisions in the ten provinces of Canada. He reported that Saskatchewan, Newfoundland and Quebec had the strongest systems, that the frequently cited Ontario system was 'acceptable but restricted' and that the federal EA system compared poorly with most provincial systems. Neither his views nor his evaluation framework are universally accepted, but Gibson (1993) agreed that the federal system was less satisfactory than that in Ontario.

Environmental assessment is a high profile process in Canada, partly because its application provides one of the most visible manifestations of the government's commitment to the environment and partly because it often provides the best available opportunity for public participation in environmental decision making. (Much of Canada possesses little in the way of enforceable land use planning outside the more populated areas (Fenge and Rees, 1987).) Further, EA has been the subject of significant jurisdictional conflicts between federal and provincial governments, though joint EA arrangements are supposed to apply (and have often worked satisfactorily). EA is a major employer in Canada: the Canadian 'EA world' probably consists of over 2000 people.

The need for EARP reform had been widely recognised for some time (FEARO, 1988b) and the court cases emphasised the urgency of change and raised EA decision making to public prominence. Accordingly, in fulfilment of a 1984 manifesto commitment (Fenge and Smith, 1986), the federal government introduced the Canadian Environmental Assessment Bill in 1990. After much public debate and substantial revision, the Bill was given Royal Assent in June 1992. Four key regulations required for the operation of the Canadian Environmental Assessment Act (CEAA) were published for comment in late 1993 and they came into effect when the Act was proclaimed in force early in 1995.[4]

The CEAA is intended to entrench in law the federal government's obligation to integrate environmental considerations in all its decisions relating to projects (but not to policies, plans and programmes to which EARP in principle applied). The Act is part of a package intended not

61

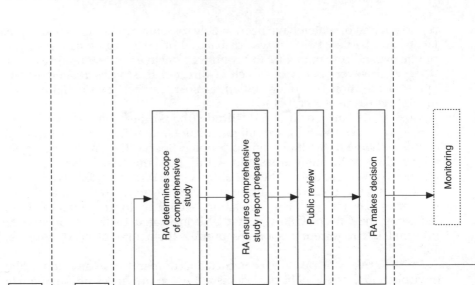
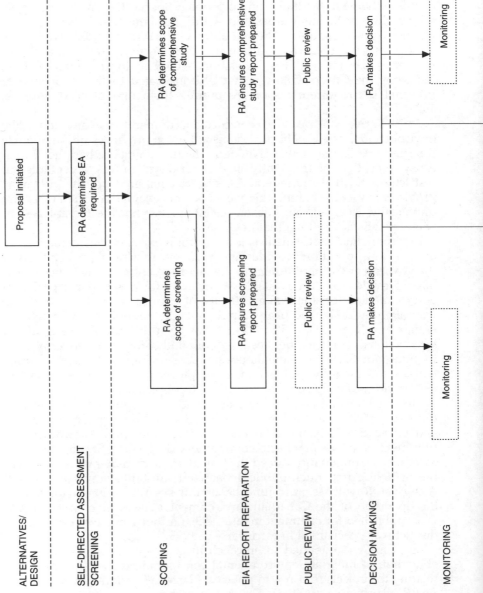

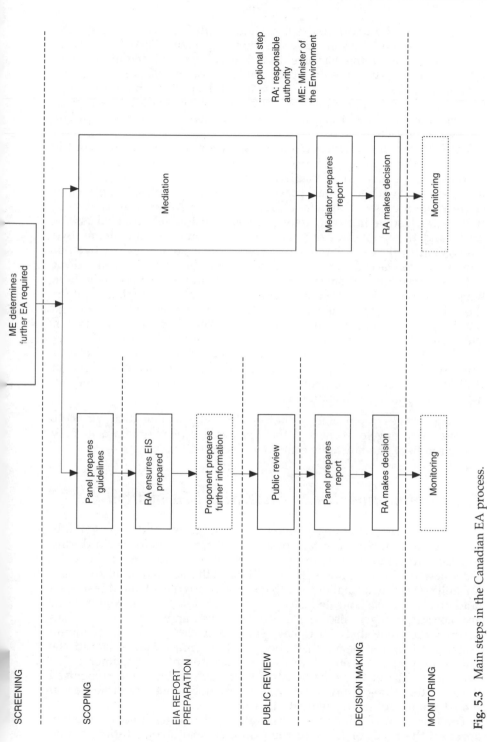

Fig. 5.3 Main steps in the Canadian EA process.

63

only to reduce the uncertainties associated with EARP but to make the EA process more 'efficient, effective, fair and open' (FEARO, 1991). A second element of the package is the non-legislated requirement to release a summary of the environmental assessment of every proposal submitted to Cabinet at the time the Cabinet decision is announced. The final element of the reform package is a participant funding programme to help individuals and organisations to involve themselves in public reviews of projects.

The EA process under CEAA bears many similarities to EARP but is much less discretionary. FEARO has been replaced by a new, more autonomous agency: the Canadian Environmental Assessment Agency (CEA Agency) with additional powers over the EA process. The principal steps in the process are shown in Fig. 5.3. The CEAA system consists of two separate, sometimes successive, procedures: the 'self-directed assessment' and the 'public review'. Each of these procedures contains two assessment tracks each with its own steps. The system is exhaustive but allows for the vast majority of federally controlled projects to be dealt with as 'screenings'. A small (but increased) number of projects will be subject to 'comprehensive study' (which can be seen as a replacement for the EARP initial environmental evaluation) and panel review and thus require an environmental impact statement. Notable positive features of the Act are the provisions relating to cumulative environmental effects, to the use of mediation,[5] and to follow-up provisions. One of the four purposes of the Act, along with consideration of environmental effects in decision making, consideration of transboundary effects and the encouragement of public participation is:

> to encourage responsible authorities to take actions that promote sustainable development and thereby achieve or maintain a healthy environment and a healthy economy (Section 4(b)).

The Commonwealth of Australia EIA system

Australia, as well as taking many of its immigrants and an essentially British system of parliamentary government from the old world during the last two centuries has, in more recent years, taken much from the older new world: the United States. Although the population of Australia (17 million) is much smaller than that of the contiguous United States, its land area is similar and its history of colonisation and its frontier ethic are, not surprisingly, also similar. By the 1960s and early 1970s, it had become apparent that Australia's state-controlled land use planning systems, though stronger than those in the United States, lacked the power to prevent some of the worst excesses of development. The Commonwealth of Australia (Australia's federal government) recognised the need for stronger environmental control and looked with interest at the advantages of the US federal EIA process.[6]

The Commonwealth and the six states and two territories could not agree on the need for EIA and the Commonwealth constitution, dating from 1901, made no mention of the environment. While many of the states (each of which has its own government) resented Commonwealth interference in their right to control their own environments, others were more enthusiastic and some states proceeded with their own informal EIA provisions. Despite several meetings of the environment ministers forming the (then) Australian Environment Council, the Commonwealth eventually recognised that uniformity was unattainable (Anderson, 1990), and passed its own legislation. The Environment Protection (Impact of Proposals) Act became law in 1974, some 14 years before the European Commission's Directive on EIA came into effect. Like the US National Environmental Policy Act (NEPA), this Act related only to federal activities (and not to those undertaken by state or local govern-ments) and was independent of existing procedures. There was, however, quite deliberately far more discretion in the Australian Commonwealth system and, consequently, far less opportunity for resort to the courts than in the United States.

More recently, there has been widespread recognition by the states of the need for environmental controls and for their harmonisation in the interests of trade and competition in Australia (Australian and New Zealand Environment and Conservation Council – ANZECC – 1991). As a result, the states have passed legislation or set procedures in place to extend the scope of EIA to their own activities. As Anderson (1990) has remarked:

> It is perhaps not going too far to say that the Impact of Proposals Act established the framework within which all subsequent EIA processes in Australia have been developed.

Some of these EIA systems (between which the similarities are more apparent than the differences) are stronger and more effective than their Commonwealth precursor (Porter, 1985; Formby, 1987; ANZECC, 1991).

The Commonwealth does not formally oversee the EIA activities of the states. There has been some duplication between Commonwealth and state EIA requirements in the past but joint working on a single assessment or Commonwealth acceptance of state EIA procedures is now usual. There have been various reviews of, and amendments to, the Commonwealth EIA system over its 18 year life and the EIA function of the Commonwealth has now been re-positioned within a new Com-monwealth Environment Protection Agency (EPA) which remains a part of the Department of the Environment, Sport and Territories.[7]

The main steps in the Commonwealth EIA system are summarised in Fig. 5.4. The Act contains permissive powers for screening, for scoping, for the preparation of draft and final environmental impact statements, for consultation and participation, for assessment of the final EIS, for this assessment to be taken into account in decision making, for the provision of reasons for decisions and for the holding of inquiries. There also exists a discretionary power to require monitoring of approved actions. EPA has a staff of about 30 to implement the requirements of the 1974 Act, which applies only to a limited number of proposals in which the

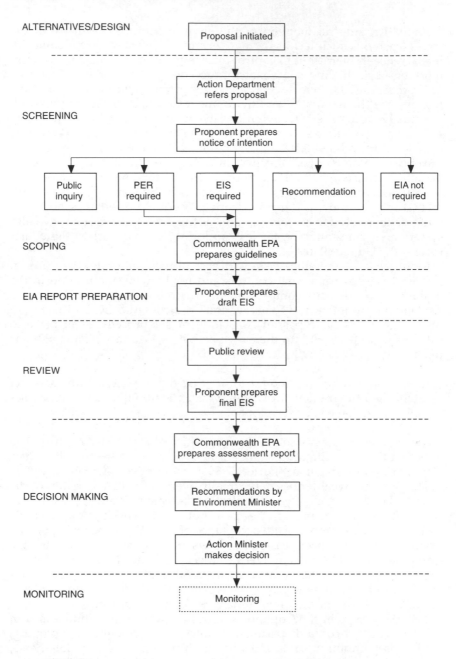

ALTERNATIVES/DESIGN — Proposal initiated

Action Department refers proposal

SCREENING — Proponent prepares notice of intention

Public inquiry | PER required | EIS required | Recommendation | EIA not required

SCOPING — Commonwealth EPA prepares guidelines

EIA REPORT PREPARATION — Proponent prepares draft EIS

Public review

REVIEW — Proponent prepares final EIS

Commonwealth EPA prepares assessment report

DECISION MAKING — Recommendations by Environment Minister

Action Minister makes decision

MONITORING — Monitoring

···· optional step
PER: public environment report
EPA: Environment Protection Agency

Fig. 5.4 Main steps in the Commonwealth of Australia EIA process.

Commonwealth has a decision-making responsibility. These proposals are often controversial.

Rather like NEPA, the Impact of Proposals Act provides a legal framework which contains little procedural detail about the various stages of the EIA process. However, it does provide for the approval of administrative procedures. Accordingly, the Australian Government issued the detailed Environment Protection (Impact of Proposals) Administrative Procedures, which were approved by the Governor-General in 1975. They were extensively revised in 1987 (Commonwealth of Australia, 1987). The Procedures set out the details of how the Act is to be administered. Various leaflets and subsidiary guidelines have also been produced (see also Wood 1993b; Anderson, 1994).

Despite a number of reviews, there has been little fundamental change to the Commonwealth EIA system beyond the introduction of scoping and, in 1987, of the less rigorous and detailed 'public environment report' procedure. Between the Act coming into force in 1974 and the end of June 1993, 132 environmentally significant proposals required the preparation of an EIS, and 24 a public environmental report, an average of just over 8 EIA reports per annum.

The New Zealand EIA system

New Zealand has undergone a revolution in environmental management.[8] Several government departments have been abolished or restructured, local government has been reorganised and environmental law has been reformed. The Resource Management Act 1991, which swept away numerous previous Acts, including the Town and Country Planning Act 1977, the Clean Air Act 1972 and the Water and Soil Conservation Act 1967, introduced environmental impact assessment as a central element in a decision-making process designed to achieve the goal of sustainable management (Box 5.1). This is EIA Mark II in New Zealand, the Mark I system

Box 5.1 Purpose of the New Zealand Resource Management Act 1991

Section 5. Purpose: (1) The purpose of this Act is to promote the sustainable management of natural and physical resources.

(2) In this Act, 'sustainable management' means managing the use, development and protection of natural and physical resources in a way, or at a rate, which enables people and communities to provide for their social, economic, and cultural wellbeing and for their health and safety while:

(a) Sustaining the potential of natural and physical resources (excluding minerals) to meet the reasonably foreseeable needs of future generations; and
(b) Safeguarding the life-supporting capacity of air, water, soil, and ecosystems; and
(c) Avoiding, remedying, or mitigating any adverse effects of activities on the environment.

---

having been introduced in 1974. EIA is now, in principle at least, almost comprehensive and flexible in that it applies, at the appropriate level of detail, to all projects and, in addition, to policies and plans prepared under the Resource Management Act provisions.

New Zealand has a slightly larger land area than the United Kingdom but, with 3.3 million people, its population density is only 5 per cent of the UK's. New Zealand local authorities, which are generally not as well staffed as those in the United Kingdom, serve average populations about four times smaller. Under the current arrangements in New Zealand, local government carries the principal responsibility for the administration of the EIA system and, in particular, the planning departments in local authorities are responsible for dealing with proponents and making recommendations on the basis of the EIA.

New Zealand introduced its Environmental Protection and Enhancement Procedures, partially modelled on the Canadian EIA procedure, in 1974 by means of a Cabinet minute. These evolved over the years and in the latest version of the Procedures (Ministry for the Environment – MfE, 1987) public agencies were required to undertake screening, agree project-specific scoping guidelines with the Ministry for the Environment, consult with various statutory, local and other authorities, publish an environmental impact report, submit it to a formal published 'audit' by the Parliamentary Commissioner for the Environment (an 'ombudsman' reporting not to the Government but to Parliament) and agree monitoring arrangements with MfE. This EIA process was subject to considerable public oversight but was largely discretionary and, in practice, related to only a limited number of public sector projects virtually all of which were approved. Wells and Fookes (1988) reported that an average of three formal environmental impact reports were formally audited (reviewed) each year between 1977 and 1989 but that hundreds of impact statements of variable size, scope and quality were being produced annually for local authorities and other bodies but not being subjected to the Procedures. In principle, but not in practice, the Procedures also applied to policies.

Considerable debate about the Procedures developed (see, for example, Morgan, 1983, 1988; Fookes, 1987b). Wells and Fookes (1988) summarised the need for reform to produce an EIA process which was 'simple but effective, comprehensive but not complex, flexible yet consistent, authoritative yet accessible'. At the same time, wide-ranging reviews of environmental administration, in which the Minister for the Environment (later Prime Minister) took a strong personal interest, and of local government, resulted in complete reorganisations of both (Memon, 1993). The various objectives of these reorganisations included the decentralisation of decision making, increased consideration of the environment at all levels of decision making, increased public involvement, and the elimination of administrative and legislative fragmentation.

New Zealand now has a basic system of regional and territorial (city and district) authorities (replacing hundreds of local and special-purpose bodies), responsible for many duties formerly undertaken by central government (Cocklin, 1989; Memon, 1993). Environmental impact assessment is one of these: EIA is now inextricably interwoven into the local authority procedures for determining applications for land use and subdivision consents and for coastal, water and discharge permits under the provisions of the Resource Management Act. The Ministry for the Environment has issued a guide to the Act (MfE, 1991b), a guide to scoping (MfE, 1992b) and several guides mentioning the environmental assessment of regional policies and plans and of district plans (MfE, 1991c, d, 1992a, 1993), a leaflet on the assessment of environmental effects (MfE, 1991a) and several other EIA leaflets.[9] It has also commissioned a more general guide on EIA (Morgan and Memon, 1993). Several commentaries on the new EIA system have been published (see, for example, Morgan *et al.*, 1991; Dixon, 1993a, b; Morgan, 1993; Montz and Dixon, 1993; Wood, 1993a).

A simplified version of the main steps in the EIA system in New Zealand (derived from Hughes (1992) and Montz and Dixon (1993)) is shown in Fig. 5.5. The Act makes broad provisions in relation to the EIA system and has devolved almost all responsibility for the administration of environmental impact assessment from central to local government. The Act provides the outline of the EIA process, but leaves much detail to be provided by individual regional authorities in their regional policy statements and regional coastal plans and by territorial authorities in their district plans.

The Resource Management Act contains provisions which effectively provide for a two-phase screening process and encourage scoping. It indicates the content requirements for an EIA report (which include alternatives), provides for public participation and consultation and requires that the report be considered in the decision. It also contains provisions relating to monitoring. In addition, the Act provides for public hearings into applications and for the call-in of requests for resource consents by the Minister for the Environment where issues of national significance are raised. There is a third-party right of appeal against LA decisions which are heard by a Planning Tribunal. Apart from the value of the precedents created by the Tribunal's findings, the EIA system is also subject to scrutiny by the Parliamentary Commissioner for the Environment.

It is too early for any review of the Resource Management Act to have been contemplated. Indeed, transitional arrangements, including the use of the Environmental Protection and Enhancement Procedures for central government projects not covered by the Act, were still in operation in 1994. At that time the Ministry of the Environment employed only a very small number of central and regional office staff on EIA. Generally, the number of central and local government officers, consultants, academics and environmental campaigners involved in EIA in New Zealand is very small (probably less that 50 full-time equivalents).

It is notable that EIA is not mentioned by name in the Resource Management Act but that the procedures and aims of EIA suffuse both its policy and plan preparation provisions and its resource consent provisions (Dixon, 1993b). In this sense, as in others, the highly sophisticated Resource Management Act makes a marked contribution to the advancement of sustainable development policy.

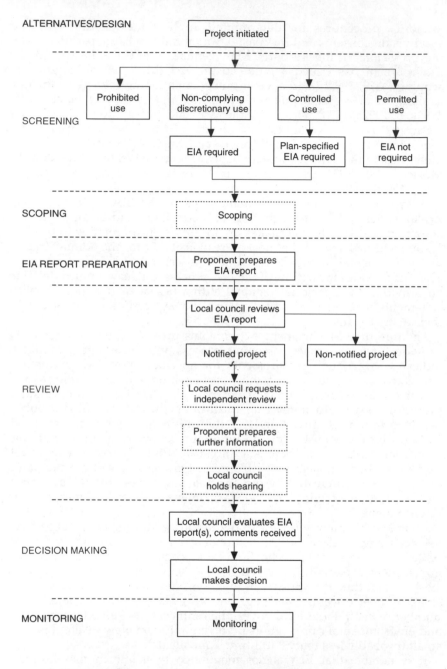

Fig. 5.5 Main steps in the New Zealand EIA process.

Notes

1. This account, and the subsequent treatment of The Netherlands EIA system in this book, is partially based on interviews. I am very grateful to Harry Boschloo, Jan Jaap de Boer, Ralph Hallo, Victor Jurgens, Maartje Nelemans, Michiel Odijk, Marja van Eck, Jeroen van Haeren, Jules Scholten, Romke Seijffers and Rob Verheem for their assistance.
2. The 1987 Decree, the Environmental Protection (General Provisions) Act 1986 and the Environmental Management Act 1994 are available in English from the Ministry of Housing, Physical Planning and the Environment (VROM).
3. This account, and the subsequent treatment of the Canadian EIA system in this book, is partially based on interviews. I am very grateful to David Barnes, Gordon Beanlands, Karen Brown, Bill Couch, Karen Finkle, Kathy Fischer, Robert Gibson, Stephen Hazell, Patrice LeBlanc, Jon O'Riordan, Stephane Parent, Ray Robinson, Barry Sadler and Graham Smith for their assistance.
4. The Canadian EA process described in Chapters 6–20 of this book is the CEAA system but EARP continues to apply to all federal proposals initiated before proclamation.
5. See, for a useful discussion, Sadler (1994b).
6. This account, and the subsequent treatment of the Australian EIA system in this book, is derived from Wood (1993b) which, in turn, is partially based on interviews. I am very grateful to Eric Anderson, John Ashe, Warren Atkinson, John Bailey, Ann-Marie Delahunt, Frank Downing, Rob Fowler, Clark Gallagher, Nick Harvey, Malcolm Hollick, Tor Hundloe, Ian Lamb, Steven Münchenberg, Brett Odgers, Helen Ross, Robin Saunders, Rob Sippe and Ian Thomas for their assistance.
7. Previously called the Department of the Arts, Sport, the Environment, Tourism and Territories and, later, the Department of the Arts, Sports, the Environment and Territories.
8. This account, and the subsequent treatment of the New Zealand EIA system in this book, is derived from Wood (1993a) which, in turn, is partially based on interviews. I am very grateful to Jim Clark, Chris Cocklin, Jenny Dixon, Bob Drury, Tom Fookes, Lindsay Gow, Paddy Gresham, Helen Hughes, Alisdair Hutchinson, Bob McClymont, Ali Memon, Richard Morgan and Barry Weeber for their assistance.
9. In addition to the Ministry for the Environment (1991a) leaflet the following have been issued in Wellington by MfE:
   (1991) *Assessment of Effects*. Resource Management Information Sheet 5.
   (1992) *A Guide for Councils: How to Respond to Resource Consent Applications During the Transitional Period*. Assessment of Environmental Effects Information Sheet 1.
   (1992) *A Guide for Applicants: How to Assess the Effects of Proposals under the Resource Management Act*. Assessment of Environmental Effects Information Sheet 2.
   (1992) *Principles and Issues Concerning the Assessment of Environmental Effects*. Assessment of Environmental Effects Information Sheet 3.

CHAPTER 6

Legal basis of EIA systems

Introduction

It is now generally accepted that EIA systems should be based upon clear specific legal provisions. The next section of this chapter explains the reasons for advancing this as an evaluation criterion for EIA systems. It then discusses the requirements for the legal basis of EIA systems in rather more detail and puts forward several more explicit criteria. These requirements are then used to assist in the comparison of the legal bases of the US, Californian, UK, Canadian, Dutch, Australian and New Zealand EIA systems.

The legal basis of EIA systems

Following the passage of the National Environmental Policy Act 1969 in the United States (NEPA), it was not uncommon for governments in other countries to maintain that the main principles of EIA were already provided by their existing legislation. The United Kingdom, for example, argued for many years that the principal elements of EIA were satisfactorily furnished by the existing town and country planning system (see Chapter 4). It is, of course, true that many excellent EIA studies have been carried out in the United Kingdom and elsewhere without the framework of a formal EIA system. Kennedy (1988b) has categorised this type of approach as 'informal–implicit'. In it, EIA is adapted to meet the needs of particular situations, an EIA report as such may not be prepared and authorities are not accountable for taking EIA into consideration in decision making.

In the formal–explicit approach, on the other hand, EIA requirements are codified in legislation or regulations, an EIA report must be prepared and authorities are accountable for considering EIA (for example, through judicial review). Kennedy (1988b, p. 258) argues that, 'generally speaking, EIA is only integrated in decision-making (that is, it only works) when it is applied in a formal–explicit way'. This view is now generally accepted, though the same arguments have been aired in relation to the environmental assessment of programmes, plans and policies in recent years (for a UK example, see Department of the Environment, 1992).

A further relevant issue in EIA is the question of how far the detailed operation of the EIA process should be prescribed in laws and regulations, and how much it should be left to the discretion of the relevant authorities. The advantages of a legally specified EIA system may be summarised (Fowler, 1985; Buckley, 1991b) as: permanence and evidence of commitment; avoidance of uncertainty; provision of a firm basis for public participation; and enforcement of acceptance of EIA.

On the other hand, the advantages of a largely discretionary EIA system, only the broad details of which are enshrined in law or regulation (Fowler, 1985) are: the desirability of voluntary compliance; the avoidance of judicial involvement; and the retention of discretion. Fowler (1985, p. 205) concludes that 'where a firm political commitment to EIA happens to exist at the time of adoption of a scheme, this is reflected in a legislative base'. He suggests that there is a gradual shift towards EIA systems 'in which both administrative and judicial supervision is seen as necessary in terms of the efficiency of the overall process'. This appears to hold generally true. Thus the Canadian Federal Government, after almost 20 years of discretionary EIA codified the system in legislation in 1992. Similarly, the New Zealand Government, after 17 years use of discretionary EIA procedures, passed legislation specifying many of the requirements of the process in 1991. It is true, of course, that some degree of discretion in the application of any EIA system must remain to meet particular circumstances (below).

A further issue is whether the EIA system should be independent of existing decision-making procedures or whether it should be integrated into them. NEPA introduced a completely new procedure which cut across existing decision making. This, not surprisingly, led to confusion and delay in the early years, to considerable duplication of control, and to a desire by other countries to avoid similar problems. This has been exacerbated by a desire to avoid both litigation and the loss of control and power over decisions to external agencies. The advantage of separation is the creation of a fresh approach which emphasises the importance of EIA, something which may not be apparent where EIA is integrated into existing procedures if prevailing attitudes among practitioners and decision-makers are not modified. The European Directive on EIA is specifically phrased, notwithstanding the US separation model, to allow member states to introduce EIA into existing decision-making procedures. Many of them (including the United Kingdom) have followed this route.

There are clearly advantages in both approaches, if they are implemented effectively. Indeed, this distinction between separate and integrated EIA systems may be somewhat arbitrary in practice, since the essential aim of EIA is that decisions on actions are made which take full account of the outcomes of the EIA process. This may, or may not, be achieved in systems utilising both approaches.

There is always a danger that, unless the various steps in the EIA process are mandatory, there will be some proponents, consultants, consultees or authorities who will fail, in certain circumstances, to discharge their responsibilities fully. For this reason, each step in the EIA process needs to be specified sufficiently in a law or in a binding regulation to provide a measure of certainty to the participants in the EIA process. The finer points involved in each stage of the process need not be spelled out in law provided that appropriate additional guidance is made available (for example, in the form of advisory guidelines). It is important, in the interests of certainty, that the specified system is adhered to, and that accepted procedures are not changed arbitrarily.

While lawyers drafting laws and regulations will always strive to make them unambiguous, other lawyers will endeavour to discover loopholes and ambiguities if it is in their clients' interests to do so. Clearly, for an EIA system to function effectively, ambiguities need to be minimised. Where they exist and cause problems in the operation of the EIA system, they should be remedied at the first available opportunity. Herein lies another advantage of specifying some details in the form of regulations or guidance, since they can then be modified without recourse to primary legislation.

Some degree of discretion in the operation of the various steps of the EIA process needs to exist since every eventuality cannot be foreshadowed in laws and regulations. In particular, flexibility is necessary to ensure that the EIA system is focused on the desired outcome of EIA, environmentally sensitive decisions, rather than on ensuring that all the procedural formalities have been completed. Such discretion takes its most extreme form in the state of Victoria, where the EIA system is now based on guideline procedures quite different from the provisions of the Act underpinning them (Wood, 1993c). However, the discretion remaining should not be sufficient to remove reasonable legal certainty, and nor should it enable any participant in the EIA process to gain undue advantage. It is for this reason that 'fast-track' solutions to EIA decision making which automatically permit a development to proceed unless the relevant authorities take appropriate EIA action are generally unsatisfactory. The discretion remaining in an effective EIA system should be broadly acceptable to all parties.

EIA is a process which applies to certain types of actions, but not to all proposals. The legal requirements relating to EIA should be clearly distinguished from those relating to other types of action so that no confusion exists between different processes. This applies equally to systems in which EIA is separated from other decision-making procedures and those in which it is integrated. The need for differentiation is strongest in integrated EIA systems since the scope for confusion, particularly among proponents, is higher than in separated systems.

In the last analysis, it may be necessary to take enforcement action against one of the participants in the EIA process. This might, for example, be against the relevant authority for not screening the proposal adequately, or for not considering the comments on the EIA report sufficiently in reaching its decision, or against the proponent for not meeting conditions attached to a permission. Such action might be taken by any of the participants in the EIA process, including the public. It is necessary, therefore, that there should be adequate opportunities for the various participants to appeal administratively or to the courts to ensure that the various obligations in the EIA process are properly discharged.

It is important that a clear outline of all the procedures involved in the EIA process be available so that proponents, developers, consultees, the public (and the relevant authorities) can gain an overview of the whole process. This outline should include the time allocated to each stage in the process (a necessary requirement to prevent it from becoming over-lengthy) and any charges involved in it. The various criteria for analysing the legal basis of EIA systems are summarised in Box 6.1. These requirements are used in conjunction with the central criterion to assist in the review of the legal basis of each of the seven EIA systems which follows.

United States of America

The US Federal EIA system is based upon the broad provisions of legislation – the National Environmental Policy Act 1969 – the brevity of which is matched by its ambiguity. The various requirements of the Act have been clarified over the years by both the courts and the Council on Environmental Quality (CEQ) Regulations, themselves based upon legal rulings. Further guidance has been issued by CEQ to clarify matters not covered fully in the Regulations.[1] While the substantive intent of NEPA, to change the nature of federal decision making, has been gradually whittled away over the years to become a largely procedural requirement, the legal basis of the US EIA system is clearly specified by it. The detailed steps in the process are specified in the Regulations, which are widely regarded as providing a model basis for an EIA system, being comprehensive, specific, clear and surprisingly readable.

There is reasonable agreement by proponents, practitioners and

Box 6.1 Evaluation criteria for the legal basis of EIA systems

---

Is the EIA system based on clear and specific legal provisions?

Is each step in the EIA process clearly specified in law or regulation?

Are the legal provisions sufficiently unambiguous in application?

Is there a degree of discretion in the provisions which is acceptable to the participants in the EIA process?

Are the EIA requirements clearly differentiated from other legal provisions?

Is each step in the EIA process enforceable through the courts or by other means?

Are time limits for the various steps in the EIA process specified?

Does a clear outline of procedures and time limits exist for the EIA system as a whole?

---

environmental groups that the Regulations leave an appropriate degree of discretion for the EIA process to be applied to the activities of the very wide range of federal agencies affected by NEPA. Generally, the various agencies (over 50) have issued guidelines or regulations to apply the CEQ Regulations specifically to their own activities and the Environmental Protection Agency (EPA, 1993) has published a sourcebook for the environmental assessment process which includes a set of computer discs. This documentation (primarily intended for agency staff) forms part of a voluminous literature on NEPA (see, for example, Environmental Law Institute, 1989; Fogleman, 1990; Mandelker, 1993b; Hildebrand and Cannon, 1993). There is no single official explanation of the EIA system as a whole beyond the EPA sourcebook and the CEQ Regulations themselves.

Each major procedural step in the EIA process can be challenged in the courts. The fact that such a litigious society as the United States generates less that 100 court cases per annum on NEPA when its provisions are applied to at least 50,000 actions must be regarded as a vindication of the Regulations. Nevertheless, while the legal provisions are generally regarded as being reasonably unambiguous, the continuing legal actions arguing that environmental assessments or EISs ought to have been prepared demonstrates the scope they leave for interpretation.

The Regulations contain provisions relating to the time limits for consultation and participation. They also provide that, subject to certain limitations, the agency must set time limits if an applicant for the proposed action requests them (Section 1501.8). This provision was the one which the business community found the most attractive (Yost, 1981). In practice, it has not been used a great deal.

The EIA requirements are clearly differentiated from other legal provisions. Indeed, this separation caused animosity, confusion and delay in the early years when NEPA was applied retrospectively to projects which were already under construction. While experience has considerably reduced these problems, some of the frustration with the complexity of NEPA and other environmental regulations which drove Congressman Hinshaw to enter the holy hypothetical saga reproduced in Box 6.2 into the record of the House of Representatives still exists (see, for example, Mandelker, 1993a).

California

The California Environmental Quality Act (CEQA) is very similar to the US National Environmental Policy Act (NEPA), though there are minor differences in terminology, in procedures and in substantive effects. CEQA contains explicit provisions for screening, for scoping, for the preparation of a draft environmental impact report (EIR), for a final EIR to be submitted, for the agency's decision to be made public and for monitoring. There are specific requirements for public participation and for mitigation of impacts. There are also provisions for different types of EIR, for tiering, for various public notices and for time limitations.

Box 6.2 'God and EPA'

In the beginning God created heaven and earth.

He was then faced with a class action lawsuit for failing to file an environmental impact statement with HEPA (Heavenly Environment Protection Agency), an angelically staffed agency dedicated to keeping the Universe pollution free.

God was granted a temporary permit for the heavenly portion of the project, but was issued a cease and desist order on the earthly part, pending further investigation by HEPA.

Upon completion of his construction permit application and environmental impact statement, God appeared before the HEPA Council to answer questions.

When asked why he began these projects in the first place, he simply replied that he liked to be creative.

This was not considered adequate reasoning and he would be required to substantiate this further.

HEPA was unable to see any practical use for earth since 'the earth was void and empty and darkness was upon the face of the deep'.

Then God said: 'Let there be light.'

He should never have brought up this point since one member of the Council was active in the Sierrangel Club and immediately protested, asking, 'how was the light to be made? Would there be strip mining? What about thermal pollution? Air pollution?' God explained the light would come from a huge ball of fire.

Nobody on the Council really understood this, but it was provisionally accepted assuming (1) there would be no smog or smoke resulting from the ball of fire, (2) a separate burning permit would be required, and (3) since continuous light would be a waste of energy it should be dark one-half of the time.

So God agreed to divide light and darkness and he would call the light Day and the darkness Night. (The Council expressed no interest with in-house semantics.)

When asked how the earth would be covered. God said: 'Let there be firmament made amidst the waters, and let it divide the waters from the waters.'

One ecologically radical Council member accused him of double talk, but the Council tabled action since God would be required first to file for a permit from ABLM (Angelic Bureau of Land Management) and further would be required to obtain water permits from appropriate agencies involved.

The Council asked if there would be only water and firmament and God said 'Let the earth bring forth the green herb, and such as may seed, and the fruit tree yielding fruit after its kind, which may have seen itself upon the earth.'

The Council agreed, as long as native seed would be used.

About future development God also said: 'Let the waters bring forth the creeping creature having life, and the fowl that may fly over the earth.'

Here again, the Council took no formal action since this would require approval of the Game and Fish Commission coordinated with the Heavenly Wildlife Federation and Audobongelic Society.

It appeared everything was in order until God stated he wanted to complete the project in six days.

At this time he was advised by the Council that his timing was completely out of the question... HEPA would require a minimum of 180 days to review the application and environmental impact statement, then there would be public hearings.

It would take 10 to 12 months before a permit could be granted.

God said: 'To Hell with it.'

Source: *United States Congressional Record* (1974).

CEQA now runs to some 70 pages and the Guidelines (or regulations) to another 240 pages. The Act has been revised almost annually by the state legislature in response to perceived problems and court decisions. (The Act is also subject to constant interpretation before the courts.) Almost every detail of the EIA procedure in California is specified in either the Act or the Guidelines in considerable detail. However, though every step in the EIA process is clearly spelt out, there is sufficient ambiguity and open-endedness remaining to allow differences in practice and continuing legal challenge. Although most of these challenges are for legitimate environmental reasons, some are made for non-environmental reasons by adjoining jurisdictions, local residents, business competitors and even unions as a result of the very broad interpretation of legal standing applied in California.

While there is considerable discretion remaining in the Act and Guidelines to allow some flexibility of operation, the fear of court action leads to over-elaboration in the production of negative declarations where projects could be excluded, in the production of EIRs where negative declarations would suffice and in the generation of very lengthy EIRs. While legal action cannot be brought until the agency makes its decision, each step in the process can then be challenged. Most challenges, as at the federal level, relate to failure to produce environmental documentation or to its inadequacy.

The EIA requirements in California are clearly differentiated from other legal provisions. There is no necessity to integrate other state and local environmental reviews with the EIR, though this is encouraged and often occurs under CEQA (Bass and Herson, 1993c). CEQA provides for time limits at each stage of the EIA process. Once an application for a private development project is accepted as complete, the lead agency has one year to prepare the EIR and approve or disapprove the project. There are also limitations upon the time during which legal action can be brought.

The Guidelines have not kept pace with the changes in CEQA: the last revision was in 1986. This is a result of staffing cuts which have taken place at the Office of Planning and Research, themselves a consequence of the lack of political will to provide leadership on environmental issues. There is, despite the use of updated discussion in the Guidelines, no clear up-to-date State of California outline of procedures and time limits for the EIA system as a whole. Ironically, this lack of up-to-date central guidance has fostered an industry in consultancy services. In addition, private sector guides to CEQA procedures (Bass and Herson, 1993b; Hernandez *et al.*, 1993; Remy *et al.*, 1993) have been published to meet the need for current information that keeps pace with statutory changes and court interpretations.

United Kingdom

The regulations incorporating environmental assessment (EA) into the town and country planning system (and other statutory procedures) were made under the European Communities Act, the provisions of which do not permit the requirements of the European Directive on environmental assessment (or of any other directive) to be exceeded. However, the Planning and Compensation Act 1991 now allows the Secretary of State for the Environment to require EA for projects needing planning permission other than those listed in the Directive (Chapter 4) and this power has now been exercised. The Town and Country Planning (Assessment of Environmental Effects) Regulations implement the provisions of the European Directive almost to the letter, though there is some disagreement about whether the content requirements for environmental statements specified by the Regulations accurately reflect the Directive. The Regulations, which were laid before Parliament prior to coming into effect on 15 July 1988, provide the legal basis for each of the steps shown in Fig. 4.1. Not only are all the main steps covered by the Regulations but time limits are specified for each of them. No mention is made in the Regulations of either scoping or monitoring.

The Regulations have not proved to be unambiguous. For example, the definition of certain types of project (for instance, urban development projects, fish farms) has proved uncertain. These ambiguities have almost all stemmed directly from the wording in the European Directive on EIA. Though EIA in the United Kingdom is largely integrated into the town and country planning system, the requirements are clearly distinct from those for normal planning applications, for example, in relation to timescales. The degree of discretion provided by the Regulations (which mirrors that in the existing land use planning system) appears to be broadly acceptable to most of the main participants.

The Regulations provide the developer with a right of appeal against a local planning authority (LPA) determination that EA is required. As the EA system is integrated into town planning procedures, normal appeal provisions against the non-determination of planning applications (and against planning decisions) apply and the Secretaries of State can call applications in for determination by central government. In principle, therefore, there is strong central control over the freedom of LPAs to determine applications to which the Regulations apply. While there is no third party right of administrative appeal in the British planning system, access to the courts is possible where the EIA requirements have not been properly discharged. In practice, there have been very few such cases to date.

Circular 15/88 (Department of the Environment – DOE, 1988b) and the *Guide to the Procedures* (DOE, 1989) together provide clear and detailed guidance on the operation of the procedures. The indicative criteria and thresholds to be used by LPAs in reaching a judgement about whether EA is to be required for Schedule 1 projects are contained in the advisory Circular, and not in the Regulations. The criteria (and the other advice contained in the Circular) can therefore be changed relatively easily and, in any event, do not have regulatory force.

It is apparent that the Regulations and accompanying guidance contain provisions which clearly and specifically define the basis of the EA system integrated into British planning procedures.

The Netherlands

The Environmental Management Act 1994 contains some 10 pages and 43 sub-sections of detailed requirements relating to the coverage of EIA, the content of the environmental impact statement (EIS), the preparation of the EIS, the evaluation of the EIS, the decision-making procedure and the post-project evaluation. The Act provides for the EIA process to be integrated into existing decision-making procedures. Together with the Notification of Intent Decree and the amended EIA Decree, the Act contains provisions relating to each step in the EIA process shown as obligatory in Fig. 5.1. The one area not provided for in the Act relates to the provision of supplementary information as a result of the review of the EIS by the EIA Commission and the public.

The Act and the Decrees have proved not to be unambiguous in application. The principal ambiguities were felt by the Evaluation Committee on EIA (ECW, 1990) to relate to the exemption provisions and to the nature of alternatives dealt with in the EIS. Notwithstanding the detailed nature of the legal requirements, numerous requests for clarification have been addressed to the Ministry of Housing, Physical Planning and Environment (VROM). Many of these have concerned the issue of whether economic and social impacts should be included in the EIS. The Ministry has taken the view that the EIS should be confined mainly to physical environmental impacts, though proponents are free to include economic and social impacts if they choose to do so. Generally, the Ministry and the EIA Commission have both acted to clarify the meanings of the legal provisions when necessary. The remaining degree of discretion in the operation of the EIA system is felt to be appropriate by most of the participants in the EIA process.

There is a general right of appeal against the decision to which the EIA process is linked in The Netherlands. This, together with the publication of virtually all the documentary material associated with the EIA process, means that there can be adequate enforcement of the various steps in the EIA process. Some of the stages in the EIA process (e.g. screening) are also enforceable through the courts though there have been less than ten cases involving EIA in The Netherlands. Cases have related to the granting of exemptions (the 'no significant impact' provision for which was deleted in 1994), to failure to carry out EIA and to the nature of the guidelines for the EIS (the content of these was deemed not to be open to appeal).

While the EIA requirements are integrated into existing decision-making procedures, they are clearly differentiated from them. EIA can be linked to licences and permits under the Environmental Management Act 1994, to local land use plans, to the waste policies in provincial environmental management plans and to other sectoral decisions (e.g. those concerning airports, motorways) as well as to policy decisions for activities for which EIA is required at the project level.

There are three time limits specified in the Act: 13 weeks are permitted from receipt of the notification of intent to the issue of the guidelines (21 weeks if the proponent and competent authority are the same). The competent authority is allowed six weeks to determine whether the EIS is acceptable, prior to its publication. Once it has been published, the EIA Commission usually has nine weeks to determine whether the EIS meets the legal requirements and the guidelines. The time period for the making of the decision is determined by the regulations relating to the type of decision concerned (e.g. licences, land use plans, etc.).

There is a large handbook on the EIA procedure (which was updated in 1994), together with an explanatory leaflet (VROM and Ministry of Agriculture, Nature Management and Fisheries, 1991). In addition, as mentioned in Chapter 5, numerous other documents on more technical aspects of EIA exist. In general, there is probably more official information about EIA in The Netherlands than in any jurisdiction outside North America. The existence of this information, like that of the very clear and specific legal provisions relating to EIA in The Netherlands, reflects the careful preparation for (and subsequent administration of) the Dutch EIA system.

Canada

The Canadian Environmental Assessment Act 1992 (CEAA) lays down the steps in the EA process in considerable detail in its 82 sections. There are separate parts of the Act dealing with interpretation, purposes, projects to be assessed, the assessment process, transboundary and related effects, access to information, administration, the Canadian Environmental Assessment Agency, review and transitional arrangements. Some government officials fear that, in endeavouring to overcome some of the shortcomings arising from the wide discretion provided by the Environmental Assessment and Review Process (EARP) Guidelines and in attempting to ensure that the courts will be able to enforce the EA process, the authors of the Act may have over-prescribed it.

The Act contains provision for numerous regulations to be made. Four sets were promulgated when the Act was proclaimed. These related to the list of acts and regulations under which an activity could trigger CEAA, a list of projects to be excluded from the provisions of the Act, an inclusion list for physical activities to be covered by the Act and a list of projects to be subject to a comprehensive study.

It is too early to state whether the provisions of the Act and the regulations are sufficiently unambiguous. It would be very surprising, given the number of provisions in the Act and the regulations, if no ambiguities were to surface. However, since the Act is intended to reduce the uncertainty of application inherent in EARP, such ambiguities should be limited in number. Indeed, the most significant amendments to CEAA during its passage through the House of Commons centred on the reduction of residual discretion (Gibson, 1993). Much of the detail of CEAA's implementation is prescribed in regulations, which can be altered relatively easily. It appears that many stakeholders in both the environmental movement and in industry regard the Act as an advance (though not always as a significant advance) on the discretionary provisions in EARP.

The Act, like EARP, specifies an EA process which is almost entirely separate from other legal provisions. Where the federal responsible authority (RA) proposes the project, or grants money or land for the project, EA is a separate process. Where the RA is involved in a regulatory role, such as issuing a licence, EA is effectively integrated into existing decision-making processes. Even here, however, the legal EA requirements are quite distinct and clearly differentiated from other provisions.

The various steps in the EA process are open to varying degrees of public participation and to challenge in the courts. There had been over 50 court cases under EARP by mid-1994 and, following the landmark challenges of 1989, the authors of CEAA were determined to reduce substantially the number of legal challenges by prescribing the EA steps carefully in the legislation. Only time will tell whether they have succeeded.

No time limits are specified in the Act. However, Section 59(a) states that regulations can be made to prescribe the time periods relating to the EA process and it is likely that appropriate regulations to limit the time taken for the various steps in the EA process will be made. (No time limits were ever set for EARP.)

A comprehensive guide to CEAA was published in 1993. This consists of a 'manager's guide', a 'practitioner's guide' and several reference guides including advice on, *inter alia*, cumulative impacts and the assessment of significance (Federal Environmental Assessment Review Office, 1993a).[2] This lengthy guide provides a clear outline of EA procedures and some indication of methods but does not mention time limits. It is intended that the guide should eventually be supplemented by separate sector-specific guides for the EA of particular types of project.

Commonwealth of Australia

EIA at the national level in Australia is clearly differentiated from other legal provisions relating, for example, to land use planning or pollution control. The Environment Protection (Impact of Proposals) Act 1974 is basically enabling legislation, leaving much of the detail of the Commonwealth procedure to be elaborated in more detailed orders. The Environment Protection (Impact of Proposals) Administrative Procedures (Commonwealth of Australia, 1987), which were laid before Parliament, can thus be altered without changing the primary legislation. There is, for example, only one time limit mentioned in the Act, that relating to the making of recommendations by the Environment Minister following receipt of the final environmental impact statement. The precise legal status of the Procedures appears to be somewhat uncertain since they are neither formal regulations nor advisory guidance.

This uncertainty also exists elsewhere in the EIA system. While the Act and the Procedures together set down provisions for each stage of the process, they leave a number of ambiguities. For example, they contain an open-ended definition of the activities to be covered (Fowler, 1982, p. 19). There has been considerable debate about the nature of enforceable obligations on decision-makers to take the EIS into account, as the Act requires. It would appear that judicial review is the only available means of compelling decision-makers to take appropriate account of the EIS (Fowler, 1982, p. 28). However, there has been but a single major challenge in the courts to the system, and that was dismissed.[3] While the general foundations of the EIA system are clearly based in law, considerable legal uncertainty about specific aspects of the EIA process therefore remains, despite the theoretical possibility of court challenge.

There thus continues to be some ambiguity in the application of the legal provisions, particularly in regard to the proposals which should be subject to EIA. The discretion remaining in the provisions appears to be unacceptable both to proponents and to environmental groups. Proponents are anxious that both the application of EIA and the detailed requirements of the process are uncertain. Environmental groups feel that too much discretion in the application of EIA resides in the hands of powerful federal ministries. Both participants thus seek greater legal certainty in the application of the EIA procedure.

While various helpful leaflets and informal guidelines relating to specific aspects of the EIA process have been prepared, no up-to-date comprehensive guidance setting down procedures and time limits for each step in the process exists for the Commonwealth EIA system.

New Zealand

The Resource Management Act 1991 contains almost all the legal provisions relating to the New Zealand EIA system. As explained in Chapter 5, this revolutionary legislation also provides the legal basis for almost all of

New Zealand's environmental protection measures and, accordingly, it runs to nearly 400 pages in length. The EIA provisions constitute only a small part of the total (around 20 of the over 400 sections) and are closely integrated with other resource planning provisions.

The term EIA is not used in the Act, which uses phrases such as 'an assessment of any actual or potential effects ... on the environment' (Section 88(4)(b)) to describe EIA. No phrase is used to describe the EIA report in the Act, despite the fact that New Zealand had 15 years' experience of environmental impact reports. Dixon (1993b, p. 244) believed the authors of the Act replaced the term 'impact' by 'effect' to signify 'a fresh approach to EIA as all proposals and plans now come under this scrutiny'.

The Act provides a broad indication of the projects to be assessed: land use and subdivision consents, discharge, water abstraction and coastal permits unless exempted by local authorities in their plans and policies. The Act encourages scoping and provides an indication of the contents of an EIA report which can be modified by regional and territorial (city and district) authorities in their policy statements and plans. Time limits are imposed upon these authorities to process applications once they have sufficient information. The results of public participation and consultation on the basis of the EIA report, and the report itself, must be considered in the decision. The Act further provides for monitoring of the impacts of approved proposals. In short, the Resource Management Act provides a clear framework for EIA in New Zealand but deliberately leaves very considerable latitude to local authorities to determine their own specific EIA requirements. Several outlines of the procedures and the tight time-table for each step in the EIA process exist in Ministry for the Environment documentation (Chapter 5) and in Morgan and Memon (1993).

The legal provisions are far from unambiguous. For example, since *all* resource consent applications, large or small, are controlled under the provisions of the Act, there is ample scope for discretion in the inter-pretation of terms such as 'major', 'minor' and 'significant', about which no further advice has been issued by the Ministry for the Environment. Montz and Dixon (1993) suggested that the way in which such terms were interpreted by planners when applications were received would be important in the successful implementation of EIA.

The provisions of the Act are enforceable by third party appeal to the Planning Tribunal and, on points of law, beyond the Tribunal to the courts. The Tribunal consists of a judge assisted by a small number of part-time members, whose judgements form the basis of planning law in New Zealand. Clearly, the decisions of the Planning Tribunal will help to provide clarification and legal precedents. The independent Parliamentary Commissioner for the Environment can involve herself or himself in a limited number of EIA issues where this is thought likely to provide useful results.

Summary

Table 6.1 summarises the extent to which the seven EIA systems are based on clear and specific legal provisions. All seven systems meet the criterion. However, the US, California and the Commonwealth of Australia acts provide a general outline and rely on detailed regulations for their implementation and the UK EIA system is based almost entirely on regulation, supported by non-statutory guidance. The New Zealand legislation relating to EIA is extraordinarily brief, has not been supported by regulations, and does not even mention the term 'EIA'. In developing countries, the legal basis of EIA is often weak or non-mandatory (Appendix). The most detailed legislation is to be found in The Netherlands and Canada, where sets of regulations relating to particular aspects of the EIA process have been issued to support the very specific EIA acts. There also exist specialised EIA agencies in The Netherlands and Canada and, to a lesser extent now, in the United States to provide advice and guidance on EIA both in general and in specific cases.

Table 6.1 The legal basis of the EIA systems

Criterion 1: Is the EIA system based on clear and specific legal provisions?		
Jurisdiction	**Criterion met?**	**Comment**
United States	Yes	National Environmental Policy Act and Regulations clearly define separate EIA system.
California	Yes	California Environmental Quality Act (CEQA) and Guidelines clearly define an EIA process separate from other legal decision-making procedures.
United Kingdom	Yes	Regulations specifically implement European Directive on EIA. EIA integrated within town and country planning system, administered by local planning authorities (LPAs).
The Netherlands	Yes	EIA Act and decrees specifically provide for clearly defined EIA process integrated into other decision-making procedures.
Canada	Yes	Canadian Environmental Assessment Act (CEAA) and regulations clearly define EA process largely separate from other decision-making procedures, together with powers of CEA Agency.
Australia	Yes	Act and procedures together outline separate EIA system. Considerable legal uncertainty and great discretion in implementation.
New Zealand	Yes	Act provides clear broad framework for EIA but allows local authorities very considerable discretion in operation.

Five of the seven EIA systems involve procedures which are quite separate from other authorisation systems. In the United Kingdom and New Zealand, on the other hand, EIA procedures (while identifiable legally) are firmly integrated into other types of consent procedure. It is not surprising that the legal requirements for EIA in these two countries are expressed much more briefly than in the acts and regulations specifying the EIA systems in the other five jurisdictions. However, given that there is no central body responsible for EIA in these countries (as there is in The Netherlands, Canada and the Commonwealth of Australia) and that there is a much more limited possibility of appeal to the courts than in the United States and California, it is apparent that there is likely to be rather more discretion in their EIA systems than in the other five.

Notes

1. Council on Environmental Quality (1981a, b); Memorandum: Guidance Regarding NEPA Regulations *48 Federal Register* 34263, 28 July 1983. These documents are reproduced in CEQ annual reports and in Bass and Herson (1993a).
2. Addendum: An updated version of this FEARO guide was published late in 1994 (Canadian Environmental Assessment Agency, 1994, *The Canadian Environmental Assessment Act: Responsible Authority's Guide*. CEAA, Hull, Quebec). *A Citizen's Guide* was also published in 1994. The CEAA guide made few substantive changes to the FEARO advice summarized in this book.
3. There have been eight cases challenging various aspects of the Commonwealth EIA process, but these have had little effect upon it.

CHAPTER 7

Coverage of EIA systems

Introduction

The coverage of EIA systems relates both to the range of actions subjected to EIA and to the range of impacts regarded as relevant. While it is generally accepted that the impacts of all environmentally significant new projects should be subject to EIA, there is little unanimity about the extension of EIA to higher tier sections such as programmes, plans and policies or about the definition of the word 'environment'.

The next section of this chapter discusses several criteria which can be employed in the analysis of EIA systems. These criteria are used to assist in the comparison of the legal bases of the US, Californian, UK, Dutch, Canadian, Australian and New Zealand EIA systems.

Coverage of actions and impacts

The National Environmental Policy Act 1969 (NEPA) applies to public actions by the federal government. These actions include the granting of permits for private actions. 'Actions' also include the making of plans and the enactment of legislation but one of the main intentions of the Act was clearly to ensure that the projects initiated by federal government were environmentally acceptable. In practice, the overwhelming emphasis of NEPA application has been upon projects, many of them public. From the outset, the word environment was defined to include social and economic impacts, as well as physical environmental impacts (e.g. pollution, effects on ecology, etc.). The coverage of later EIA systems has seldom followed the precedent set by NEPA.

If the objective of EIA is to ensure that the environmental impacts of significant actions should be assessed prior to implementation, there appears to be little point in distinguishing between public and private actions. There are sufficient examples of both well- and poorly-considered projects by public and private proponents in most countries to counter any argument that, say, private but not public proponents should be subject to EIA. That private proponents have not, in general, been subject to EIA at federal level in the United States is largely a consequence of a system of government which limits federal intervention in state and local matters.

It is clearly important that no significant types of project, whether public or private, should be exempt from EIA unless there is an overwhelming reason for this (for example, national security considerations). It is not uncommon, for example, to exempt specific legislation passed by national governments from EIA requirements, as the European Directive on EIA permits. Clearly, the unjustified use of such legislation could bypass EIA requirements. This has, in practice, taken place in a number of countries on occasion. There may be certain classes of project which, though their environmental impacts are significant, are normally exempt from EIA. For example, the construction of city tower (high-rise) blocks and the conversion of natural areas to intensive agriculture has frequently escaped EIA in many countries.

It is also important that the impacts arising at different stages of the project are assessed. Thus, impacts arising at exploration, construction, operation, modification and decommissioning stages should be considered. Further, impacts under both normal operating and potential accident conditions need to be evaluated.

Following from this is the issue of whether all the types of project which, in principle, are subject to the legal provisions of the EIA system are actually assessed, or whether some actions are effectively exempted in practice. Such avoidance of EIA requirements may be a result either of accident, or of the difficulty of undertaking EIA, or of the exercise of power by those assessed, or of setting criteria for applicability too high. (The screening of environmentally sensitive projects is discussed in Chapter 9.)

The argument that the EIA of, say, the construction of a road, takes place too late in the decision-making process to influence crucial choices between different types of transport system and hence their environmental impacts is well established (Wathern, 1988c; Wood and Djeddour, 1992; Therivel *et al.*, 1992). The same argument applies to projects such as housing schemes where the cumulative impacts of several projects can only be adequately covered at the plan-making stage (Wood, 1988a). The need for EIA at strategic level, i.e. at programme, plan and policy tiers, is widely accepted but few jurisdictions have implemented effective strategic environmental assessment (SEA) provisions, though many permit at least some SEA to take place. The infrequent use of SEA provisions in many jurisdictions may be a result of both perceived methodological difficulties and reluctance to cede power over decision making. (SEA is dealt with in some detail in Chapter 19.)

The definition of 'environment' in EIA has been treated differently in different jurisdictions. The European Directive on EIA eschews consideration of social and economic impacts whereas the EIA systems in many other parts of the world, including many developing countries, evaluate impacts other than those upon the physical environment. It is inevitable that in any reasonably democratic decision-making procedure economic and social factors will strongly influence the outcome as a result of the political process. It was the neglect of the physical environment in decision making which was the original stimulus for EIA in the United States and this was the reason why the European Directive on EIA was narrowly focused; it was felt that the balance needed to be redressed.

The difference in approach between Europe and the United States can probably be best explained by settlement patterns. Europe is much more densely populated than the United States and Canada and propinquity has led to the need for relatively stringent controls over new development. As a consequence, EIA in the United States and Canada developed partially to fill a vacuum created by the absence of a strong planning system. A further factor is the tradition of distrust of government in the United States which has resulted in greater citizen participation and more open government than in Europe (Wood, 1989b). Since the public finds the distinction between the treatment of the different types of impact of a given proposal to be artificial, there has inevitably been pressure to treat such impacts comprehensively. In the circumstances, consideration of the social and economic impacts of development within the EIA framework was inevitable. A similar situation has arisen in many other jurisdictions, for example in Australia and New Zealand where EIA has responded to the need to give explicit consideration to the needs of pre-European peoples.

In the last analysis, the issue of whether or not EIA covers impacts other than those on the physical environment is probably not crucial, especially as the distinction between them is often a narrow one in practice. It is, however, important that *all* impacts on the physical environment are encompassed by the EIA system. Thus, impacts on the various environmental media (e.g. the air), on living receptors (e.g. people, plants) and on the built environment (e.g. buildings) should be considered. Further, indirect impacts arising from other types of induced activity (e.g. ancillary service development) and the interrelatedness of environmental impacts (e.g. emissions of sulphur dioxide affecting the acidity of freshwater) and cumulative impacts (Cocklin *et al.*, 1992) need to be assessed. The danger of double-counting impacts should be constantly guarded against.

The various criteria which can be used in considering the coverage of EIA systems are summarised in Box 7.1. These are used to assist in the review of the coverage of each of the seven EIA systems which follows.

United States of America

The National Environmental Policy Act 1969 applies only to federal actions, but not to state actions or to most private projects except where they require a federal permit. Section 102(2)(C) of NEPA (Box 2.1) states that all agencies of federal government must:

Box 7.1   Evaluation criteria for the coverage of actions and impacts

---

Must the relevant environmental impacts of all significant actions be assessed?

Does the EIA system apply to all public and private environmentally significant projects?

Are the provisions applied in practice to all the actions covered in principle?

Are all significant environmental impacts covered by the EIA system?

... include in every recommendation or report on proposals for legislation and other major federal actions significantly affecting the quality of the human environment, a detailed statement ...

The meaning of each phrase has been picked over by countless court deliberations. Bass and Herson (1993a, p. 27) state that actions typically consist of:

- Adoption of official policies, rules and regulations
- Adoption of plans
- Adoption of program[me]s
- Approval of specific projects, including private undertakings approved by agency permit or regulatory decision.

Examples of activities which may be subject to NEPA include discharges to wetlands and federal land management activities such as mining, oil and gas development, highway and airport construction, port development and navigation projects, timber harvesting, etc.

The very large number of environmental assessments prepared each year bears eloquent testimony to the fact that NEPA is somewhat broader in application than it might at first appear. While there continue to be legal arguments about whether an environmental assessment (EA) is required in particular cases, or whether an environmental impact statement (EIS) rather than an EA should be prepared in certain circumstances, it appears to be true that the EIA provisions enshrined in NEPA are generally applied in practice to almost all the actions to which it is addressed.

If there has been an area of under-application it has been in relation to programmes, plans and policies. NEPA was always intended to permit strategic environmental assessment but, until relatively recently, EIA was largely confined to projects. However, while many programmes and plans are now being subjected to NEPA (Chapter 19) the EIA of policies and legislative proposals appears still to be a relatively neglected area.

The types of environmental impacts covered by NEPA are broad. The impacts which should be included in an EIS have been summarised by Bass and Herson (1993a, p. 69) as:

- Direct effects
- Indirect effects
  - Reasonably foreseeable consequences

- Growth-inducing effects
- Changes in land use patterns, population density, or growth rate
- Cumulative effects
- Conflicts with land use plans, policies, or controls
- Other types of effects
  - Unavoidable effects
  - The relationship between short-term uses of the environment versus long-term productivity
  - Irreversible or irretrievable commitments of resources
  - Energy requirements and conservation potential
  - Natural or depletable resource requirements
  - Effect on urban quality
  - Effect on historical and cultural quality
  - Socioeconomic and market effects.

In practice, as a result of numerous legal challenges, these types of impact are generally covered where they are likely to be significant. A 1994 executive order on environmental justice (No. 12898) requires the coverage of environmental effects, including human health, economic and social effects of federal actions, including effects on minority and low-income communities, in EISs. Recently, there has been debate about how the coverage of issues such as acid precipitation, global climate change and global loss of biodiversity should be included in EISs.

To summarise, the federal EIA system is partial in its application to environmentally significant actions but comprehensive in its coverage of federal actions and environmental impacts. It is thus, perhaps, typical of the fragmentation of US environmental policy which it was designed to address.

California

The California Environmental Quality Act (CEQA) applies to all public and private projects not subject to an exemption. 'Project' is, however, much broader in meaning than might be expected. In California:

'Project' means the whole of an action which has a potential for resulting in a physical change in the environment, directly or ulti-mately, that is:
- An activity directly undertaken as a public agency, including:
  - Public Works construction activities
  - Clearing or grading of land
  - Improvements to existing public structures
  - Enactment and amendment of zoning ordinances
  - Adoption and amendment of local general plans.
- An activity which is supported in whole or part through public agency contracts, grants, subsidies, loans, or other assistance from a public agency.

- An activity involving the public agency issuance of a lease, permit, license, certificate, or other entitlement for use by a public agency.
(Bass and Herson, 1993b, p. 18)

Various government activities fall outside the meaning of 'project'. Others are treated as statutory exemptions (for example, 'ministerial' decisions involving non-discretionary permits such as building permits). Yet others are treated as categorical exemptions (much like the NEPA categorical exclusions). Each public agency may establish its own list of specific activities falling within the 29 classes established by the Resources Agency (e.g. minor modification of existing facilities) but very few have done so.

While there is some discussion about the lists of categorical exemptions established by certain agencies, there is general agreement that CEQA is applied to almost all significant projects. In practice, projects are actually subject to CEQA because of the vigilance of the public in bringing or threatening legal actions where some omission has occurred. Further, there are certain exemptions from the categorical exemptions (Chapter 9).

Olshansky (1991) reported that almost 37 per cent of local EIRs relate to residential projects, 19 per cent to mixed use projects, 11 per cent to retail and office developments, 8 per cent to infrastructure projects, 5 per cent to industrial developments and the remainder to a variety of other projects.

CEQA applies to many policies, plans and programmes and to a number of non-project actions such as the use of pesticides. There is specific provision for 'program' EIRs. A 'program' EIR is prepared for an agency programme or series of actions that can be characterised as one large project. 'Program' EIRs are also prepared for agency plans, policies, or regulatory programmes (Guidelines, sections 15165, 15168). A general plan EIR can be a type of 'program' EIR. There are also provisions for staged, subsequent and supplemented EIRs, for addenda to EIRs and, since 1994, for 'master' and focused EIRs (Chapter 19). Despite the use of the word 'project', CEQA is therefore broad in its application. There is provision for tiering, i.e. the EIA of later projects in policy, plan or program EIRs (Guidelines, section 15385), but this is little used because of agency concerns over potential exposure to lawsuits. Master EIRs provide a five year CEQA exemption period for projects covered by the EIR and are intended to streamline the CEQA process. However, they appear to add little, beyond additional complication, to previous provisions.

The coverage of physical environmental impacts under CEQA is broad. Appendix I to the Guidelines consists of a checklist which covers earth, air, water, plant and animal life, noise, light and glare, land use, natural resources, risk of upset (accident), population, housing, transportation, public services, energy, utilities, public health, aesthetics, recreation and cultural resources. Further, it is a requirement that significant environmental impacts, including direct, indirect, short and long term, and unavoidable impacts from all stages of a project must be covered (Guidelines, section 15126). Growth-inducing and cumulative impacts must be included as must archaeological and historic features. Economic and social effects should only be analysed where they are related to a physical change in the environment.

In practice, the evaluation of social or economic effects is generally treated as optional, but tends to be less comprehensive than that in EISs prepared under NEPA. Generally, practice indicates that most significant environmental impacts are considered in the project EIA process. However, the impacts of plans and programmes are often evaluated and less comprehensively evaluated and cumulative, long-range and indirect impacts are poorly covered in most EIRs.

United Kingdom

Environmental assessment (EA) applies to all projects listed in the European Directive on EIA, subject to the use of screening criteria, no matter under which legislation they fall. This list is lengthy but not comprehensive. As mentioned in Chapter 6, EA has been incorporated into the town and country planning system (and other statutory procedures) by means of regulations made under the European Communities Act which do not allow the requirements of the Directive to be exceeded. However, the Planning and Compensation Act 1991 now allows the Secretary of State for the Environment to require EA of other planning projects. This power was exercised in relation to private motorways, motorway service areas, wind generators and coast protection works in 1994. (Previous proposals to include water treatment plants, trout farms and golf courses were not implemented.)

The UK EIA system is not confined to projects approved under the town and country planning procedures. Together with the other project approval systems into which EIA requirements have been integrated, the Planning Regulations provide for the assessment of most types of project. Exceptions relate to certain types of agricultural project, including some afforestation schemes, and to classified defence projects. Formal arrangements have been put in place for the EIA of projects approved by specific act of Parliament.

Nearly all types of public and private projects are thus subject to assessment. However, whether a particular project is assessed depends upon the screening criteria and thresholds which apply to the project type. It also depends on the application of those criteria by local planning authorities (LPAs). Practice is varied (Chapter 9).

The UK Planning Regulations and other EIA regulations do not apply to programmes, plans and policies (but see Chapter 19). As in the European Directive, the Regulations define the word 'environment' to mean the physical environment: human beings, flora, fauna, soil, water, climate, the landscape, the interaction between any of these, material assets, and the cultural heritage. The social and economic environment is not overtly included in this definition. It is, however, open to LPAs to consider these matters in reaching a planning decision if they choose to do so. The definition of effects adopted in the United Kingdom includes both direct and indirect effects, but while 'secondary, cumulative, short-, medium-, and long-term, permanent and temporary, positive and negative effects' may be described in an environmental statement (Planning Regulations, Schedule 3, para. 3), this is not mandatory.

In summary, the coverage of environmentally significant types of projects requiring planning permission is, in principle, comprehensive but the mandatory coverage of types of environmental impacts cannot be regarded as complete.

The Netherlands

The EIA provisions of the Environmental Management Act 1994 (Wm) apply to significant policy plans and programmes involving locational decisions, as well as to projects. In practice, 30–40 non-project EIAs had been carried out by the end of 1993, on subjects such as waste management, electricity supply, water supply, minerals extraction and land use plans. Most of these EIAs either involve site selection for specific projects or alterations to an existing land use plan as a result of a proposed project. There have also been one or two EIAs for central government policy plans but EIA can often be avoided in these cases.

The EIA system applies to all the projects specified in the EIA Decree. The 1987 Decree, amended in 1992 and 1994, contains a list of projects for which EIA is mandatory. This now corresponds to Annex I of the European Directive (Chapter 3), but for some time The Netherlands had failed to apply EIA unconditionally to all Annex I projects (for example, instead of applying to all oil refineries, it applied to refineries with a capacity of more than 1 million tonnes p.a.).

Part C of the Schedule to the EIA Decree contains a list of activities and decisions for which EIA is required, and a list of criteria to be applied. This list did not cover all the projects set down in Annex II of the Directive, either because they were not considered to have harmful effects or were thought unlikely to arise. This discrepancy between the Dutch provisions and the requirements of the Directive was eliminated by revisions to the Regulations in 1994.

The Dutch EIA system applies to both public and private projects though most of the EIAs undertaken have been for private developments. The provision for exemptions from EIA which is contained in the Act has been used only occasionally (Fig. 5.2) and was due to be partially repealed in 1994. There have been cases where EIA has not been undertaken because of over-liberal interpretation of the guidelines and thresholds but, in general, EIA is carried out if the project is listed in Part C and its size exceeds the relevant threshold. As mentioned in Chapter 6, the courts have required EIA where these criteria were met.

The Act no longer contains a definition of the 'environment' to be covered in EIA, though Section 7.1(2) specifies that energy, resource, waste disposal and traffic impacts are included. The delineation of topics tends to be undertaken by the EIA Commission in the scoping process.

However, it was originally assumed that indirect and cumulative impacts should be included and this has become the EIA Commission's practice. There is, as mentioned in Chapter 6, some confusion over the relevance of social and economic impacts to EIA, with the Ministry of Housing, Physical Planning and the Environment anxious to confine coverage mainly to impacts on the physical environment.

Canada

The Canadian Environmental Assessment Act (CEAA) requires EA where the federal responsible authority (RA):

- proposes the project
- grants financial assistance to the project
- sells or leases land to enable a project to be carried out
- grants a permit or licence for a prescribed project.

'Project' is broadly defined to include construction, operation, modification, decommissioning or other undertaking (and certain physical activities not related to physical works such as dredging). Not only are public projects covered by the Act but most private projects require federal money or land or a federal permit and are thus also covered by the provisions of CEAA. Each of the provincial and territorial governments has its own EA system and many of those private projects not subject to CEAA are caught by their requirements (Chapter 5).

As mentioned in Chapter 5, the Environmental Assessment and Review Process (EARP) covered, in principle at least, policies, plans and programmes as well as projects. CEAA covers only projects, though the Canadian Government has committed itself to the environmental assessment of policies, plans and programmes going before Cabinet (Federal Environmental Assessment Review Office, 1993b). It is quite possible that if the implementation of this commitment proves disappointing there will be pressure for legislated requirements for strategic environmental assessment.

Because the Canadian environmental assessment process relies heavily on the RAs for its implementation, there must be some question as to whether the comprehensive provisions of CEAA will be applied fully in practice, despite the power of the Minister of the Environment to demand additional EA. As Gibson (1993, p. 18) has stated, 'responsible authorities are frequently also proponents and/or advocates of proposed undertakings' and may thus be reluctant to require full EA. There is no action-forcing mechanism in the Act (beyond the public participation and access to information provisions) to ensure compliance by certain departments which have, in the past, ignored EARP with impunity (Smith, 1993, p. 130). Under EARP the number of initial assessment decisions increased markedly when the courts ruled that departments were bound to implement the Guidelines Order they had considered to be discretionary. This probably bodes well for practical compliance at the self-assessment stage of the process, though (as ever) the nature of EA practice may well be variable.

The coverage of environmental effects in CEAA is potentially broad. Not only must direct changes to the biophysical environment be covered, but also 'effects in several socio-economic and cultural areas that flow directly from the environmental effects of the project' (Federal Environmental Assessment Review Office, 1993a). However, the requirement to link effects on:

• human health
• socio-economic conditions
• physical and cultural heritage
• traditional aboriginal uses

directly with project environmental effects can be very restrictive. In addition, it is a requirement that EA should include coverage of:

> the environmental effects of malfunctions or accidents that may occur in connection with the project and any cumulative environmental effects that are likely to result from the project in combination with other projects or activities that have been or will be carried out (Section 16 (1)(a))

together with an evaluation of their significance.

Canada has long been concerned with cumulative environmental effects (Peterson *et al.*, 1987; Sonntag *et al.*, 1987), perhaps because so much of its environment is still in a relatively undisturbed condition. Cumulative environmental effects from projects are those (frequently minor) effects which, when combined (perhaps synergistically) with the effects of other projects, become additively important (Chapter 19). This exemplary concern with cumulative impacts may be complicated by artificial jurisdictional limitations on the coverage of EA as well as by methodological uncertainties. In the past, if an EA had been triggered by federal regulatory involvement in a project, the consideration of impacts under EARP had to be restricted to areas of federal jurisdiction rather than covering the full range of impacts permitted for federal-interest projects. This lack of completeness in the coverage of the impacts of certain projects may also apply under CEAA, weakening the comprehensiveness of the Act. In practice, completeness of coverage may be achieved by application of provincial EA provisions but the potential scope for confusion is considerable.

Commonwealth of Australia

The intended coverage of the Australian Environment Protection (Impact of Proposals) Act 1974 is shown in Box 7.2. In debating the bill in Parliament, the responsible Minister stated that 'we will not be limiting its scope in terms of the type of proposal that could be the subject of a statement' (Cass, 1974, p. 4082). However, as shown in Chapter 19, the

Box 7.2 Coverage of Australian EIA system

The object of this Act is to ensure, to the greatest extent that is practicable, that matters affecting the environment to a significant extent are fully examined and taken into account in and in relation to:

(a) the formulation of proposals;
(b) the carrying out of works and other projects;
(c) the negotiation, operation and enforcement of agreements and arrangements (including agreements and arrangements with, and with authorities of, the States);
(d) the making of, or the participation in the making of, decisions and recommendations; and
(e) the incurring of expenditure

by, or on behalf of, the Australian Government and authorities of Australia, either alone or in association with any other government, authority, body or person.

Source: Environment Protection (Impact of Proposals) Act 1974, Section 5(1).

coverage of the Act has, in practice, been almost entirely restricted to projects.

Generally, the Act has been applied in Australia to proposals which affect the environment to a significant extent and fall into one or more of the following categories:

- activities and projects carried out by Commonwealth departments and authorities, including defence projects, railways, national highways, airports, postal and telecommunication facilities and developments on Commonwealth land
- grants to state governments for programmes, such as Bicentennial road programmes and international standard sporting facilities
- proposals which require Commonwealth approval to export primary products which currently include uranium, coal, mineral sands, aluminium, oil, gas and woodchips
- proposals involving foreign investment, particularly in mining and manufacturing, real estate development and tourist developments.

The Act therefore applies to Commonwealth proposals, and to some state and private projects. As in the United States, the potential coverage of the EIA system is broad but excludes many environmentally significant state and private proposals, many of which are subject to state EIA processes.

In practice, it has proved easy for Commonwealth departments to avoid the EIA procedures. Although memoranda of agreement have been signed between the Commonwealth and many federal departments and most of the states, much discretion remains; so much so that only one special parliamentary bill to exempt a proposal from Commonwealth EIA has ever been necessary (the designation of World Heritage properties).

The 'environment' is broadly defined in Section 3 of the Act as including 'all aspects of the surroundings of human beings, whether affecting human beings as individuals or in social groupings'. In practice, while the Act has generally been interpreted to cover all physical environmental impacts, there has been considerable debate about the extent to which social impacts are covered. In practice, many EISs now deal with social impacts and the trend is clearly towards fuller treatment. There have been several demands that the Australian Commonwealth EIA system should more formally deal with social and cultural impacts, recognising the fact that about 50 per cent of public submissions in relation to EISs have dealt with these (Formby, 1987; Ecologically Sustainable Development Working Groups – ESDWG, 1991; Fowler, 1991). The 1992 Intergovernmental Agreement on the Environment resolved the issue:

> The parties agree that impact assessment in relation to a project, program or policy should include, where appropriate, assessment of environmental, cultural, economic, social and health factors.
> (Commonwealth Environment Protection Agency, 1992, Schedule 3)

It has also been suggested that ecologically sustainable development criteria be explicitly included in the preparation of guidelines (ESDWG Chairs, 1992). Overall, then, while the coverage of actions, in practice, leaves much to be desired, the coverage of environmental impacts is potentially comprehensive.

New Zealand

The Resource Management Act 1991 covers local government actions comprehensively. It applies not only to projects but to policies and plans proposed under the provisions of the Act (Chapter 19). The Act applies to almost every proposed project, as nearly all projects require a resource consent in New Zealand. The EIA provisions thus apply to land use and subdivision consents (which used to be dealt with under the Town and Country Planning Act) and to discharge, water abstraction and coastal permits (which were previously dealt with under a variety of legislative provisions). Some central government projects with major environmental effects are still dealt with under the Environmental Protection and Enhancement Procedures where they fall outside the provisions of the Act (Chapter 5).

The extent to which the Act applies in practice to consents with minor environmental effects depends largely on the screening procedures adopted by local councils. These procedures include the designation of certain types of development as permitted uses (thus not requiring EIA) and as controlled uses (thus requiring a limited form of EIA). They also include the identification of persons likely to be affected by the proposed development who should be informed about it (Chapter 9).

The definition of the term 'environment' adopted is also very broad as it includes 'ecosystems and their constituent parts, including people and

Box 7.3 Content of a New Zealand EIA report and matters to be considered in its preparation

---

Resource Management Act Fourth Schedule

1. **Matters that should be included in an assessment of effects on the environment** – Subject to the provisions of any policy statement or plan, an assessment of effects on the environment for the purposes of section 88(6)(b) should include:
   a) A description of the proposal;
   b) Where it is likely that an activity will result in any significant adverse effect on the environment, a description of any possible alternative locations or methods for undertaking the activity;
   c) Where an application is made for a discharge permit, a demonstration of how the proposed option is the best practicable option;
   d) An assessment of the actual or potential effect on the environment of the proposed activity;
   e) Where the activity includes the use of hazardous substances and installations, an assessment of any risks to the environment which are likely to arise from such use;
   f) Where the activity includes the discharge of any contaminant, a description of:
      i) The nature of the discharge and the sensitivity of the proposed receiving environment to adverse effects; and
      ii) Any possible alternative methods of discharge, including discharge into any other receiving environment.
   g) A description of the mitigation measures (safeguards and contingency plans where relevant) to be undertaken to help prevent or reduce the actual or potential effect;
   h) An identification of those persons interested in or affected by the proposal, the consultation undertaken, and any response to the views of those consulted;
   i) Where the scale or significance of the activity's effect are such that monitoring is required, a description of how, once the proposal is approved, effects will be monitored and by whom.

2. **Matters that should be considered when preparing an assessment of effects on the environment** – Subject to provisions of any policy statement or plan, any person preparing an assessment of the effects on the environment should consider the following matters:
   a) Any effect on those in the neighbourhood and, where relevant, the wider community including any socio-economic and cultural effects;
   b) Any physical effect on the locality, including any landscape and visual effects;
   c) Any effect on ecosystems, including effects on plants or animals and any physical disturbance of habitats in the vicinity;
   d) Any effect on natural and physical resources having aesthetic, recreational, scientific, historical, or cultural, or other special value for present or future generations;
   e) Any discharge of contaminants into the environment, including any reasonable emission of noise and options for the treatment and disposal of contaminants;
   f) Any risk to the neighbourhood, the wider community, or the environment through natural hazards or the use of hazardous substances or hazardous installations.

communities; and all natural and physical resources; and amenity values' together with relevant 'social, economic, aesthetic and cultural conditions' (Section 2). EIA in New Zealand thus encompasses social impact assessment and, in particular, consideration of Maori cultural and community impacts. The comprehensive interpretation of effects is evident from the 'matters to be considered' listed in Box 7.3.

Finally, the word 'effect' is also given a broad meaning in Section 3 of the Act and includes: positive or adverse, temporary or permanent, past, present or future and cumulative effects regardless of their scale, intensity, duration or frequency. It also includes risk. Cocklin *et al.* (1992) believed that the Act, by encouraging the territorial authorities to integrate EIA within the plan-making process, provides the basis for effective cumulative effects assessment, something which has previously been lacking in New Zealand, as elsewhere.

It is apparent that the Mark II New Zealand EIA system, in common with many other systems of long standing, employs considerably broader definitions of actions, of the environment and of effects than, for example, the Mark I UK EIA system.

Summary

The coverage of the seven EIA systems is shown in Table 7.1. The coverage of impacts and projects in the EIA systems in California, The Netherlands and New Zealand is, at least in principle, comprehensive. However, those in the United States, the United Kingdom and the Commonwealth of Australia only partially meet the criterion and that in Canada does not do so. The EIA systems in most developing countries would fail to meet the project coverage criterion, though most would cover different types of environmental impacts (see Appendix).

It is no coincidence that the United States, Canada and Australia are federal countries. The reason for the failure of their systems to cover all significant projects is largely constitutional. The jurisdiction of the federal government only extends to certain projects – the remainder are subject to state or local control, as in the comprehensive Californian EIA system. Even so, many major public and private projects are covered.

The problem may be that certain states, provinces or territories may possess less comprehensive EIA systems than that of California and thus that some environmentally significant projects escape scrutiny altogether. This could also happen in, say, The Netherlands if the thresholds and criteria are set too high. The coverage of projects in many developing countries tends to be very selective, often being dependent on the requirements of the development assistance agencies.

This jurisdictional boundary problem is also the reason why the treatment of certain impacts is limited in Canada. Equally, the UK EIA system leaves some discretion in the coverage of, for example, cumulative, indirect, economic and social impacts to the proponent

Table 7.1   The coverage of the EIA systems

Criterion 2: Must the relevant environmental impacts of all significant actions
                be assessed?

Jurisdiction	Criterion met?	Comment
United States	Impacts: Yes Actions: No	Applies only to federal, not state or most private, projects: comprehensive coverage of impacts of significant federal actions (including some non-project actions).
California	Yes	Coverage of both impacts and projects is comprehensive. CEQA applies also to plans and programmes.
United Kingdom	Impacts: No Actions: Yes	Comprehensive coverage of projects approved under town and country planning process. Some discretion in impact coverage.
The Netherlands	Yes	Covers highly significant projects and certain policies, plans and programmes. Indirect and cumulative environmental impacts covered, but not legally specified.
Canada	No	Artificial limitations on otherwise comprehensive coverage of impacts of certain projects possible. Restricted to federal projects and projects requiring federal finance, land or permit.
Australia	Impacts: Yes Actions: No	Coverage of impacts potentially comprehensive: includes social, economic and cultural impacts. In practice, coverage of actions confined to certain projects.
New Zealand	Yes	Act provides for all local authority approved policies, plans and projects to be subject to EIA covering physical environment, social and economic impacts.

and the LPA. In practice, the coverage of impacts on the physical environment in EIA reports in both countries tends to be reasonably comprehensive.

CHAPTER 8

Consideration of alternatives in EIA systems

Introduction

The consideration of alternatives has been described as 'the heart of the environmental impact statement in the US' (Council on Environmental Quality, 1978, Regulation 1502.14). It has also proved to be a contentious area in EIA. It is, for example, not a mandatory requirement of the European EIA Directive that alternatives to the proposed project be considered in the EIA report. This chapter discusses why the treatment of alternatives in EIA is important and advances several evaluation criteria to assist in the review of EIA systems. These criteria are then employed in the analysis and comparison of the EIA systems in the United States, California, the United Kingdom, The Netherlands, Canada, the Commonwealth of Australia and New Zealand.

Consideration of alternatives in EIA systems

The consideration of the alternatives to an action is the first step in the EIA process (Fig. 1.1). The proponent of an action has a set of aims to be met which can normally be satisfied in a number of alternative ways, each of which has different effects upon the environment. For example, electricity provision to a newly developing suburb might involve some mixture of the construction of new generating capacity, the import of electricity from another region and stringent energy conservation measures. Should it be decided to construct a new generation facility, further alternatives regarding the type of generator (e.g. gas-fired, wind-powered, etc.), location and site layout and design will exist. The choice of an alternative which minimises the environmental impact of the action should be an important determinant of any decision to proceed.

It is at the design stage, before any commitment to any particular action has been made, that it is easiest and cheapest to choose the alternative which best reduces the environmental impacts of an action. Later in the

EIA process it may be necessary to consider another alternative if unforeseen impacts are predicted to arise from that chosen. In order for choices between alternatives to be made, the designer needs to have access to environmental expertise and/or to simple evaluative tools (Brown, 1992). For example, regular meetings between designers and environmental professionals together with a specific, if brief, evaluation of the impacts of different alternatives (including the option of not proceeding with the action) can assist in making environmentally appropriate choices. Easy-to-use, if unsophisticated, methods such as shadows to show the effects of noise from roads in alternative locations can be employed helpfully at this very early stage. Informal consultations with decision-makers, environmental authorities and representatives of the communities affected may be very helpful but, because of their potential sensitivity at such an early stage in the siting process, need to be handled with great care (Lee, 1989).

Once the decisions regarding broad approach and location have been made, more detailed design of the action can take place (Fig. 1.1). Here, where more resources are committed to the action, it is equally important that the avoidance and/or mitigation of environmental impacts continues to be considered. The same techniques of meetings with environmental professionals, specific evaluation and, if appropriate, consultation, together with the use of simple assessment methods, apply as the range of design alternatives narrows and the preferred design emerges. This step in the EIA process, which involves more detailed environmental evaluation is much easier both to accomplish and to demonstrate if the environmental impacts of alternative ways of achieving aims and of alternative locations have been evaluated earlier.

It is, of course, all too easy to profess that a thorough evaluation of the environmental effects of alternatives has been carried out and that the environmental consequences of the detailed design have been fully considered when the reality is different. However, in practice, it is very difficult to assert that the environment has been fully considered in detailed design if the alternative chosen for further elaboration is manifestly more damaging to the environment than some of those rejected. It is for this reason that the analysis of alternatives is so important in the EIA process.

The Council on Environmental Quality surveyed federal agencies in the United States in 1991 to determine how extensively alternatives were actually considered and to what extent this consideration was influencing decisions. The outcome was interesting:

> The results of the survey indicated that when alternatives are not fully considered in the NEPA process, litigation is more likely, and agencies are less likely to achieve the original goals of the project in an efficient, economical manner.
>
> (Bear and Blaug, 1991, p. 18)

While many proponents will be as good as their word in considering alternatives, the public nature of the EIA process demands that the environmental evaluations undertaken at the alternatives/design stage be demonstrated in documentary form. Analysis of alternatives should not only be done but be seen to be done.

One method of providing an early check that the environmental effects of alternatives really have been fully considered is their inclusion in preliminary documents produced prior to the EIA report (e.g. in the notice of intent in the United States). Such documentation should show clear evidence of the mitigation/avoidance of environmental impacts in the initial action designs. For the reasons stated above, this evidence will usually be in the form of an evaluation of the environmental consequences of the alternatives considered. If such evidence is not forthcoming, the proponent can be encouraged to return to this evaluation and, if necessary, to redesign the proposal before too many resources are expended. If documentation does not have to be produced until the scoping stage, or even later in the EIA process, the proponent's commitment to the design will be greater and the chance of cost-effective amelioration may be reduced.

While less satisfactory than the early submission of public documentation, any requirement to discuss the action with the decision-making and/or environmental authorities prior to submission of the EIA report (e.g. at the scoping stage or, preferably, at the screening stage of the EIA process) will involve the inspection of design documents. This will provide an opportunity to check that the most environmentally appropriate alternative and design meeting the proponent's aims has been chosen and, if it has not, to require that further iteration of the design process takes place.

As a final vital check that the environmental consequences of alternative approaches, locations and designs to meet the proponent's aims have been considered, the EIA report should contain evidence to this effect. If the documentation proves to be inadequate it may be possible for the proponent to supply supplementary information but in some instances reconsideration of the proposal may be necessary.

Since the range of possible alternatives to an action may be legion, a choice will need to be made as to the alternatives to be detailed in the EIA report (and other documentation). This choice is usually made on a case-by-case basis, the standard test being that of reasonableness. In the United States, non-feasible, remote or speculative alternatives need not be analysed but sufficient information to permit a reasoned choice of alternatives so far as environmental aspects are concerned, including the no-action alternative, must be provided (Fogleman, 1990, p. 132).

The existence of published advice on the treatment of the environmental impacts of alternatives in the EIA process will be beneficial not only to developers but to consultants, decision-making authorities, environmental authorities, consultees and the public. The various criteria which can be used in evaluating the treatment of alternatives are summarised in Box 8.1. These criteria are employed to assist in the analysis of the treatment of alternatives in each of the seven EIA systems which follow.

Box 8.1   Evaluation criteria for the consideration of alternatives in action design

---

Must evidence of the consideration, by the proponent, of the environmental impacts of reasonable alternative actions be demonstrated in the EIA process?

Must clear evidence of the consideration of the environmental impacts of alternatives be apparent in preliminary EIA documentation?

Must the realistic consideration of the impacts of reasonable alternatives, including the no-action alternative, be evident in the EIA report?

Does published guidance on the treatment of the impacts of reasonable alternatives exist?

Does the treatment of alternatives take place effectively and efficiently?

United States of America

The treatment of alternatives lies at the heart of the US EIA system. The National Environmental Policy Act 1969 (NEPA) specifically refers to the coverage of alternatives to the proposed action (Section 102(2)(C)(iii)). This is evident throughout the EIA process, commencing with the environmental assessment (EA). The EA, which is supposed not to exceed 15 pages in length, must briefly discuss the feasible alternatives to the proposed action and their environmental impacts (unless there are no 'unresolved conflicts concerning alternative uses of available resources') (Regulations, Section 1508.9(b)). In practice, EAs frequently considerably exceed 15 pages in length but often include adequate discussion of the environmental impacts of both the proposal and of the alternatives to it.

The evaluation of alternatives in the environmental impact statement (EIS) is governed by the so-called 'rule of reason' under which an EIS must consider, analyse and compare a reasonable range of options that could accomplish the agency's objectives. An explanation of why alternatives were eliminated should be included. The Regulations (Section 1502.14) state that the range of alternatives to be considered should include:

- alternative ways of meeting the objective
- the no-action alternative
- alternatives outside the lead agency's jurisdiction.

The Regulations further require that rigorous evaluation and comparison are required, that the preferred alternative must be identified and that measures to mitigate the environmental impacts of alternatives must be described. Further, the environmentally preferable alternative must be identified in the record of decision for a proposal.

The Regulations, together with the treatment of alternatives in the first seven of 'NEPA's forty most asked questions' (Council on Environmental Quality, 1981a), provide guidance on the consideration of reasonable alternatives. However, considerable scope for uncertainty remains, given the infinite number of alternatives to an action which may be feasible. The question of alternatives has exercised the courts in the United States on a large number of occasions.

The Council on Environmental Quality is right to claim that the treatment of alternatives is 'the heart of the environmental impact statement' (Regulations, Section 1502.14). There can be no doubt that the US EIA process requires the demonstration, by the proponent, that the environmental impacts of alternative actions have been considered.

California

The treatment of alternatives is given considerable prominence in the Californian EIA system. However, although alternatives must be evaluated in considerable detail, this is a less onerous requirement than that of the federal system under the National Environmental Protection Act (NEPA) which demands that all alternatives must be evaluated equally in an EIS. Unlike the 'environmental assessment' provisions, there is no requirement for discussion of alternatives in the 'initial study' prepared under the Californian Environmental Quality Act (CEQA), or in the 'negative declaration' which usually results from it. Since most projects are judged to have no significant impacts or to be capable of mitigation, lack of discussion of alternatives is common. In any event, alternative approaches are sometimes employed as mitigation measures to ensure that a 'negative declaration' is made.

It is a requirement that responsible agencies (those which must be consulted) should respond to the 'notice of preparation' by indicating the reasonable alternatives they require to be explored in the draft environmental impact report (EIR) (Guidelines, Section 15082). Public consultation to help in the identification of alternatives is also recommended (Guidelines, Section 15083). Practice on the identification of alternatives in scoping varies, with some agencies and (frequently) the public failing to respond substantively. The lead agency may make an initial determination as to which alternatives are feasible and merit in-depth consideration and which do not (Bass and Herson, 1993b, p. 63).

The EIR must describe a range of reasonable alternatives to the project or project location that could attain the objectives for the project, and evaluate their comparative environmental merits. It must focus on the avoidance or reduction of impacts and identify the environmentally superior alternative and explain why alternatives other than the proposed project were rejected (Guidelines, Section 15126). Alternatives must include the 'no-project' option in which a description of the maintenance of existing environmental conditions is used as a baseline for comparing the impacts of alternatives.

Alternatives that are remote or speculative, or whose effects cannot be reasonably predicted, need not be considered. Nevertheless, the range of alternatives evaluated must be full and a clear justification of choices must be provided. For example, reasonable alternatives to the project location must be discussed when appropriate, especially where the proposed project site would result in significant unavoidable impacts.

Box 8.2   Feasibility of alternative sites in Californian EIRs

---

Alternative sites are feasible when:

- the project is proposed by a public agency
- the developer owns or controls feasible alternative sites
- the developer has the ability to purchase or lease other sites
- the developer otherwise has access to suitable alternative sites
- two or more developers are seeking approval from a local agency for the same type of development at different locations, or
- other circumstances necessitate review of alternative sites.

Source: Adapted from Bass and Herson (1993b, p. 65).

While alternative sites are more likely to be feasible for public projects, the Californian courts have determined that they may be feasible for private projects in certain circumstances (Bass and Herson, 1993b, p. 64). These are shown in Box 8.2.

In practice, alternatives to both projects and sites are discussed in EIRs. While the treatment of alternatives is generally considered adequate, this often amounts to a necessary elaboration of the limited variants to the preferred project, rather than to a genuine attempt to choose the environmentally preferable alternative. Unsurprisingly, given the less onerous regulatory requirements (above), the environmental evaluation of alternatives in EIRs is usually less detailed than in EISs for federal projects.

United Kingdom

The desirability of the integration of environmental factors in the choice of alternative and in initial design is a fundamental reason for making the proponent responsible for producing the EIA report in the United Kingdom. As the Department of the Environment (DOE) has put it:

> From the *developer's* point of view, the preparation of an environmental statement in parallel with project design provides a useful framework within which environmental considerations and design development can interact. Environmental analysis may indicate ways in which the project can be modified to anticipate possible adverse effects, for example, through the identification of a better practicable environmental option, or by considering alternative processes.
>
> (DOE, 1989, p. 3, emphasis in original)

Since preliminary EIA documentation does not have to be submitted to the local planning authority (LPA) or to any environmental authority, and since discussion between the proponent and these bodies is not required in the UK EIA system, the environmental statement (ES) provides the only formal check on the treatment of alternatives.

It is, therefore, perhaps surprising that the UK Planning Regulations do not require that alternatives must be discussed in the ES. In this, and in their permissive approach to the treatment of alternatives, they reflect the European Directive on EIA.

Schedule 3 to the Regulations states that an ES may include, 'by way of explanation or amplification':

> (in outline) the main alternatives (if any) studied ... and an indication of the main reasons for choosing the development proposed, taking into account the environmental effects (para. 3).

Alternatives are also included in the checklist of matters to be considered for inclusion in an ES in the DOE Guide to the UK EIA process. It is suggested here that the 'main alternative sites and processes considered, where appropriate, and reasons for final choice' may be relevant (DOE, 1989, Appendix 4, 1.4).

The lack of regulatory weight given to the treatment of the environmental impacts of alternatives is reflected in practice. About one-fifth of a sample of 24 LPAs questioned in 1990 felt that the treatment of alternative locations and of alternative processes and/or designs in the ESs they had received was unsatisfactory (Wood and Jones, 1991). Jones *et al.* (1991) also found that the treatment of alternatives in ESs was frequently unsatisfactory.

The Netherlands

The Dutch EIA system lays considerable emphasis on the treatment of alternatives. In The Netherlands alternatives may include measures which would be described as mitigatory elsewhere (e.g. technical controls over pollution). Although the Notification of Intent Decree does not specify that the proponent must include mention of alternatives in the notification, some information on alternatives is almost always provided. This provides the EIA Commission with an indication of the alternatives the developer is likely to consider in the EIS. The EIA Commission then suggests, in its guideline recommendations, the alternatives which it believes it is reasonable to study. These alternatives may include different sites, processes, designs or mitigation methods and are normally adopted by the competent authority.

Section 7.10 of the Environmental Management Act 1994 (Wm) specifies the minimum contents of an EIS and lays great emphasis on the coverage of reasonable alternatives. Alternatives to the proposed activity and their environmental consequences must be described both in the EIS and in the non-technical summary. Further, a comparison between the environmental impacts of the proposed development and of the alternatives considered is required. Section 7.10 effectively requires that the 'no action' alternative be described in the EIS because the future environment in the absence of the proposal must be described. Sometimes, if it is viable, the no-action alternative is specifically described. Finally, 'the alternative in which the best available possibilities for protection of the environment are applied' must also be described in the EIS (Section 7.10(3)).

There is a published guide specifically dealing with the treatment of the environmental impacts of alternatives and some general guidance is available in the EIA handbook (Chapter 5). The EIA Commission provides project-by-project guidance on alternatives in its guideline recommendations. A considerable proportion of the guidelines for each project is normally devoted to the treatment of alternatives.

Following receipt of the EIS, the competent authority is obliged to state the grounds on which the decision is based (Chapter 13). It is therefore compelled to indicate how it has taken into account the effects of the proposed activity, and of its alternatives, on the environment.

There are different views about how well alternatives are treated in The Netherlands. Many Government and business participants in the EIA process believe that the treatment of alternatives is broadly satisfactory whereas environmental groups believe that the analysis of alternatives is frequently the weakest part of an EIS. Alternatives appear often to involve changes at the margin, rather than radically different approaches. The Evaluation Committee on EIA (ECW, 1990, p. 7) supported this view:

> There is a statutory requirement to specify the most environmentally sound alternative to the proposed activity. If the alternative is not worked out in sufficient detail – which often happens – it is impossible to obtain a satisfactory impression of the most viable alternative.

It suggested a reformulation of the Act to clarify the definition of the environmentally most benign alternative in order to force proponents to think more deeply about alternative ways and means of implementing their proposals. This change was implemented in 1994: *inter alia*, consideration of compensation measures is now required. The EIA Commission (1994) has said that there are shortcomings in the treatment of alternatives in some EISs.

Canada

There is considerable emphasis on the treatment of alternatives in the various documents referred to in the Canadian Environmental Assessment Act (CEAA). There is, however, no requirement for the consideration of alternatives and their environmental effects to be included in screening reports. Section 16(1)(e) of the Act states that any other matter relevant to the screening may be included if the responsible authority (RA) requires it, including 'alternatives to the project'. The Guide to the Act states that the RA may require consideration of:

> alternative means of carrying out the project: a description of the alternative means, the environmental effects of any such alternative means and a rationale explaining why the alternatives were rejected.
> (Federal Environmental Assessment Review Office, 1993a, p. 103)

It is mandatory to consider 'alternative means' in comprehensive study, mediation and panel reports. Section 16(2)(b) requires that such reports include consideration of:

alternative means of carrying out the projects that are technically and economically feasible and the environmental effects of any such alternative means.

The consideration of 'alternatives to' the project remains discretionary in these reports but inclusion may be required by the Minister of the Environment in consultation with the RA. Box 8.3 summarises the distinction between 'alternative means' and 'alternatives to' in the Canadian system. This table represents the full extent of the coverage of alternatives in the Guide to CEAA, beyond stating that the rationale explaining why the alternatives were rejected should be included (above).

Gibson (1992, 1993) felt that the discretionary provision relating to 'alternatives to' the project is likely to be used rarely and falls short of the comparative evaluation of alternatives necessary to force effective integration of environmental concerns into the design stage for new proposals. However, Canadian practice in relation to alternatives has been generally satisfactory at the panel review stage. Thus, in relation to the expansion of Toronto Airport, the Panel Guidelines (FEARO, 1990) required consideration of the environmental effects of various alternatives:

1. Improvements in productivity achieved with the existing configuration.

Box 8.3 The Canadian Environmental Assessment Act provisions for alternatives

---

Alternatives

The CEAA distinguishes between 'alternative means' and 'alternatives to':

'Alternative means' of carrying out the project are methods of a similar technical character or methods that are functionally the same. 'Alternative means' with respect to a nuclear power plant, for example, includes selecting a different location, building several smaller plants and expanding an existing nuclear plant. 'Alternative means' that are technically and economically feasible must be considered in a comprehensive study, mediation, and panel review, but are discretionary under a screening.

In contrast, 'alternatives to' the project are functionally different ways of achieving the same end. For example, 'alternatives to' the nuclear power plant include importing power, building a hydro-electric dam, conserving energy, and obtaining the energy through renewable sources. Consideration of 'alternatives to' the project is at the discretion of the RA in screening, or of the Minister in consultation with the RA in a comprehensive study, mediation or panel review.

Source: Federal Environmental Assessment Review Office (1993a, p. 106).

2. Diversion of air traffic to other regional airports.
3. Addition of only one (not two) runway.
4. Development of a second major airport.
5. Development of a replacement major airport.
6. A combination of some or all of these alternatives.

The provision of this 'alternative means' information under EARP clearly involved considerable effort but was broadly achieved. In general, Canadian practice in relation to the treatment of alternatives under EARP has been mixed, with some EA reports providing broad-ranging coverage. However, consideration of alternatives has been negligible in initial assessments. There is a danger that this will continue to be the case, with most screening reports failing to mention alternatives but with largely satisfactory treatment of 'alternative means' (but not of 'alternatives to') in other types of report.

Commonwealth of Australia

The treatment of alternatives is an important element in the Australian EIA system. Thus, in the notice of intention prepared for screening purposes (Chapter 9), a list of the alternatives considered must be prepared. Equally, Box 8.4, which lists the content requirements for Commonwealth environmental impact statements, indicates the stress placed on the no-action alternative, 'feasible and prudent' alternatives and the reasons for choosing the preferred alternative. The requirements for the treatment of alternatives in public environment reports are similar.

No specific guidance on the treatment of alternatives exists and, in practice, their coverage in EISs tends to be weak. In particular, the no-action alternative is frequently glossed over in EISs. Overall, there is little evidence that the environmental impacts of meaningful alternatives are adequately discussed in Australian EISs.

New Zealand

The consideration of alternatives in the EIA report is expected in New Zealand. The Fourth Schedule to the Resource Management Act 1991 (Box 7.3) requires proponents of proposals likely to have a significant impact to provide information on 'any possible alternative locations or methods for undertaking the activity'. In addition, where the discharge of a pollutant is involved, the proponent must describe 'any possible alternative methods of discharge, including discharge into any other receiving environment'. This emphasis on the best practicable environmental option is interesting, since the Act provides probably the first instance of the integration of legislative provisions for EIA and for achieving this option anywhere in the world.

Where a project is deemed by the local council to have major effects and is notified, the council may:

Box 8.4   Content requirements for alternatives in Australian EISs

To the extent appropriate in the circumstances of the case, an environmental impact statement shall:

(a) state the objectives of the proposed action;
(b) analyse the need for the proposed action;
(c) indicate the consequences of not taking the proposed action;
(d) contain a description of the proposed action;
(e) include information and technical data adequate to permit a careful assessment of the impact on the environment of the proposed action;
(f) examine any feasible and prudent alternative to the proposed action;
(g) describe the environment that is likely to be affected by the proposed action and by any feasible and prudent alternative to the proposed action;
(h) assess the potential impact on the environment of the proposed action and of any feasible and prudent alternative to the proposed action, including, in particular, the primary, secondary, short-term, long-term, adverse and beneficial effects on the environment of the proposed action and of any feasible and prudent alternative to the proposed action;
(i) outline the reasons for the choice of the proposed action;
(j) describe, and assess the effectiveness of, any safeguards or standards for the protection of the environment intended to be adopted or applied in respect of the proposed action, including the means of implementing, and the monitoring arrangements to be adopted in respect of, such safeguards or standards; and
(k) cite any sources of information relied upon in, and outline any consultations during, the preparation of the environmental impact statement.

Source: Commonwealth of Australia (1987, Para. 4.1).

Require an explanation of:
(i) any possible alternative locations or methods for undertaking the activity and the applicant's reasons for making the proposed choice (Section 92(2)(a)).

This provision clearly invites councils to ensure that alternatives are fully explored. It is noteworthy, however, that the no-action alternative is not specifically mentioned.

The Act states only that information on alternatives and on other matters in the Fourth Schedule should be included 'subject to the provisions of any policy statement or plan'. However, since Section 88(6)(b) of the Act states that the assessment 'shall be prepared in accordance with the Fourth Schedule' it is apparent that a proponent ought to provide this information where an EIA report is required unless the local or regional council has provided a different specification. It is, of course, open to regional and district councils to include more stringent requirements relating to the treatment of alternatives in their policy statements or plans.

There is no published guidance on the treatment of the impacts of alternatives in the New Zealand EIA system. Indeed, it is notable that none of the guidance commissioned (Morgan and Memon, 1993) or published by the Ministry for the Environment (Chapter 5) stresses alternatives. It is thus apparent that the treatment of alternatives has not been given priority in the implementation of the New Zealand EIA system. This is reflected in practice: alternatives have not generally been treated satisfactorily in EIA reports, with the exception of those dealing with new road proposals.

Summary

The consideration of alternatives in the seven EIA systems is shown in Table 8.1. It can be seen that the 'alternatives' criterion is met, to a greater

Table 8.1   The consideration of alternatives in the EIA systems

Criterion 3:   Must evidence of the consideration, by the proponent, of the environmental impacts of reasonable alternative actions be demonstrated in the EIA process?

Jurisdiction	Criterion met?	Comment
United States	Yes	Treatment of alternatives required in almost every environmental assessment and lies at 'heart of environmental impact statement' (EIS).
California	Yes	Full range of alternatives must be evaluated in environmental impact report (EIR) and a clear justification of choice must be provided.
United Kingdom	No	No regulatory requirement. Regulations permit consideration of alternatives and guidance advises it. Practice varies.
The Netherlands	Yes	Alternatives, including the 'no-action' and the environmentally preferable alternatives, must be considered in scoping, the EIA report and the decision.
Canada	Yes	Treatment of 'alternative means' in comprehensive study, mediation and panel reports but not in screening reports; discretionary provision for treatment of 'alternatives to' in all EA reports. Practice varies.
Australia	Yes	Alternatives must be listed in notice of intention and fully treated in EIA reports. Practice often inadequate.
New Zealand	Yes	EIA report should contain discussion of alternative locations and methods. Practice is often weak.

or lesser extent, in the United States, California, The Netherlands, Canada, Australia and New Zealand. It is, nevertheless, true that the treatment of alternatives in these jurisdictions often leaves a great deal to be desired. If practice in the United States and California is at the leading edge, practice in the treatment of alternatives in the Australian federal EIA system and in New Zealand is frequently unsatisfactory, and that in some cases in Canada and The Netherlands has been properly criticised. The treatment of alternatives in the EIA systems in developing countries is often weak.

The only EIA system which does not always require the treatment of alternatives in EIA reports is that in the United Kingdom. The consideration of alternatives is, in effect, discretionary in Britain and the official guidance strongly advises that alternatives be described in environmental statements. In practice, some UK ESs contain adequate discussion of a reasonable range of alternatives to the proposed action but this is totally absent from others.

Practice in relation to the treatment of the 'no-action' alternative and the environmentally preferable alternative could undoubtedly be improved in all the jurisdictions and in developing countries (see Appendix) to enable a better informed comparison to be made with the proposed action and hence to assist in reducing the severity of its environmental impacts. The first step to such improvement in the United Kingdom (and in many developing countries) must be the introduction of a formal requirement that the environmental impacts of alternatives to every proposed project be considered.

CHAPTER 9

Screening of actions

Introduction

The determination of whether or not an EIA report is to be prepared for a particular action normally hinges upon the question of the significance of its environmental impacts. Two broad approaches to the establishment of significance may be identified in EIA systems:

- the compilation of lists of actions and of thresholds and criteria to determine which should be assessed
- the establishment of a procedure for the discretionary determination of which actions should be assessed.

In practice, most EIA systems adopt a hybrid approach involving lists, thresholds and the use of discretion. In some, different types of EIA (with different documentary and participation requirements) are employed for actions with different levels of significance.

This chapter discusses the issue of significance of impacts of actions and puts forward several evaluation criteria intended to assist in the analysis of EIA systems. The EIA systems in the United States, California, the United Kingdom, The Netherlands, Canada, the Commonwealth of Australia and New Zealand are then reviewed utilising these criteria.

Treatment of the screening of significant actions in EIA systems

It is clearly important that some form of screening takes place in any EIA system. Without it, large numbers of actions would be assessed unnecessarily and/or actions with significant adverse impacts would not be assessed. The question of significance has created difficulty in the United States from the outset and has been the most frequent cause of litigation on EIA over the years (see Chapter 2). The US courts have generally ruled that an environmental impact statement was necessary when significant effects were present and where its preparation was reasonable under the circumstances. This type of legal test could, at least in principle, be applied in other jurisdictions. Interestingly, the courts in the United States have rejected specific size or monetary factors as a guide to determining the significance of an action (Bear, 1989).

The Council on Environmental Quality Regulations set forth ten general criteria for the determination of significance on a case-by-case basis (Box 9.1). These 'intensity' criteria are to be applied within the societal and environmental 'context' in which the action would occur. Some of these criteria have been adopted or modified by other jurisdictions.

As mentioned in Chapter 3, the European Directive on EIA lists a limited number of Annex I projects for which assessment is mandatory (save in very exceptional circumstances). It then goes on to specify other (Annex II) projects for which criteria and thresholds may be applied to determine whether or not they should be assessed. These criteria should reflect the nature, size and location of the proposal. Other jurisdictions (for example, the Commonwealth of Australia) have not published lists of criteria or lists of proposals but have established discretionary screening procedures.

The United States possesses a *de facto* simplified form of EIA report: the environmental assessment. The use of an additional level of assessment was considered but rejected by the European Commission (Chapter 3) but has been adopted, in a modified form, by other jurisdictions such as the Commonwealth of Australia.

Box 9.1   US Council on Environmental Quality significance criteria

---

1. Is the impact adverse or beneficial?

2. Does the action affect public health or safety?

3. Is the action located in a unique geographic area?

4. Are the effects likely to be highly controversial?

5. Does the proposed action pose highly uncertain or unique or unknown risks?

6. Does the action establish a precedent for future actions with significant effects, or represent a decision in principle about future considerations?

7. Is the action related to other activities with individually insignificant but cumulatively significant impacts?

8. To what degree may the action affect designated or listed and protected sites?

9. To what degree may the action adversely affect endangered or threatened species and habitats?

10. Could the action contravene other environmental legislation?

Source: Amended from CEQ Regulations 40 CFR 1508.27(b), by Land Use Consultants (1992).

In practice, there is a variety of procedures for screening actions which mix the approaches described and which permit different levels of scrutiny and of consultation and participation. Whichever approach is adopted, it is clearly important that, if screening is to be operated effectively, the proponent should be required to submit information to assist the decision-maker and/or the relevant environmental authorities in determining whether EIA is necessary in any particular case. The environmental assessment nominally performs this function in the United States.

Since proponents require as much certainty as possible in determining whether assessment is likely to be required, clear and detailed information about actions, criteria, thresholds and screening procedures generally should be available. Such guidance is helpful not only to the proponent but to all the other participants in the EIA process.

To instill confidence in the screening process, an identifiable decision should be made by a publicly accountable body and the reasons for making it should be on the public record. Further involvement by the public and by environmental authorities would require a formal period of public participation and, beyond that, a third party right of appeal against screening decisions. This would obviously include the right of the proponent to appeal.

Whatever screening procedure and level of participation is adopted, it is obviously necessary that it works effectively and efficiently. That is, in general, those actions with significant impacts (and only those actions) should be assessed and decisions should be made within a specified period of time without undue expenditure by any of the participants in the EIA process.

The criteria which can be advanced to analyse the treatment of screening are summarised in Box 9.2. These criteria for dealing with the

Box 9.2   Evaluation criteria for the screening of actions

Must screening of actions for environmental significance take place?

Is there a legal test of whether the action is likely to affect the environment significantly?

Is there a clear specification of the type of action to be subject to EIA?

Do clear criteria/thresholds exist (e.g. size, location)?

Do different types of EIA exist for different types of actions?

Must documentation be submitted by the proponent to assist in screening?

Is information about actions, criteria, thresholds and screening procedures readily accessible?

Is the screening decision made by a publicly accountable body?

Does consultation and participation take place during screening?

Is there a right of appeal against screening decisions?

Does screening function effectively and efficiently?

---

issue of significance of action impacts are now used in reviewing the way this stage is handled in each of the seven EIA systems.

United States of America

There is no formal legal test in the US EIA system of whether the proposed action is likely to affect the environment significantly. Nor is there a set of criteria or thresholds in existence which permit precision in determining whether an environmental impact statement (EIS) must be prepared. Rather, a three-step process applies under the provisions of the National Environmental Policy Act (NEPA):

1. Determine whether NEPA applies to the proposed action (includes determination of whether action is 'categorically excluded').
2. Determine whether the proposed action may 'significantly affect the quality of the human environment' (usually involves preparation of an environmental assessment – EA).
3. Prepare an EIS if significant impacts are anticipated.

In practice, the EA is bypassed for certain types of proposed actions for which agencies have published EIS criteria in their own NEPA regulations.

The EA is a supposedly concise public document which is intended to serve three purposes:

- provide sufficient evidence and analysis to determine whether an EIS is required
- support an agency's compliance with NEPA when no EIS is required
- facilitate preparation of an EIS when one is required.

It must discuss the need for the proposed action, reasonable alternatives (Chapter 8), the probable environmental impacts, and the agencies and persons consulted. Agencies must provide notice of the availability of EAs. Scoping of EAs is not a requirement but is undertaken by some agencies. Other federal agencies and the public are supposed to be involved in the preparation of the EA (Regulations, Section 1501.4(b)).

The decision to proceed to an EIS or to prepare a finding of no significant impact (FONSI) is taken by the lead agency. This public document must succinctly state the reasons for deciding that the action will not have significant effects on the human environment and summarise or attach the EA. Many federal agencies are now preparing mitigated FONSIs, i.e. reducing all the significant impacts of the proposed action to less than significant levels. Although the different federal agencies have different rules about consultation and public participation there have been many legal challenges to EAs and FONSIs, indicating considerable dissatisfaction with the screening process.

This is not surprising. Some 40,000–50,000 EAs are prepared annually generally by federal agencies (rather than by contractors or applicants), but few agencies use EAs to determine whether an EIS is needed (Blaug,

1993). Some EAs are undoubtedly EISs in disguise, perhaps to try to avoid the public scrutiny and possible delay involved in EIS preparation (Bear, 1989). Despite the length of EAs (about 75 per cent exceed the recommended 15 pages) and their cost (frequently up to $US100,000) the majority of federal agencies do not, in practice, involve the public (Blaug, 1993). Generally, since around 100 EAs are prepared for each EIS, this indicates serious shortcomings in the US EIA system.

California

The screening process in California is lengthy. It consists of a preliminary review to determine whether a project is subject to CEQA, the preparation of an initial study, and a decision about whether the project is likely to have a significant environmental impact. If it is, an environmental impact report must be prepared. If it is not, then a negative declaration or mitigated negative declaration is prepared. The preliminary review begins with the filing of application forms and a detailed project description and an acceptance by the lead agency that the application is complete.

With very few exceptions 'statutory exemptions' are based on types of activity (e.g. temporary water transfers of up to one year's duration). However, most 'categorical exemptions' are based on criteria which generally have to be set by the lead agency (e.g. subdivision of certain properties in urban areas into four or fewer parcels). There are provisions for exceptions to the categorical exemption list if, for example, the project would cause substantial adverse changes in significant historic resources. Where a lead agency decides that a project is exempt from the California Environmental Quality Act (CEQA) it may file a notice of exemption providing its reasons and must notify the applicant.

If the project is not exempt, an initial study must be undertaken. This is a brief document, similar to the environmental assessment (EA) under the US National Environmental Policy Act, the contents of which include:

– project description
– environmental setting
– potential environmental impacts
– mitigation measures for any significant effect
– consistency with plans and policies
– names of preparers.

(Bass and Herson, 1993b, p. 29)

The initial study may use a checklist format and an example of a checklist is provided in Appendix I to the Guidelines. However, an explanation must be used to support the completion of the checklist (which provides boxes to be ticked).

If, on the basis of substantial evidence from the completed initial study, the lead agency believes that a fair argument exists that the project may have a significant impact on the environment, an EIR must be prepared.

If, on the other hand, there is no such evidence, or if the proponent agrees to project revisions that clearly mitigate any effects, a negative declaration may be issued. To help the agency make this decision a set of mandatory findings of significance are set out in the Act (CEQA, Section 21083). Thus, for example, an impact is considered significant if it will cause 'substantial adverse effects on human beings, either directly or indirectly'. There is, in addition, a list of effects which are normally considered to be significant (Guidelines, Appendix G). One example of these impacts is that the project would 'have a substantial, demonstrable negative aesthetic effect'. To help in making these decisions, agencies are encouraged to consult with other agencies or develop their own thresholds of significance. Some have attempted to provide more practical indicators than those in the Guidelines by adopting quantitative thresholds (see, for example, Bass and Herson, 1993b, Appendix 5).

The negative declaration must be made available to the public for review. It must contain a description of the project, the proposed finding of no significant effect, attach a copy of the initial study and state the proposed mitigation measures.

The Californian screening procedure generally works reasonably well and the quality of documentation has improved. Most initial studies do not meet the advice in the Guidelines that they should be less than 15 pages long (but few exceed 30 pages). Unlike the NEPA EAs they are used to determine whether an EIR should be prepared. Like EAs, initial studies are also environmental impact assessment documents in their own right. The two main problems with the screening procedure are the lack of involvement by other agencies when consulted, principally as a result of staff shortages, and the openness to subsequent legal challenge of the screening decision. Consequently, there is no certainty that the screening stage is actually complete since significant impacts may come to light later in the EIA process.

United Kingdom

As described in Chapter 4, the UK planning regulations contain two lists of projects. For Schedule 1 projects (for example, oil refineries, large power stations, and toxic waste disposal sites), for which environmental assessment (EA) is mandatory, it is normally clear whether a particular project requires EA. For the longer list of Schedule 2 projects, whether a project will require an environmental assessment depends on the likely significance of its environmental effects. These, in turn, will depend on the nature and scale of the project, the location and the complexity or adversity of effects (Department of the Environment – DOE, 1989). Because these criteria are necessarily general, the UK advisory Circular (DOE, 1988b) sets out quantified thresholds and indicative criteria for different categories of Schedule 2 projects. An example of these criteria and thresholds for certain mineral operations is presented in Box 9.3.

The UK system is binary: either a project is subject to EA or it is not. There is no provision for 'simplified' EA, though, inevitably, the environ-

Box 9.3 Indicative UK screening criteria and thresholds for mineral projects

Whether or not mineral workings would have significant environmental effects so as to require EA will depend upon such factors as the sensitivity of the location, size, working methods, the proposals for disposing of waste, the nature and extent of processing and ancillary operations and arrangements for transporting minerals away from the site. The duration of the proposed workings is also a factor to be taken into account.

It is established mineral planning policy that minerals applications in national parks and areas of outstanding natural beauty should be subject to the most rigorous examination, and this should generally include an environmental assessment.

All new deep mines, apart from small mines, may merit an environmental assessment. For open-cast coal mines and sand and gravel workings, sites of more than 50 ha may well require an environmental assessment and significantly smaller sites could require an environmental assessment if they are in a sensitive area or if subjected to particularly obtrusive operations.

Whether *rock quarries* or *clay operations* or other mineral workings require EA will depend on the location and the scale and type of the activities proposed.

For *oil and gas extraction* the main considerations will be the volume of oil or gas to be produced, the arrangements for transporting it from the site and the sensitivity of the area affected. Where production is expected to be substantial (300 tonnes or more per day) or the site concerned is sensitive to disturbance from normal operations, EA may be necessary. *Exploratory deep drilling* would not normally require EA unless the site is in a sensitive location or unless the site is unusually sensitive to limited disturbance occurring over the short period involved. It would not be appropriate to require EA for exploratory activity simply because it might lead to production of oil or gas.

Source: Department of the Environment (1988b, Appendix A (emphasis in original)).

mental statements for certain projects (especially those with potentially complex impacts such as toxic waste incinerators) tend to present a much fuller treatment than those for others (e.g. afforestation projects). Many minor developments which fail to be approved by local planning authorities (LPAs) within the town and country planning system are not, of course, subject to EA.

Where there is any doubt about the need for EA, the developer is advised to consult the LPA, which is an elected body, to obtain an informal view or a formal 'opinion'. The LPA may, in turn, refer to the statutory consultees for advice (DOE, 1988b). Where a formal opinion is sought by the proponent, information about the nature, purpose and possible effects of the proposal on the environment must be provided. If the LPA, on receipt of this information or of a planning application unaccompanied by an environmental statement (ES), determines that one

121

is required, there is provision for the developer to appeal to the Secretary of State for the Environment against this screening decision.

The Secretaries of State can issue a 'direction' that an EA be undertaken and submitted to the LPA even in the absence of an appeal by the developer. Public pressure or the opinion of statutory consultees could thus succeed (no doubt on very rare occasions) at central government level in obtaining EA, even if it fails to convince the LPA of the need. There is, however, no formal third party right of appeal against screening decisions. The types of projects potentially subject to EA, the criteria and thresholds and the procedures to be followed are clearly set out in the Planning Regulations in the Circular (DOE, 1988b) and in the *Guide to the Procedures* (DOE, 1989), supplemented by Circular 7/94 (DOE, 1994a).

The formal procedures for obtaining opinions and directions have been used very little in practice. Of a sample of 24 ESs received by LPAs, just over half were initiated by a request from the LPA, with just under a third being volunteered by the developer without previously notifying the LPA. Statutory consultees were sometimes involved in the screening process but the public did not participate. In a quite separate sample of 24 LPAs which had not received ESs, 30 cases (in 12 LPAs) were discovered where EA could have been appropriate had the advice in the Circular been followed. In the 24 cases in the second sample, half the LPAs appeared not to have considered the question of whether or not to request EA, the 'decision' often being taken by junior planners having little knowledge of EA (Wood and Jones, 1991).

There was widespread agreement among the 36 LPAs questioned that further guidance (including examples) was needed on how to define 'significant environmental effect' for Schedule 2 projects. The interpretation of the criteria in the Circular by LPAs tended to vary and several were thought to be ambiguous (Wood and Jones, 1991).

Overall, it can be seen that, while screening of environmentally significant actions does take place, practice is variable and is not always effective.

The Netherlands

As explained in Chapter 7, Part C of the Schedule to the Environmental Impact Assessment Decree contains a list of activities for which EIA is mandatory. This Schedule contains three columns: in the first the activities for which EIA is required are set down; in the second column threshold values beyond which EIA must be applied are given; and in the third column the decisions are specified which involve mandatory EIA (Commission of the European Communities, 1993c). The thresholds are usually based on area (e.g. an industrial estate of more than 100 ha) or weight (e.g. an industrial waste incinerator for 25,000 tonnes p.a. or more). Little discretion is seemingly allowed: implementation of EIA is mandatory if the extent of the proposed activity exceeds the threshold. (Proponents may, if they choose, voluntarily undertake EIA.)

This list of activities was insufficient to meet the requirements of both Annex I and Annex II of the European Directive on EIA (Chapter 3) and, in addition, many of the threshold values are considered by the participants in the EIA process to be arbitrary and to be set too high. Further, a number of problems were experienced as a result of the lack of clarity of certain descriptions. Most of these difficulties have now been remedied by amendment to the Regulations to include all Annex I projects, by the provision of extra information or the making of supplementary policy or occasionally by judicial opinion (Commission of the European Communities, 1993c). In effect, The Netherlands has operated a highly sophisticated EIA system for a highly selective set of projects. Many environmentally significant proposals have eluded the EIA net in the past.

A new screening procedure was to be implemented to supplement the list in Part C in 1994, partly as a result of the recommendations of the Evaluation Committee on EIA (1990). Part D of the Schedule to the EIA Decree was to contain a further list of activities and thresholds which fully meets the requirements of the EIA Directive. This Part D list was to be utilised with accompanying criteria relating to the nature of the area, whether impacts are likely to be cumulative, the complexity of the project, etc. The new screening procedure was to be applied in each individual case by the competent authority.

There is no provision for different types of EIA but the project-specific scoping guidelines issued by the competent authority provide strong advice on the coverage of each EIA. The notification of intent was not, previously, used for screening purposes but documentation is now submitted by the proponent to assist in screening under the new procedures. The notification of intent (which should be brief) now covers the purpose and goal of the proposal, a general description, the alternatives to be considered, a description of the area and the impacts to be considered. Information on screening was to be available not only in the EIA Decree but in the new EIA handbook (Chapter 5).

While there is consultation of select bodies (e.g. the provincial environmental inspectorate) during screening there is no public participation. There is a third party right of appeal to the competent authority at the decision stage of the EIA process which may relate to screening decisions. On balance, while too many projects have escaped the EIA process in the past, screening has worked well for the limited range of projects covered in The Netherlands.

Canada

The Canadian screening situation is complicated by two factors. First, the EA system has two separate, sometimes successive, procedures, each with its own screening steps. It is convenient to discuss the screening step for each procedure separately (a pattern that will frequently be followed in the steps discussed in succeeding chapters). Second, one track in the self-directed assessment is called a 'screening', an important purpose of which is to determine whether additional EA is required.

In the *self-directed assessment* the responsible authority (RA) determines, utilising project information provided by itself or by the proponent (usually an application), the exclusion list and the inclusion list, whether EA is required. If EA is required, the RA determines whether this will be in the form of a screening or of a comprehensive study by referring to the comprehensive study list (which consists of types of projects, sizes of projects and certain locational considerations). There is no legal test of significance but decisions are potentially subject to judicial review in the federal courts. It is likely that the prescription of projects subject to comprehensive study will lead to around 25–50 reports being produced each year (far more than the number of initial environmental evaluations undertaken under the Environmental Assessment and Review Process (EARP)). An additional category of EA document is the 'class screening' report which acts as a model for future projects. This report presents a comprehensive discussion of the environmental effects of a class or type of project and identifies known mitigation measures.

There is no legal requirement for the proponent to provide a document to assist in screening nor for any form of public participation or administrative appeal. It is too early to say whether screening in the self-directed assessment procedure will function effectively and efficiently. Certainly, it is expected to operate more effectively than the initial assessment under EARP. The discretion provided by EARP caused much confusion and some agencies failed to carry out screening before the 1989 court cases made EARP mandatory (Fenge and Smith, 1986). Subsequently, however, practice has improved and public participation in screening is now widespread.

The decision to require further EA (*public review*) beyond the self-directed assessment, is to be taken by the Minister of the Environment, often at the request of the RA. The Minister can take this decision following the RA's decision upon completion of the screening report or comprehensive study report, while self-directed assessment is taking place or, occasionally, before self-directed assessment has commenced. The criteria for the decision are the likely significant adverse effects of the project, and/or the existence of public concerns.

The Minister, after seeking the advice of the RA and the Canadian Environmental Assessment Agency, may decide that mediation will take place either instead of a panel review or will be applied during the course of a panel review. The criteria for determining the appropriateness of mediation include:

- What are the potential sources of uncertainty or disagreement? For example, do the disputes involve fundamental opposition to the proposed project, technical issues, the determination of environmental effects and their significance, or the effectiveness of mitigation measures?
- Are these disagreements negotiable? Is there room for compromise and consensus?

- Who are the main parties involved?
- Do the parties agree on the areas of uncertainty or disagreement?
- Are there representatives who can speak on behalf of the interests?
- Are the parties willing to participate in mediation?
     (Federal Environmental Assessment Review Office, 1993a, p. 135)

Unlike the screening step in the self-directed assessment procedure, there are no clearly specified types of projects which should be subject to public review. The Minister has the benefit of the screening and comprehensive study reports and of the RA's decision in making the screening decision. The fact that so few projects are subject to formal public review in Canada makes screening a highly political decision and the Minister will clearly be subject to public and political pressure (the decision will sometimes be taken by Cabinet). There is no formal consultation or participation beyond that provided in the self-directed assessment procedure. However, the Minister could be challenged in the courts, as in the past under EARP, if a public review is not required. It appears that the screening decision will involve the same difficulties as under EARP and that it will, perhaps inevitably, sometimes be challenged and sometimes be inefficient (i.e. it will take time).

Commonwealth of Australia

The Commonwealth of Australia Environment Protection (Impact of Proposals) Act is triggered not by the government agency responsible for the environment (the Commonwealth Environment Protection Agency – EPA) or by the Environment Minister (as it is in most of the state EIA systems) but by the so-called 'Action Minister' (the Commonwealth Minister responsible for the action or decision). The Action Minister must determine whether or not a proposal is environmentally significant. If it is, it must notify EPA and designate a proponent, which can be either a Commonwealth department or authority (if the proposal is a Commonwealth development) or a private company (if the proposal is a private sector development requiring Commonwealth approval). The proponent is then required to provide preliminary information on the proposal to EPA, usually in the form of a document called a 'notice of intention'.

This notice is similar to the US environmental assessment and its intended purpose is broadly similar. It should contain a brief summary of the proposal, illustrated as appropriate, a list of the alternatives considered, a description of the environment affected, an indication of the potential environmental impact and a description of mitigation measures to be taken to reduce this (Commonwealth of Australia, 1987, Procedures, para. 2.2). While there are no provisions relating to the making public of notices of intention, they are normally available to interested parties on request.

Following the submission of the notice of intention, EPA can determine that no further environmental impact assessment is necessary, though it may well suggest that certain environmental recommendations are met (Fig. 5.4). The Procedures contain several criteria to assist in determining the need for further assessment, for example 'the endangering, or further endangering, of any species of fauna or flora' and 'any environmental assessment action taken relevant to the proposed action by any State' (para. 3.1.2). If further assessment is deemed necessary, the Minister can choose one of three types:

1. A public environment report (PER) which provides a more selective treatment of the environmental implications of a proposal than does an environmental impact statement.
2. An environmental impact statement.
3. A commission of inquiry (which may require an EIS).

Since the Act came into force, over 3000 environmentally significant proposals have been referred to the Environment Department or EPA, many informally. As mentioned in Chapter 5, by the end of 1993, 132 of these required the preparation of an EIS, and 24 required the preparation of a PER (introduced in 1987) – only 4 inquiries had been required. There is no appeal against the determination of whether assessment will take place, and at what level. However, the decision to require an EIS or PER must be publicised and the Minister must provide, if requested to do so, reasons for not requiring further assessment. There has been a perceptible trend towards the preparation of notices of intention containing detailed impact mitigation measures by consultants, leading to a reduction in the need for EISs, as in the United States.

The lack of a departmental power of direction and of clear screening criteria and information leaves the EIA system open to political discretion and this, in turn, has led to demands that the EPA or the Environment Minister should trigger the Act (Formby, 1987; Australian and New Zealand Environment and Conservation Council, 1991; Fowler, 1991). However, while this has been firmly resisted by other departments and by industry in the past, business and conservationists alike agree on the need for greater legal certainty (Buckley, 1991b; Kinhill and Phillips Fox, 1991), and a reduction in the level of discretion which may have been an over-reaction to the legal challenges made to the US National Environmental Policy Act (Fowler, 1982). This demand has been met by a Prime Ministerial commitment:

To reduce uncertainty, national criteria will be developed for those classes and types of proposals that will normally attract the application of formal environmental impact assessment procedures. This will guide industry on the likely assessment requirements even before project proposals are developed.

(Keating, 1992, p. 85)

New Zealand

Since, in principle, EIA applies to all resource consent applications, screening is obviously very important in eliminating minor or irrelevant projects from further consideration. The Resource Management Act 1991 delegates this task, like most other EIA responsibilities, to the regional and district authorities. Not only does the Ministry for the Environment have no direct role in screening but it has not issued specific guidance on screening or recommended screening criteria. Figure 5.5 shows that various options exist at the initial screening stage, and indicates the two-phase nature of screening.

The role of mandatory regional policy statements and plans and district plans (which must accord with regional policy statements and plans) in setting out the criteria for determining whether resource consents require EIA in New Zealand is crucial. In particular, rules may be included in district plans which prohibit certain uses and permit others without the need for resource consent. Similarly, controlled uses, and the nature of any assessment required for these, can be specified. Finally, non-complying and discretionary uses, for which EIA is always required, can be set down in rules. By defining the nature of uses, the limits on the size of controlled uses or permitted uses and/or the nature of their effects, district plans effectively provide the means of screening projects subject to EIA. However, many minor projects requiring regional consents will still require EIA if they are not controlled or permitted uses.

All applicants for discretionary or non-conforming uses are required to complete a prescribed form and attach an assessment of the environmental impacts of their proposal 'in accordance with the Fourth Schedule' to the Act, however minor. The relevant plans are likely to specify more restricted EIA requirements in the case of controlled activities. This type of arrangement continues practice under the previous requirements when large numbers of informal EIA reports were submitted (Chapter 5).

Local authorities (usually districts) may require further EIA information (for example, on alternatives – Chapter 8) when the proposal is notified. The decision on whether to notify the application is taken by the local council on the basis of the EIA report submitted. If it determines that the project is likely to have major effects it is effectively making a screening decision and requiring a fuller EIA of the application.

There is no provision for different types of EIA report. Rather, the Resource Management Act 1991 specifies that any assessment:

shall be in such detail as corresponds with the scale and significance of the actual or potential effects that the activity may have on the environment (Section 88(6)(a))

provided that the requirements of the Fourth Schedule (Box 7.3) are met. There is thus an expectation of a continuum of report sizes, from half a page to several hundred pages (Hughes, 1992). This has transpired, as Morgan (1993) has reported.

The New Zealand EIA system therefore relies on the judgement of the region or district in determining whether EIA is necessary for non-complying, discretionary uses on the one hand and for controlled uses on the other. There is public participation and consultation in the preparation of policy statements and plans (and thus in determining the project EIA criteria employed). People likely to be affected by a proposed activity have to give their written consent before non-notification is permitted (i.e. they are involved, to some extent, in project screening decisions). There is a right of appeal to the Planning Tribunal on screening as on other issues.

As Morgan (1988) has pointed out, 'there is a danger that different authorities will adopt different screening criteria, leading to differences in the treatment of similar projects'. This is particularly true during the transitional period before policies and plans are fully in place. Morgan (1993) has indicated that, in practice, there are considerable variations in the provision of guidance to applicants by councils and in local authority screening decisions. Many councils appear to have found considerable difficulty in establishing screening procedures.

Summary

Table 9.1 summarises the treatment of screening in the seven EIA systems. It is inconceivable that any EIA system could be operated without some form of screening, so it is not surprising that six of the seven systems are adjudged to meet the screening criterion. The EIA systems use a variety of approaches, criteria and thresholds for screening. It is, perhaps, more surprising that the Commonwealth of Australia fails to meet the screening criterion. The reason for this is that the screening is done by the proponent government department without real control by the environmental agency. This lack of environmental agency control marks it out from the other six systems. A similar situation arises in many developing countries where the main criterion for the inclusion or exclusion of a project from EIA appears to be the insistence of the development assistance agency that EIA be undertaken, rather than any environmental imperative (see Appendix).

It is notable that five of the seven jurisdictions make use of more than one type of EIA document. In the United States the environmental assessment is nominally a screening document but, in practice, it is an EIA report in its own right for thousands of projects each year. The initial study in California serves a similar purpose. Some of the documents in the two-stage Canadian screening process (and especially screening reports) are not dissimilar. Screening in Australia results in different types of EIA report and in New Zealand in EIA reports of varying length and complexity.

Only The Netherlands and the United Kingdom have a single type of EIA report. It is, perhaps, no coincidence that both have strong land use planning systems under which the environmental impacts of the less significant projects can be assessed. Even here, The Netherlands tends to

Table 9.1 The treatment of screening in the EIA systems

Criterion 4: Must screening of actions for environmental significance take place?

Jurisdiction	Criterion met?	Comment
United States	Yes	Use of categorical exclusions, inclusion criteria, and (rarely, in practice) environmental assessments to determine significance of impacts.
California	Yes	Initial study must be prepared and published to determine whether significant impacts are likely, based upon Guidelines checklist.
United Kingdom	Yes	Use of lists of projects, indicative criteria and thresholds in screening by LPAs varies.
The Netherlands	Yes	Lists of activities, thresholds and criteria in EIA Decree allow competent authorities little discretion.
Canada	Yes	Screening by responsible authority using lists of projects results in 'screening' or comprehensive study. Further screening by Minister can lead to panel review or mediation.
Australia	No	Provisions for screening and less detailed EIA reports exist but absence of Environment Minister power of direction, and of clear criteria, lead to uncertainty.
New Zealand	Yes	Local authorities must specify types of, and criteria for, actions subject to EIA in their policies and plans.

employ more restrictive thresholds and criteria than the United Kingdom, resulting in proportionately fewer EIA reports. Screening practice in developing countries varies but many employ only a single type of EIA report.

CHAPTER 10

Scoping of impacts

Introduction

Scoping is the name applied to the process of determining the range of issues to be addressed in the EIA report and for identifying the significant issues related to a proposed action (Bear, 1989). Scoping was not an original requirement of the US National Environmental Policy Act (NEPA) but was added in 1978 in response to the encyclopedic nature of many environmental impact statements (EISs). Scoping was intended to ensure that more focused EISs were prepared and, incidentally, has assisted in increasing coordination between proponents in the EIA process and in the agreeing of action-specific timetables. It has proved to be a successful innovation. As Bear (1989, p 10065) has stated: 'A well designed scoping process can have an extremely positive ripple effect throughout the rest of the NEPA process.'

While there is no provision for scoping in the European Directive on EIA, many other jurisdictions have adopted scoping procedures. This chapter describes approaches to scoping and suggests criteria for use in the evaluation of EIA systems. These criteria are then utilised in the review of the EIA systems in the United States, California, the United Kingdom, The Netherlands, Canada, the Commonwealth of Australia and New Zealand.

Determination of the scope of the EIA report

Scoping is intended to focus the EIA on the most important issues, while ensuring that indirect and secondary effects are not overlooked and eliminating irrelevant impacts. In many older EIA systems it is a requirement that the public participate in scoping so that proponents and decision-makers are made aware of public concerns early in the EIA process. Consultation of environmental authorities can also reveal useful insights and, further, may ensure that coordination between them is more

Box 10.1   Steps to be considered in scoping

---

1. Develop a communications plan (decide who to talk to and when).

2. Assemble information that will be the starting point of discussion.

3. Make the information available to those whose views are to be obtained.

4. Find out what issues people are concerned about. (Make a long list.)

5. Look at the issues from a technical or scientific perspective in preparation for further study.

6. Organise information according to issues including grouping, combining and setting priorities (make the long list into a shorter list).

7. Develop a strategy for addressing and resolving each key issue, including information requirements and terms of reference for further studies.

Source: Ministry for the Environment (1992b, pp. 9, 10).

likely. Scoping has been defined as 'determining key concerns through open communication at an early enough stage to influence the planning' (Ministry for the Environment, 1992b, p. 12).

Scoping is part of a cyclical process. There are several steps involved, one representation of which is shown in Box 10.1. An important starting point is the preliminary identification of impacts. Generally, reference should be made to published guidance documents, whether of a general or generic (i.e. concerned with the specific type of action) nature or to previously prepared action-specific guidelines or EIA reports. These last may be helpful either because they relate to actions similar to that proposed or to actions in locations similar to that proposed. Such guidance or guidelines will normally furnish a checklist of impacts to be considered.

Other systematic methods of ensuring comprehensiveness in the iden- tification of impacts are the use of matrices or flow diagrams (see Bisset, 1988). The set of impacts identified as a result of using these approaches will include many which are irrelevant or insignificant and may still exclude some which are potentially important. Consultation with decision-making and environmental authorities, with interest groups such as local voluntary conservation groups, and with the local commu- nity should assist in ensuring that all potentially significant impacts are identified. This scoping process may or may not include meetings. However, public meetings are the best way of ensuring open dialogue about the significance of impacts but questionnaires, surveys and, in certain cases, the support of community-led scoping can be helpful (Beanlands, 1988).

Like screening decisions, scoping decisions frequently hinge on the issue of significance. In the last analysis these decisions often have to be made by individuals with the appropriate levels of knowledge and expertise who are able to say from past experience: what significant

131

effects are likely to arise; how they are likely to impact on the environment; and what steps might be taken to deal with them. Different mechanisms are used in different jurisdictions to ensure that the skills of such individuals are brought to bear on the action in question. These vary from the use of appropriate consultants by the proponent, through informal and formal consultations of 'interested parties at all levels of government, and all interested private citizens and organisations' (Bear, 1989, p. 10064), to the use of specialist panels and of representative consultation committees.

The preparation of action-specific scoping guidelines may not be a formal requirement in some jurisdictions, whereas in others the proponent will be required to prepare and publicise such guidelines and, in yet others, the decision-making body or environmental authority may be responsible for their preparation (Ministry of Housing, Physical Planning and Environment, 1981).

The most basic means of ensuring that some form of scoping takes place is to require the proponent to consult the decision-maker and/or environmental authorities prior to submission of the EIA report. Such consultation provides an opportunity for the opinions of the relevant authorities about the scope of the EIA to be expressed. So that consultation is not only required but is seen to take place, some record of the discussion should be available for public inspection. This record should, preferably, demonstrate that the various relevant authorities have indeed expressed their views about the scope of the EIA and that these views have been considered by the proponent.

Whether or not such consultation takes place, the proponent may be required to address a general or generic set of impacts in the EIA. That such a requirement has been met can easily be demonstrated in the EIA report. Clearly, the use of guidance specific to the type of action proposed is likely to be more helpful than the use of guidance equally applicable to all types of action or to certain classes of action. (The consideration of a general set of impacts, provided it is reasonably comprehensive is, nevertheless, very considerably preferable to the absence of any such requirement.) It is now generally accepted that the preparation of action-specific (rather than generic or general) guidelines by either the proponent or by the decision-making or environmental authorities renders the EIA process markedly more effective. A danger in such an approach, particularly where public consultation takes place, is that irrelevant impacts will be incorporated in the guidelines. Some mechanism for eliminating such impacts, and hence permitting the EIA to focus on the relevant issues, is necessary. Such a mechanism needs to be seen to be operated equitably.

The scoping stage in the EIA process will clearly be discharged best if guidance on scoping procedures and methods is available to the proponent and to the other participants. In the original American terminology, public participation is a necessary element of 'scoping', but this is not a requirement in all jurisdictions. There is, none the less, considerable unanimity of view that scoping is most effective where public participation and consultation are undertaken. Clearly, there should be a public

Box 10.2   Evaluation criteria for the scoping of impacts

Must scoping of the environmental impacts of actions take place and specific guidelines be produced?

Must the proponent consult the environmental authority early in the EIA process?

Is scoping mandatory in each case?

Must a general or generic set of impacts be addressed in the EIA?

Must action-specific scoping guidelines be prepared?

Are irrelevant impacts screened out?

Does published guidance on scoping procedures and methods exist?

Is consultation and participation required in scoping?

Is there a right of appeal against scoping decisions?

Does scoping function efficiently and effectively?

---

record of the outcome of scoping in this case. A further check that participation is effective in influencing the scope of EIA would be a third party right of appeal against scoping decisions.

Whatever scoping procedure and level of participation is adopted, it is obviously necessary that it works effectively and efficiently. Box 10.2 summarises this, and the other criteria which can be employed to review the treatment of scoping. These criteria for ways of establishing the coverage of the EIA for a particular proposal are now utilised in analysing the scoping process in each of the seven EIA systems.

United States of America

The National Environmental Policy Act 1969 (NEPA) requires a formal scoping process for each EIS. The first formal step in EIS preparation is the publication of a notice of intent which must contain a description of the agency's proposed scoping process, including any scoping meetings (which are recommended but are not a requirement of the Regulations). The open scoping process is intended to obtain the views of other agencies and the public regarding the scope of the EIS. The Regulations (Section 1501.7(a),(b)) state that the objectives of scoping include:

- determining which significant issues should be analysed in depth in the EIS
- identifying and eliminating issues which are insignificant or which have been dealt with elsewhere
- allocating responsibilities among agencies
- identifying relevant environmental review procedures, documents and consultation requirements
- setting page and time limits.

133

There is no prescribed list of impacts which must be included in EISs beyond the specification that direct, indirect, 'connected', 'similar' and 'cumulative' actions must be considered, together with a consideration of alternatives and mitigation measures (Bass and Herson, 1993a). A record of the scoping process must be kept since the Regulations (Section 1502.9(a)) state that draft EISs 'shall be prepared in accordance with the scope decided upon in the scoping process'. The Council on Environmental Quality (1981b) has issued guidance on scoping which not only advocates the use of public meetings and other methods of ensuring participation but suggests that a scoping report be prepared. This should be a record of the decisions made during the scoping process and contain a summary of the issues to be evaluated in the EIS and of the views of those participating in the scoping process.

There is no formal right of appeal against scoping decisions but it is customary for additional impacts to be addressed in the draft EIS if it later becomes apparent that they are likely to be significant. On the whole, the formalised scoping process works well, providing an agreed list of contents to be covered in the draft EIS. There is, however, a tendency to include impacts of questionable significance, rather than to exclude them, partly because of the fear of legal challenge. Further, while some public scoping meetings are very well attended, others fail to attract a single participant. Notwithstanding its shortcomings, scoping is generally regarded as a valuable addition to the EIS preparation process in the United States and is also sometimes used in connection with the preparation of environmental assessments.

California

Scoping is mandatory under the California Environmental Quality Act. The lead agency, having prepared an initial study, must publish a notice of preparation and send it to all the responsible agencies, to the State Clearinghouse (in most cases) and to parties requesting notice in writing. The notice of preparation must describe the project and describe the environmental effects to be evaluated in the environmental impact report (EIR). The various agencies are obliged to respond within 30 days and the lead agency must include the information requested in the EIR (Guidelines, Section 15082).

While the mandatory notice of preparation is sent to other agencies, there is no requirement to consult the public unless this is requested. Nevertheless, the Guidelines (Section 15083) suggest that consulting individuals or organisations will be helpful and define this involvement of the public as 'scoping':

> Scoping has been helpful to agencies in identifying the range of actions, alternatives, mitigation measures and significant effects to be analysed in depth in an EIR and in eliminating from detailed study issues found not to be important.

Upon request lead agencies must organise meetings with the relevant agencies, or with the public, as a tool for obtaining information about the scope and content of an EIR. The various means for determining the scope of the EIR are thus the initial study, the notice of preparation, consultation with designated agencies, optional public consultation and optional meetings (Bass and Herson, 1993b, p. 50).

In practice, scoping appears to be successful in defining the range of issues to be covered in the EIR but not in eliminating irrelevant issues. Generally, for most non-controversial projects, neither the public nor the designated responsible agencies tend to provide a significant contribution to the scoping process, which relies mostly upon the initial study. There have, however, been examples of cases where scoping meetings have been held and where many public concerns have been integrated into the EIA process from the onset. There is no requirement for a public record of scoping decisions to be made but the various responses are kept on file and are thus open to inspection. Practice varies considerably from agency to agency (Bass and Herson, 1993c) but while the scoping of the impacts of large projects usually works well, it is generally less successful than under NEPA where public participation requirements are formalised.

United Kingdom

There is no formal requirement for the proponent to consult the local planning authority (LPA) prior to submission of the EIA report, or to undertake any form of scoping, in the United Kingdom. However, the Department of the Environment (DOE) has strongly advised developers to consult LPAs about the coverage of environmental statements (ESs): 'developers and authorities should discuss the *scope* of an environmental statement before its preparation is begun' (DOE, 1989, para. 24, emphasis in original). Consultation of statutory consultees and, in some instances, of the public during scoping is also recommended:

> While developers are under no obligation to publicise their proposals before submitting a planning application, consultation with local amenity groups and with the general public can be useful in identifying key environmental issues, and may put the developer in a better position to modify the project in ways which would mitigate adverse effects and recognise local environmental concerns.
>
> (DOE, 1989, para. 30)

The 'specified information', or the statutory minimum content of an ES, consists of a description of the environment and the project, the data necessary to identify and assess the main effects, a description of the likely significant effects, a description of mitigation measures, and a non-technical summary (Planning Regulations, Schedule 3, para. 2). The Regulations also indicate the additional information which may be provided 'by way of explanation or amplification of any specified information'. This includes, as well as a discussion of alternatives (Chapter 8), the likely significant secondary, cumulative, short-, medium-

and long-term, permanent, temporary, positive and negative effects of the proposal, the forecasting methods used and an explanation of the difficulties encountered (Planning Regulations, Schedule 3, para. 3). The UK *Guide to the Procedures* contains a six page checklist of issues that might need to be covered by an ES, including the risk of accidents (DOE, 1989, Appendix 4). Like the supplementary information listed in Schedule 3 to the Planning Regulations this checklist has no statutory standing. No generic sets of impacts for particular types of projects have been issued by DOE.

In a study of a sample of 24 cases where ESs were submitted, almost two-thirds of developers or their consultants undertook early voluntary consultations with the LPA. LPAs were generally able to influence the scope of the ES during these discussions. Discussions with the statutory consultees, other bodies or the public, prior to submission of the ES, took place less frequently (in about one-third of cases). Where they did occur, developers and consultants found them to be of great value (Wood and Jones, 1991).

The statutory consultees believed they were able to influence the scope of the ES in about one-fifth of the cases by suggestions to the LPA, and in rather more cases by suggestions to the developer. They mostly felt that the notice taken of their suggestions by both the LPA and the developer/consultant was considerable. On the other hand, suggestions by voluntary groups to either the LPA or the developer/consultant were thought by those groups to have virtually no influence on the scope of the ES (Wood and Jones, 1991).

The Netherlands

Scoping has been a requirement of the Dutch EIA system since its introduction in 1987. The Environmental Protection Act 1993 (Section 7.15) makes it a requirement that project-specific guidelines be prepared for each EIA. The notification of intent prepared by the proponent alerts the competent authority that an EIA is to be undertaken and that guidelines are to be prepared. In turn, the competent authority must publish the notification of intent and alert the EIA Commission, which must produce its recommendations on guidelines within nine weeks of publication of the notification. As mentioned in Chapter 6, the guidelines must be issued by the competent authority in 13 weeks (unless the authority is also the proponent).

The EIA Commission sets up a small group of independent persons from its panel known to have relevant expertise and provides administrative support. This group takes into account the results of consultation with the various relevant agencies and the representations of the public about the content of the EIS which are coordinated by the competent authority. The statutory consultees normally comment mainly on the characteristics of the proposed site and public response, which varies greatly, may focus more on the merits of the proposal than the scope of the EIA (Commission of the European Communities, 1993c).

The EIA Commission then makes recommendations on the environmental effects to be described, the objectives of the proposal, the relevant planning policies, environmental designations and environmental standards for the area, the environmental aspects that should be addressed, the specific local conditions which must be described and the alternatives that should be given attention. The EIA Commission may also make recommendations about the methods to be employed but this is not usual.

The EIA Commission's advice, which is made public, provides the competent authority with a draft of the scoping guidelines. Indeed, guidelines written by the EIA Commission are often (but by no means always) simply adopted by the competent authority. During 1993, the EIA Commission produced 101 sets of guidelines (EIA Commission, 1994). Generally, previous guidelines for similar projects provide a model for the action-specific guidelines, but no general guidelines have been prepared, despite the similarities between, say, guidelines for various waste disposal sites or for motorways. (These similarities led to a Dutch initiative to formulate generic guidelines for the EIA of motorways in 1994.)

Advice on scoping is contained in the EIA handbook (Chapter 5) and there is a specific document on scoping and guidelines (in English) (Ministry of Housing, Physical Planning and the Environment, 1981). The scoping guidelines tend to be somewhat general (despite lengths of 20–30 pages or more) and not to eliminate potentially irrelevant topics. However, guidelines very rarely neglect relevant impacts, largely because of the expertise of the increasingly influential EIA Commission. In general, scoping works well. The main problem appears to be making the competent authorities take full responsibility for, and commit themselves to, the guidelines issued in their name.

Canada

There are two distinct scoping steps, one in the *self-directed assessment* and one in the public review. It is the responsibility of the responsible authority (RA) to determine the scope of the screening or comprehensive study. The *Guide to the Canadian Environmental Assessment Act* (CEAA) suggests that there are four aspects to scoping:

- determining the scope of the EA
- determining the scope of the factors to be considered
- determining the parties involved in the project and their interests and concerns
- determining the appropriate level of effort and analysis, given the project, the likely nature of its environmental effects and the public's interests and concerns.

    (Federal Environmental Assessment Review Office, 1993a, p. 75)

The proponent must consult the RA (if the two are not one and the same) to enable the mandatory scoping stage to take place.

The EA must address the environmental effects specified in the Act (Box 10.3) and informal, action-specific, guidelines may be prepared but there is no requirement for them to be published. The intention is that only potentially significant effects will be considered in the EA and that the appropriate level of analysis should be determined. As in the preparation of the EA reports, the amount of effort devoted to scoping will depend on the scale of the EA and will therefore be much greater for a comprehensive study than for most screenings. The Guide provides guidance on scoping procedures and the general guidance to be prepared on different project types should provide further help to those involved (FEARO, 1993a). There is no requirement for consultation and participation and no right of appeal (save through the courts) against scoping

Box 10.3   Factors to be considered in Canadian EA reports

All EA reports related to screenings (including class screenings), comprehensive studies, mediations or panel reviews must address the following factors:

- the environmental effects of the project, including:

  - the environmental effects of malfunctions or accidents that may occur in connection with the project;
  - any cumulative environmental effects that are likely to result from the project in combination with other projects or activities that have been or will be carried out;

- the significance of the environmental effects;
- public comments, if applicable or received in accordance with the regulations;
- technically and economically feasible measures that would mitigate any significant adverse environmental effects of the project;
- any other matter relevant to the assessment that the responsible authority (RA) may require, such as the need for and alternatives to the project.

The RA must also identify those environmental effects (including any directly related to human health, heritage, socio-economic conditions and other factors) relevant to the assessment requiring further investigation.

In addition to the above factors, a comprehensive study, mediation or panel report must also consider:

- the purpose of the project;
- technically and economically feasible alternative means of carrying out the project as well as the environmental effects of these alternative means;
- effects on the capacity of those renewable resources likely to be significantly affected by the project to meet present and future needs;
- the need for, and the requirements of, any follow-up programme.

Source: Amended from Federal Environmental Assessment Review Office (1993a, p. 81).

decisions. However, as at the screening stage, it is in the interests of the RA to consult with the public likely to be affected and seek to overcome their concerns. While the scoping step is novel in the 'screening' process, there is some experience of scoping in the preparation of initial environmental evaluations under the Environmental Assessment and Review Process (EARP). There is thus considerable informal experience and it is anticipated that scoping under CEAA should work reasonably efficiently and effectively.

The scoping stage in *panel review* is well established in relation to the preparation of EISs under EARP. (Scoping does not formally apply in the mediation procedure.) Under EARP, once the panel has been appointed and given its terms of reference by the Minister, it normally hears the views of those involved, regarding the scope of the review, prepares draft project-specific guidelines, and revises them in the light of comments. These guidelines have frequently been demanding of the proponent under the EARP. Those for the expansion of Toronto Airport EIS, for example, ran to some 25 pages (Federal Environmental Assessment Review Office, 1990) and covered requirements for:

- summary
- project setting
- project need
- alternatives
- project description
- environmental impacts
- socio-economic impacts
- monitoring
- mitigation and compensation measures
- safety
- public consultation programme
- appendices.

The procedure under CEAA is expected to follow this pattern.

There is now considerable experience of scoping for EISs in Canada and, in general, the panel scoping procedure has worked efficiently and effectively. Public involvement in scoping is one of the reasons that confidence in panel review in Canada has generally been high. Scoping practice should improve with the prior preparation of the screening or comprehensive study report which will form the basis for the EIS (and for mediation).

Commonwealth of Australia

Scoping is a formal requirement of the Australian EIA process. The broad contents of environmental impact statements (Box 8.4) and public environment reports (which are somewhat simplified EISs) are specified. However, the Australian procedures also require the preparation, by the Commonwealth Environment Protection Agency (EPA), of guidelines on 'the matters to be dealt with, and the extent to which those matters are to be dealt with, by an environmental impact statement or a public environment report' (Commonwealth of Australia, 1987, para. 4.4).

In practice, these project-specific guidelines are agreed between the proponent and EPA and are frequently very detailed (e.g. they ran to some 27 pages in the Sydney Airport Third Runway EIA). While there is no requirement to do so, the Procedures indicate that other bodies and the public can be consulted on the content of the guidelines. Agency consultation is quite common and public participation sometimes takes place (e.g. in the Third Runway case). The Ecologically Sustainable Development Working Group Chairs (1992) have recommended that this practice be extended. It is usual for the guidelines to be reproduced in the draft EIS. Project-specific guidelines have been criticised for being insufficiently focused and for failure to consider fully the 'no-go' option. In practice, guidelines tend to be encyclopedic and scoping does not eliminate irrelevant or unimportant impacts from consideration. There is no guidance on methods of, or procedures for, scoping, beyond reference to guidelines or previous projects. Notwithstanding these difficulties, the utility of project-specific scoping guidelines is universally accepted.

It is, perhaps, significant that the Australian scoping requirements were not written into the original administrative procedures, but are now used by all the states as well as by the Commonwealth as a means of improving the quality of EIA reports (Australian and New Zealand Environment and Conservation Council, 1991). The Government has suggested that further scoping and conflict resolution techniques should be developed 'to simplify application of the legislation' (Department of the Arts, Sport, the Environment, Tourism and Territories, 1991, p. 16). It is now also committed to the preparation of national industry scoping guidelines for certain types of project, as the Prime Minister indicated: 'industry-wide guidelines will be developed which can be readily applied to specific projects' (Keating, 1992, p. 85).

New Zealand

The Resource Management Act not only specifies that the details of any assessment should correspond to the scale of the effects of the action but that it must be prepared in accordance with the Fourth Schedule (Chapter 9). The Fourth Schedule was a late addition to the Act, inserted following pressure from environmental groups and the Parliamentary Commissioner for the Environment, who were concerned that too much discretion in establishing assessment procedures was being left to local authorities. It is intended to bring together the requirements of the Act relating to the contents of an assessment of effects in the form of a checklist. Anxiety has been expressed that some local authorities may use the Fourth Schedule as an inflexible listing of required information, rather than as a guide to obtaining information about relevant significant impacts (Hughes, 1992).

While the Fourth Schedule is intended to provide a guide as to which information might be furnished by a developer in each case, it is notable that the requirements include the treatment of alternatives, the risk of accidents and monitoring, together with a strong suggestion that early consultation should take place. Scoping is not mandatory. This omission is surprising in view of the role of scoping in the Environmental Protection and Enhancement Procedures (Ministry for the Environment (MfE), 1987), the stressing of scoping in recent guidance (MfE, 1991a) and the published guidance devoted to scoping (MfE, 1992b). For notified projects (Chapter 8) the local council may 'require an explanation of (ii) the consultation undertaken by the applicant' (Section 92(2)(a)). In effect, an applicant is being instructed that it would be unwise to neglect consultation about the proposal and the scoping of its effects.

It would be almost impossible for the consent authority to take appropriate account of Maori interests, as required by the Act, without early consultation with Maori representatives (i.e. scoping) (Hughes, 1992). Plans may give guidance as to scoping and consultation requirements (MfE, 1991a). Consultation and participation is therefore not obligatory in EIA report preparation, nor is the preparation of scoping guidelines (despite the experience of scoping gained over the years for larger projects in New Zealand), but it is very strongly recommended. Morgan (1993) has reported mixed experience with scoping. Some councils have given a clear invitation to applicants to discuss proposed actions with them, whereas others have left applicants to make the initial approach or relied upon the content requirements in the Fourth Schedule. As in other jurisdictions, applicants and local authorities have generally found early discussions to be very helpful.

Summary

The treatment of scoping in the seven jurisdictions varies, with five of the EIA systems meeting the evaluation criterion (Table 10.1). Scoping is a formal requirement for full EIA reports in the United States and California. It is a general requirement for EIA reports in The Netherlands, Canada and Australia. While not a formal requirement in New Zealand, scoping is very strongly encouraged in the Resource Management Act 1991 and local authorities can set up their own scoping procedures. In each of these jurisdictions, scoping involves the preparation of action-specific guidelines and must incorporate some environmental agency and public participation.

Only the United Kingdom makes no reference to scoping in its legal provisions, though even here scoping is strongly advised. In practice, scoping in the form of discussion between the proponent and the LPA frequently takes place though records of these discussions are often not made public. Consultation of environmental authorities and the public in the United Kingdom during scoping is less common. Scoping in developing countries tends only to take place if this is a requirement of the development assistance agency (the World Bank now requires scoping) (see Appendix).

It is now widely accepted that scoping helps to ensure that the relevant

Table 10.1 The treatment of scoping in the EIA systems

Criterion 5: Must scoping of the environmental impacts of actions take place and specific guidelines be produced?

Jurisdiction	Criterion met?	Comment
United States	Yes	Public scoping is used to produce specific guidelines for EISs. Scoping is sometimes used in environmental assessments.
California	Yes	Scoping is mandatory: notice of preparation must be circulated and made available but no requirement to consult public on action-specific guidelines.
United Kingdom	No	Not a statutory requirement but strongly advised. Frequently takes place but practice varies.
The Netherlands	Yes	Public scoping process, involving EIA Commission, produces action-specific guidelines for EISs.
Canada	Yes	Unpublished scoping guidelines must be produced for self-directed assessments: action-specific EIS guidelines issued by panels following consultation.
Australia	Yes	Project-specific scoping guidelines are agreed between proponent and Commonwealth Environment Protection Agency (EPA).
New Zealand	Partially	Scoping is not obligatory, but is very strongly encouraged in Act. Practice varies.

environmental impacts are covered in EIA reports (if not that scoping helps to eliminate irrelevant impacts). It has also been found in the more mature EIA systems that scoping ensures that the various parties can participate early in the EIA process. It can therefore only be a matter of time before New Zealand (which already has considerable experience of scoping for many projects from its Mark I EIA system) and the United Kingdom (which, although its EIA system is relatively new, also has some experience of scoping) adopt formal requirements for scoping in EIA. The same is true, though the timescale may be longer, of the adoption of scoping in many developing countries.

CHAPTER 11

EIA report preparation

Introduction

If the treatment of the environmental impacts of alternatives is at the heart of the environmental impact statement (Chapter 8), the EIA report is itself at the heart of the EIA process. There can be no meaningful EIA without the preparation of a report documenting the findings relating to the predicted impacts of the proposal upon the environment. Despite an enormous literature on EIA methods, very few jurisdictions specify how the findings presented in EIA reports should be derived. They do, however, normally specify the minimum content of the EIA report and frequently indicate procedures which must be followed in the preparation of the report (e.g. the making available of information by the relevant authorities).

The next section of this chapter is concerned with ensuring that content requirements for EIA reports are achieved in EIA systems. The chapter then discusses EIA report preparation requirements in EIA systems and puts forward a set of evaluation criteria. These criteria are then used to assist in the review of EIA report preparation procedures and practice in the United States, California, the United Kingdom, The Netherlands, Canada, the Commonwealth of Australia and New Zealand.

Content of EIA reports

Virtually every EIA system possesses a requirement that EIA reports must describe the proposed actions and the environment affected, forecast the significant impacts likely to result from the implementation of the action, and present a non-technical summary. They also generally provide that EIA reports contain other material, such as treatment of alternatives (Chapter 10) and mitigation measures (Chapter 15). The preparation of this information requires the use of a wide variety of methods and techniques.[1]

The EIA process is cyclical (Fig. 1.1) and the nature of the action is continually refined as its design progresses. Design work is costly and there is therefore a temptation for the proponent to prepare EIA reports on the basis of designs which are insufficiently detailed to allow forecasts to be prepared with accuracy. The decision-making and environmental authorities, however, should be seeking a realistic estimate of impacts which may necessitate more detailed design (and more expense) than the proponent originally contemplated. Whatever degree of detail is finally determined to be appropriate, the EIA report represents no more than a record of the impacts forecast to arise from the project as developed at a particular point in time, like a photograph.

A description of the salient features of the proposal forms a required part of any EIA report. This should be succinct and comprehensible while conveying the information necessary to permit the prediction of significant impacts. It will usually be necessary to provide information about the main features of the action giving rise to impacts. These usually arise from the physical presence of a project (e.g. the external appearance of the various elements), from the use of resources (e.g. quantities of freshwater or groundwater) or from the generation and disposal of wastes (e.g. gaseous and particulate emissions). Different impacts may arise at different points in the life cycle of an action (e.g. in the case of a mineral project: exploration, construction, operation (including accident conditions), modification, decommissioning and restoration) and each needs to be described.

Various checklists designed to assist in project description have been advanced (e.g. that in Box 11.1). Such information can be provided in a variety of forms: written text, tables, process diagrams, flow charts, maps, sketches, photo-montage, etc. It can be obtained by utilising design data, published emission data (e.g. for air pollutants), published accident data, advice from expert authorities (e.g. air pollution controllers), consultancy advice, EIA reports for similar proposals and visits to the sites of similar projects.

Only by carefully and systematically describing the initial or 'baseline' environmental conditions is it possible to present an accurate and convincing picture of the likely effects that the development will have on its environment. It is very important to devote sufficient effort to this part of the EIA process, as the accuracy and plausibility of much of the remainder of the EIA report depend upon it. Wherever possible, existing data should be utilised to indicate the principal physical features (e.g. geology), existing and proposed land use; the main air, water and land quality characteristics, existing vegetation and wildlife and existing land use and other policies, plans and standards for the area (Lee, 1989). Such data will often be held by various environmental authorities and need to be readily available to proponents.

Additional information may need to be gathered by observation and measurement but its purpose and the need for it should be very carefully considered. In addition to having a clear objective in EIA report preparation, any specific pre-project baseline studies should also provide the basis for post-project monitoring (Beanlands, 1988). Data on the existing environment should, of course, be collected early enough to use it as an input into the design process. Only information directly relevant to the forecasting of impacts should be included in the EIA report and, even then, much of it may be most appropriately presented in the form of appendices.

Information on the likely magnitude of the impacts of the proposed action on the environment should be presented in the EIA report in as

Box 11.1 A checklist for project description

NATURE AND PURPOSE OF THE DEVELOPMENT

- Function of the proposal, with economic and operational context
- Demand and need for the development (if appropriate)
- Alternatives considered (if appropriate)

CHARACTERISTICS OF THE PROPOSED SITE

- Location
- Size
- Summary of topography, landscape & natural or manmade features

CHARACTERISTICS OF THE PROPOSED DEVELOPMENT

- Size
- Site layout
- Shape
- Character
- Landscape proposals (including grading)
- Car parking
- Entrances and exits
- Access to public transport
- Provision for pedestrians and cyclists
- Provision for utilities
- Any other relevant information (including emissions to air, water and land)

PHASING OF THE DEVELOPMENT

Construction phase
- Nature and phasing of construction
- Frequency, duration and location of intrusive operations
- Timing, location and extent of mitigation measures
- Use and transport of raw materials
- Number of workers or visitors

Operational phase
- Processes, raw materials
- Emissions (air, water, noise, vibration, lighting, etc.)
- Number of employees or other users
- Traffic generation

Likely expansion or secondary development
- To be covered so far as the effects of such development can be anticipated at the time the ES is prepared

---

Source: Department of the Environment (1994b, Table 7).

precise, objective and value-free manner as possible. Clearly, it is necessary to distinguish between the nature, extent and magnitude of an
impact (e.g. forecast dust levels will vary with distance from the source
and disappear when emissions cease). The forecasts of impact magnitude
also need to take full account of forecast changes in baseline conditions in
the absence of the action and of the effect of mitigation measures
(Beanlands and Duinker, 1983).

The timescale and probability of occurrence of predictions should also
be stated (Tomlinson and Atkinson, 1987b). Forecasting (or prediction)
techniques[2] include formal mathematical models (e.g. plume diffusion
models), physical models (e.g. wind tunnel simulators), laboratory models (e.g. exposure of plants to pollutants), computer simulation, analogy
with similar projects, photo-montage, etc. In general, the simplest and
least data-demanding forecasting techniques should be employed. In
order to permit external verification (Chapter 12) and auditing (Chapter
14) the limitations of the data and methods employed, together with the
confidence which can be placed in the forecasts generated should be
stated. Uncertainty in these forecasts arises not only from the probabi-
listic nature of many impacts but from inaccurate information from
subsequent changes to the design of the project and from simplifications
inherent in models and errors in their use (Ministry of Housing, Physical
Planning and Environment, 1985). Such uncertainty may be very important in decision making (Chapter 13) especially where a choice has to be
made between closely matched alternatives.

Whereas forecasting the magnitude of impacts is a matter of determining the quantitative effects, the significance of an impact is a matter
requiring value judgement. Determining the magnitude of an impact may
provide some indication of its possible significance, but there is not
necessarily a direct relationship between the two factors. For example, a
road traffic noise level of 69 dB (A) (L_{10} 18 hour) in a forest may have a
significance quite different from the same noise level in an urban area.
Because some value judgements will be inevitable in presenting forecasts
in the EIA report it is essential that statements about significance are
clearly distinguished from those about magnitude (Hyman and Stiftel,
1988).

Wherever possible emotive phraseology should be avoided and a
consistent vocabulary should be employed to describe the significance of
impacts in EIA reports. The basis on which value judgements are made
should be clearly explained since, while there may be agreement about
the magnitude of impacts, different participants in the EIA process are
unlikely to agree about their significance.

Numerous methods of dealing with the significance of impacts have
been identified (Thompson, 1990). Some formal methods have incorporated scoring and weighting in which an attempt has been made to
quantify significance and these have rightly been criticised for internalising value judgements (Hollick, 1981a; Bisset, 1988). There exist few
agreed criteria for defining significance. One approach is shown in Table
11.1.

Where thresholds or standards are exceeded (e.g. ambient noise stan-

Table 11.1 Criteria for assessing the relative significance of environmental impacts

Factor	Low significance	High significance
Nature/form	Low priority	High priority
Aggregation	Non-cumulative, discrete	Cumulative, compounded (cross-impacts)
Direction	Stable/static	Improving/worsening
Magnitude	Small	Big
Probability	Low	High
Rate	Slow	Fast
Timing/duration	Short, infrequent	Long, continuous, frequent
Area/geographic limits	Small, contained	Large, uncontained
Reversibility	Reversible	Non-reversible
Scope for amelioration	Easy, inexpensive certain	Difficult, expensive uncertain
Compliance with applicable codes, standards and norms	Compliant	Non-compliant
Unknown factors	All key factors known, predictable	Key factors unknown, unpredictable

NB The relative significance (i.e. low or high) of each factor indicated in the table is based on the assumption of *'all other things being equal'*. For example, *all other things being equal*, the smaller or more contained an environmental effect is, the less significant it is. Similarly, *all other things being equal*, the more difficult, expensive or uncertain it is to mitigate negative effects, the more significant these negative effects are.

Source: Federal Environmental Assessment Review Office (1994, emphasis in original).

dards) significance is clearly established. The same is true in certain disciplinary areas where, for example, species loss or health damage is significant. Here or elsewhere, the use of consultation methods is often indicated, e.g. the harnessing of expert opinions (scientific or professional judgement). The organisation of a panel of professionals, perhaps operating on an iterative basis (the Delphi approach), can be very helpful in establishing agreement about the significance of impacts. Analogy with similar actions is also frequently employed. The use of public opinion, again perhaps by means of a representative panel, can be of great assistance in establishing significance (Hyman and Stiftel, 1988).

The non-technical summary is frequently used to disseminate the findings of the EIA report to the general public at low cost. Since it is often the only document to be examined and since it is frequently read separately from the EIA report, it is important that it should be clear, concise, objective and well written. The summary should accurately reflect the text of the EIA report and present the main conclusions about alternatives, mitigation, etc., and explain how they were reached. A summary table setting out the main forecast impacts and the significance is often a valuable means of detailing many of the main findings of the EIA report. Thus the Department of the Environment (DOE, 1994b) advocates assessing the geographical level of importance of the issue considered (international, national, regional or county-wide, district-wide and local), then the significance of the impact (major, minor and not significant) and finally the nature of the impact (adverse/beneficial, cumulative, short/long term, permanent/temporary, reversible/irreversible, and direct/indirect). DOE suggests that this information, together with an estimate of uncertainty, should be summarised in a table.

The non-technical summary is better prepared by assembling summaries of the various sections of the EIA report as they are written and then editing them into a consistent whole rather than hurriedly as the final step in EIA report preparation. While public relations and document design expertise may be helpful in the production of the summary, any tendency to distort the document by failing to reflect the presentation in the main EIA report should be resisted.

Preparation of EIA reports

While different EIA systems have different EIA report content requirements, it is clearly important that such provisions be specified precisely. It is also important that procedures be put in place so that proponents can gain access to information about the environment (and the action) held by decision making and environmental authorities. A further requirement, given the need for checks and balances in any system driven by the proponent (especially where the proponent is responsible for EIA report preparation) is that checks to reduce the likelihood of inadequate or biased EIA reports exist.

It is ironical, given the volume of literature on EIA methods (above) that few jurisdictions specify the methods or techniques to be employed in EIA report preparation.

Various methods of checking the content of the EIA report before it is published exist. Perhaps the simplest involves review by the responsible authority before approval to release the EIA report is given. This type of check exists in most Australian EIA systems. Another is the swift review of form and content used by the US Environmental Protection Agency prior to formally acknowledging receipt. A further approach is to involve a consultative group in the preparation of the various parts of the EIA report and to reach agreement on content section by section as in Victoria (Wood, 1993c). Some jurisdictions, however, rely on the diffusion of best practice and sanctions later in the EIA process (e.g. preparation and submission of further information) as checks on the quality of EIA reports. This is the situation in the United Kingdom. It is clear that diffusion of best practice will be much speedier where adequate checks on EIA report quality exist.

Many EIA reports are prepared, in whole or in part, by consultants. Needless to say, the standard of competence among consultancies varies and, especially where EIA has been introduced only recently, some may lack the appropriate range of professional skills. While, over time, reputations for competence in EIA report preparation will be established, there is a constant danger of consultancies being selected by price or suffering from bias (for example, where lucrative design contracts may follow approval of the proposal) unless some form of accreditation of EIA consultants or code of practice is introduced. Some jurisdictions have formally introduced such requirements (e.g. Flanders in Belgium). Elsewhere, several professional associations have either set up voluntary accreditation schemes or are actively considering doing so. In others, voluntary codes of practice are in operation (e.g. Australia) or under consideration. There are non-mandatory registration schemes in existence in the United Kingdom.

A more radical approach to quality assurance in EIA report preparation is to require that it be prepared not by the proponent but by the decision-maker or the lead agency. This is the original NEPA model for permit applications to federal agencies but it takes its most developed form in California. Here, the relevant agency often asks for tenders for environmental impact report preparation from an approved list of consultants. While the proponent pays, the client is the relevant agency.

As at the other stages in the EIA process, the existence of clear and readily accessible guidelines on EIA report preparation, content and form is advantageous. Not only is such guidance helpful to proponents and consultants in preparing the EIA report, but it is helpful to decision-making authorities, environmental authorities, interest groups and the public in reviewing EIA reports.

Again, as at other stages in the EIA process, the involvement of consultees and the public in EIA report preparation (not necessarily by means of consultative committees) will lead to improved quality or at least to improved acceptability. This criterion and others which can be advanced to assist in the analysis of the treatment of EIA report preparation in EIA systems are summarised in Box 11.2. The various criteria are used in the review of EIA report preparation in each of the seven EIA systems which now follows.

United States of America

The required contents of an EIS, which are specified in the Council on Environmental Quality's (CEQ's) National Environmental Policy Act (NEPA) Regulations can be summarised as follows:

Box 11.2   Evaluation criteria for the preparation of EIA reports

---

Must EIA reports meet prescribed content requirements and do checks to prevent the release of inadequate EIA reports exist?

Must EIA reports describe actions, environments affected, forecast impacts, indicate significance and contain a non-technical summary?

Must information held by the relevant authorities about the environment or type of action be made available to the proponent?

Does published guidance on EIA report preparation exist?

Must specified EIA methods or techniques be employed?

Does accreditation of EIA consultants exist?

Do checks on the content, form, objectivity and accuracy of the information presented occur before publication of the EIA report?

Is consultation and participation required in EIA report preparation?

Does EIA report preparation function efficiently and effectively?

- cover sheet
- summary (not normally exceeding 15 pages in length)
- table of contents
- statement of purpose and need
- alternatives, including the proposed action
- affected environment
- environmental consequences, including mitigation measures
- list of preparers
- list of agencies and organisations consulted
- list of all federal permits
- appendices
- index.

The environmental consequences section of an EIS is intended to form the scientific and analytical basis for the comparison of alternatives. The discussion of environmental consequences must include the environmental effects of the alternatives, including ecological, aesthetic, historical, cultural, economic, social and health impacts (Bass and Herson, 1993a, p. 68). Conflicts between the proposed action and any relevant land use plans, policies and controls for the area must be included as 'environmental effects'.

An EIS may contain appendices which present relevant background material of an analytical nature (Regulations, Section 1502.18) but the total length of an EIS should not exceed 150 pages or, in the case of an unusually complex proposal, 300 pages. In practice, these page limits are frequently exceeded. There is no accreditation of EIA consultants but, while consultants are increasingly employed, many federal agencies possess sufficient competent in-house staff to prepare EISs. It is not uncommon for a 'draft' draft EIS to be prepared, prior to publication of the document, for internal review purposes. This preliminary EIS may occasionally be circulated to a limited number of consultees (including the Environmental Protection Agency). The draft EIS is, of course, subject to formal consultee and public scrutiny before the final EIS is prepared (Chapter 15).

The final EIS should include all the substantive comments on the draft EIS (or summaries of them). It must also include responses to the comments received during the review of the draft EIS (Regulations, Section 1503.4(b)). The final EIS may contain modified proposed actions or alternatives, it may develop and evaluate new alternatives, it may supplement, improve or modify analyses, include any necessary corrections or contain an explanation of why no further response is necessary.

The advice on EIS preparation contained in the Regulations and in the 'Forty questions' (CEQ, 1981a) requires the documents to be analytical (and not encyclopedic), to focus on significant impacts, to emphasise alternatives, to avoid *post hoc* rationalisation, to be interdisciplinary (i.e. to ensure the integrated use of the natural sciences, social sciences, and the design arts (Regulations, Section 1502.6)), to be concise, to be written in plain language and to use appropriate graphics. There is no requirement to use specified EIA methods or techniques.

These checks also apply, to a lesser extent, to the preparation of environmental assessments.

California

The content of environmental impact reports (EIRs) is specified by the California Environmental Quality Act (CEQA) Guidelines:

- table of contents or index
- summary (no longer than 15 pages)
- project description
- environmental setting
- significant environmental impacts
    direct
    indirect
    short-term
    long-term
    cumulative
    unavoidable
- alternatives
- mitigation measures
- growth-inducing impacts
- organisations and persons consulted and preparers (Guidelines, Article 9).

(There are additional requirements for plan and programme EIRs (Guidelines, Sections 15126, 15127).) EIRs should normally be no longer than 150 pages or, for unusually complex projects, 300 pages (Guidelines, Section 15141).

The notice of preparation is intended to ensure that the interests of

relevant agencies are taken into account in the EIR. There is no requirement that information held by these agencies must be made available to
the EIR preparers. In practice, however, this is the case as agencies must
specify the content of the environmental information they require to be
included in the draft EIR. Obtaining relevant information may sometimes
be difficult. There is no provision for direct public involvement in draft
EIR preparation.

Because CEQA applies to a wide variety of project types, the Guidelines cannot provide specific guidance on the methods to be employed in
writing an EIR. They do provide general advice on evaluating cumulative
impacts and on identifying the significance of impacts. The EIR should
define the threshold of significance and explain the criteria used to
determine whether an impact is above or below that threshold (Bass and
Herson, 1993b, p. 72). However, in practice this is often not done. EIRs,
according to the Guidelines, should:

• emphasise significant impacts
• forecast foreseeable impacts
• explain disagreements among experts
• adopt a level of specificity appropriate to the project
• cite all documents used in its preparation
• incorporate documents by reference and summary.

Unfortunately, many EIRs are encyclopedic and greatly exceed the
recommended page limit of 150 pages.

The lead agency may prepare the draft EIR by using its own staff, by
using consultants, by accepting a draft prepared by the applicant or the
applicant's consultant or by arranging that the applicant pay for consultants to write the EIR under contract to the agency. In each case the
agency can charge fees to cover its costs. In practice, most EIRs are
prepared by environmental consultants retained by the lead agency,
though about 15 per cent of consultants are retained directly by the
applicant (Olshansky, 1992).

Regardless of who prepares the draft EIR, the lead agency is legally
responsible for the draft's adequacy and objectivity. Accordingly, the
lead agency must exercise its independent judgement as to the adequacy
of the 'draft' draft EIR. The agency must check its coverage and content
before releasing it for consultation (Guidelines, Section 15084). Nevertheless, Olshansky (1992) reported that 20 per cent of respondents stated
that the consultant or applicant was ultimately responsible for content.
There is no accreditation of EIA consultants but, since consultants are
normally selected, by both rotation and reputation on the basis of
competitive tenders, there is virtually an approved list and the quality of
their EIA work is generally regarded as reasonable.

United Kingdom

There is no prescribed form of an environmental statement (ES), except
that it must contain a description of the environment and the project, the

data necessary to identify and assess the main effects, 'a description of the likely significant effects, direct and indirect, on the environment of the development', a description of mitigation measures and a non-technical summary (Planning Regulations, Schedule 3, para. 2).

While the developer is responsible for the content of the ES finally submitted and for the assessment methods employed, the Regulations enable the developer to collect relevant existing information from the statutory consultees (for example, English Nature) who are under a duty to provide it. Where a local planning authority (LPA) is informed in writing that an ES is in preparation, it must notify the statutory consultees so that they can be ready to provide the developer with information if requested to do so.

In nearly one-half of a sample of cases where an ES was submitted the LPA is known to have notified the statutory consultees that an ES was being prepared. In just over one-third of cases the LPA could not notify the statutory consultees because it was unaware of the preparation of the ES which was eventually submitted without any prior consultation. In the cases studied, more than two-thirds of developers and consultants preparing ESs appear to have sought relevant information from statutory consultees or other public bodies (Wood and Jones, 1991). It is apparent, however, that the intention of Regulations, to ensure that relevant existing information about the environment is used appropriately in EIA, is not yet being fully met.

Neither the Department of the Environment (DOE) Circular on EIA (DOE, 1988b) nor the *Guide to the Procedures* (DOE, 1989) specifies EIA methods or techniques. The Guide does, however, advocate a flexible approach, adjusted to individual circumstances:

> The assessment techniques used, and the degree of detail in which any particular subject is treated in an environmental statement, will depend on the character of the proposal, the environment which it is likely to affect, and the information available. While a careful study of the proposed location will generally be needed (including environmental survey information), original scientific research will not normally be necessary.

> (DOE, 1989, para. 31)

There is no formal check on the content, form and accuracy of the EIA report prior to its release because, as mentioned in Chapter 10, there is no requirement for the proponent to consult the relevant authorities prior to submission of the EIA report, though this is strongly advised. There is no formal accreditation of EIA consultants in the United Kingdom, though the Institute of Environmental Assessment operates a voluntary scheme. Neither the DOE Circular nor the DOE Guide provides detailed guidance on EIA preparation.

A manual on the EIA of major developments was commissioned by the UK Government some years ago (Clark *et al.*, 1981). This is now dated and is to be replaced by more general guidance on ES preparation (Her Majesty's Government, 1990) a draft version of which was released for consultation in July 1994. As mentioned in Chapter 4, several other guides on EA, which include some advice on ES preparation, have been published.

The Netherlands

Section 7.10 of the Environmental Management Act 1994 specifies the minimum contents of an environmental impact statement. As well as the detailed discussion of alternatives mentioned in Chapter 8, these include a description of relevant environmental plans and standards, of the methodology utilised, and of the shortcomings caused by lack of information and a non-technical summary. There must also be, of course, descriptions of the proposed action, of the environment affected and an analysis of the consequences of the action. There is no specific mention of mitigation measures (though these are subsumed within alternatives) or of the significance of environmental impacts.

Information held by the competent authority must be made available to the proponent who prepares the EIS in accordance with that authority's guidelines for the action. In practice, there is considerable discussion between the proponent and the competent authority during the preparation of the EIS and data are made freely available. The guidelines may sometimes specify the type of method to be used; for example, a specific type of noise model, though most guidelines leave the choice of method to the proponent. The general handbook on EIA and its replacement (Chapter 5) contain information on methods, as do the specific guidance documents for each type of impact. There is no shortage of written advice on EIA methods in The Netherlands! However, much of this guidance is not widely used and some of it is now dated. No accreditation scheme for consultants exists. The quality of consultants varies from the highly competent to the woefully inexperienced.

As shown in Fig. 5.1 the competent authority evaluates the acceptability of the EIS before it is made available for public review. This check ensures that the requirements of the Act and the recommendations in the guidelines are met. It had become customary for the EIA Commission to be involved unofficially in this process and the Evaluation Committee on EIA (1990) recommended formalisation of this activity, which does not involve any consultation or participation. However, it was widely felt that responsibility for this function should actually (rather than nominally) be assumed by the competent authority. Accordingly, from 1993, the EIA Commission (1994, p. 5) has declared that it will only become involved in this pre-review process if the EIS deals with an entirely new type of activity or with some new requirement introduced by the EIA Commission during scoping. Six weeks are permitted for the competent authority to decide if the proponent should provide further information. If such a request is made, it must be made public.

While a few proponents still see the EIS merely as a paper exercise, their number is diminishing rapidly. On the whole (and with exceptions) EIA report preparation is thought to take place reasonably effectively in The Netherlands. There is a noticeable improvement in the efficiency of preparation and in quality when proponents or their consultants are preparing a second or subsequent EIS rather than their first.

Canada

The content requirements for EIA reports are set out in Box 10.3. Curiously, in view of the comprehensive nature of these requirements, no non-technical summary is specified (though this will be required in guidelines for EA reports which are to be issued). There is a provision in the Canadian Environmental Assessment Act (CEAA, Section 12 (3)) for 'expert federal departments' to make information available upon request by the responsible authority (RA), the mediator or the review panel. Similar arrangements appear to have worked reasonably satisfactorily under the Environmental Assessment and Review Process (EARP) so there is every expectation that this will continue to be the case. It will be necessary to make checks on the information presented in comprehensive study reports and EISs[3] before they are released, because they are normally prepared by proponents but are the responsibility of the RA. Draft EISs have often been submitted for comment by the Federal Environmental Assessment Review Office (FEARO) under EARP. There is no mandatory consultation and participation in the actual preparation of EIA reports in Canada (though in practice this is often achieved), nor are consultants accredited.

The Guide to CEAA provides considerable advice on EA reports. It contains sections on assessing environmental effects, mitigating environmental effects, determining the significance of adverse environmental effects and preparing the EA report. These sections are to be supplemented by reference guides on:

- public registry
- cumulative environmental effects
- significance[4]
- follow-up
- sustainable use of natural resources
- health conditions
- socio-economic conditions
- heritage
- use of land and resources for traditional purposes
- role of the public.

(FEARO also intends that there will be supplementary good practice handbooks on cumulative effects assessment, public involvement, etc.)

These guides are principally procedural but provide references to the literature and sometimes include suggested methods. In the main body of the Guide it is suggested, for example, that project–environment interactions can be identified by such means as overlay maps, matrix tables and expert groups (FEARO, 1993a).

The determination of the significance of environmental effects is especially important in the Canadian EA system since this directly affects whether the responsible authority can take a decision on the project (for the self-directed assessment) or whether further review is needed through mediation or a panel review. The conclusions of the EA report (FEARO, 1993a) must state whether:

- the environmental effects are adverse
- the adverse effects are significant
- the significant adverse environmental effects are likely.

This determination is supposed to be supported by rigorous analysis, and not be based on public concerns. Criteria for determining the significance of effects include magnitude, geographical extent, duration and frequency, irreversibility, ecological context. An example of the factors used in determining whether or not environmental effects are adverse is shown in Box 11.3. Under EARP, the length of environmental screening reports varies from a page to many hundreds of pages, but the

Box 11.3 Canadian criteria for defining adverse effects

Factors used in determining Whether or Not
Environmental Effects are Adverse

Environmental changes:

- negative effects on the health of biota including plants, animals and fish;
- threat to rare or endangered species;
- reductions in species diversity or disruption of food webs;
- loss of, or damage to habitats, including habitat fragmentation;
- discharges or release of persistent and/or toxic chemicals, microbiological agents, nutrients (e.g. nitrogen, phosphorus), radiation or thermal energy (e.g. cooling wastewater);
- population declines, particularly in top predator, large, or long-lived species;
- the removal of resource materials (e.g. peat, coal) from the environment;
- transformation of natural landscapes;
- obstruction of migration, or passage of wildlife;
- negative effects on the quality and/or quantity of the biophysical environment (e.g. surface water, groundwater, soil, land and air).

Effects on people resulting from environmental changes:

- negative effects on human health, well-being or quality of life;
- increase in unemployment or shrinkage in the economy;
- reduction of the quality or quantity of recreational opportunities or amenities;
- detrimental change in the current use of lands and resources for traditional purposes by aboriginal persons;
- negative effects on historical, archaeological, palaeontological, or architectural resources;
- decreased aesthetic appeal or changes in visual amenities (e.g. views);
- loss of, or damage to commercial species or resources;
- foreclosure of future resource use or production;
- loss of, or damage to valued, rare, or endangered species or their habitats.

Source: Federal Environmental Assessment Review Office (1993a, p. 100).

vast majority have been very brief (screening reports are expected to continue to be brief under CEAA). As Rees (1980) has stated, however, the screening stage has been, in practice, essentially undocumented under EARP, a situation which has improved and should improve further under CEAA. Initial environmental evaluations (IEEs) have been very similar in length to EISs under EARP, normally running to several hundred pages. There is little reason to expect that comprehensive study reports under CEAA (which perform a broadly similar function to IEEs) will differ greatly from EISs.

Commonwealth of Australia

Environmental impact statements and public environment reports (PERs) must meet somewhat non-specific prescribed content requirements (Box 8.4). These requirements include a 'clear and concise summary' (Commonwealth of Australia, 1987, Procedures, para. 5.2). It is customary for the detailed content to be specified in the project-specific guidelines agreed between the proponent and the Commonwealth Environment Protection Agency (EPA).

The Administration Procedures (Procedures, para. 4.5) require that 'The proponent shall consult with the Department throughout the preparation of the environmental impact statement'. This consultation is a crucial element in ensuring that the draft EIS addresses the issues listed in the project-specific guidelines in an adequate manner. In practice, EPA (on behalf of the Environment Department) normally receives draft chapters and a draft of the whole EIS from the proponent. In effect, EPA's right to approve the advertisement of the draft EIS (Procedures, paras 6.3.1,2), allows it to threaten to withhold permission to release the draft EIS until it is satisfied with its quality. While not based strictly on the Administrative Procedures, this mechanism normally provides an effective latent weapon.

Permissive arrangements for the consultation of other agencies and bodies during the preparation of the EIS or PER are laid down in the Procedures (para. 4.6). In practice, this generally works well, as liaison between the proponent and EPA and other agencies (but not with public groups) is close. Despite the absence of formal guidance on EIA report preparation, the consultation process means that EPA is usually moderately content with the EIS or PER when it is released, without the preparation process taking an excessive amount of time. There is no specification of particular assessment methods to be utilised nor is there accreditation of the consultants who are employed to prepare most EIA reports.

New Zealand

As mentioned in Chapter 9, it is apparent that the document reporting on the assessment of effects (which is, confusingly, not named in the Resource Management Act) may vary in length from half a page to hundreds of pages, depending on the circumstances. The contents of the document are outlined in the Fourth Schedule to the Act. It is perhaps surprising that the Act is silent on the need for a non-technical summary (a requirement in almost every other EIA system). Provision of such a summary is, however, suggested by the Ministry for the Environment in its advice on scoping (1992b, p. 26).

The Ministry for the Environment (MfE) has suggested that the assessment information to be provided should be indicated clearly in regional or district plans and that further help can be obtained by discussing the proposal with councils to gain their initial reaction. Because, in many situations, the level of expected impact will be low and the mitigation measures easily defined, the guidance suggests that the applicant should be able to furnish the necessary information. However, where the proposal may have 'significant effects on the environment' applicants may decide they need 'professional advice and hire consultants to undertake the assessment' for them (MfE, 1991a). (There is no accreditation of EIA consultants in New Zealand.) These cases, which should generally be flagged in the regional or district plan, will usually trigger the requirement that they be notified and that information on alternatives and consultation be provided (Chapters 8 and 10). It is clearly in a developer's interests to provide such information at the outset, to avoid the delays involved in furnishing it upon request once the competent authority has received the application.

There is now guidance available to applicants on the preparation of EIA reports in the form of a 'practical guide' (Morgan and Memon, 1993), a scoping guide (MfE, 1992b) and various pamphlets (Chapter 5). However, although methodologies such as checklists, matrices and evaluation systems are described in the 'practical guide', none is specifically recommended. In addition, several seminars and workshops for practitioners have been organised, with MfE providing advice in the form of papers at many of them (see, for example, Morgan et al., 1991).

The need for guidance to developers (especially of small-scale projects where the retention of consultants is unjustified) on how an EIA report should be prepared has been pressing as regional policy statements and plans and district plans were still in the very early stages of preparation in mid-1994. In addition to a hypothetical gold mine assessment presented as an example of how to undertake an EIA (Morgan and Memon, 1993), MfE (1992a) has issued a fictional case study to provide brief model advice to developers in their approach to the assessment of the environmental effects of a tourism development. Such guidance is very important, given that there are no formal requirements for environmental information to be made available to the applicant or for any check to be made on the quality of the EIA report before it is made public.

However, given the discretion that local councils have in specifying EIA requirements, applicants will face severe difficulties until the relevant policies and plans provide the necessary indication as to the appropriate content of EIA reports. Dixon (1993b) noted that, in practice, the lack of EIA expertise among both local authority staff and applicants has caused considerable difficulties and that the Act's requirements, for example in relation to cumulative impact assessment, are not being met. Morgan (1993) reported that applicants needed project-specific advice in undertaking their assessments, and that this was seldom being provided by councils.

Summary

The performance of the seven systems against the EIA report content criterion is shown in Table 11.2. While, in practice, their performance varies substantially within as well as between jurisdictions, the United States, California, The Netherlands, Canada and Australia meet the criterion. United Kingdom ESs must meet the content requirements specified in the Planning Regulations but no checks are made to prevent the release of inadequate EIA reports. The EIA system in New Zealand contains no formal provision as to the content of EIA reports, nor are checks made on their content prior to release. Many developing countries broadly specify the contents of EIA reports but few checks are made before EIA reports are released (see Appendix).

It is probably no coincidence that neither the United Kingdom nor the New Zealand EIA system requires compulsory scoping. Scoping guide-lines provide a useful set of criteria against which to judge the coverage of

Table 11.2 The treatment of EIA report preparation in the EIA systems

Criterion 6:	Must EIA reports meet prescribed content requirements and do checks to prevent the release of inadequate EIA reports exist?	
Jurisdiction	**Criterion met?**	**Comment**
United States	Yes	Draft EISs are subject to formal checks on required contents prior to publication.
California	Yes	CEQA specifies content of EIR and checks are made by lead agency.
United Kingdom	Content: yes Checks: no	Regulations prescribe content but no formal requirement for proponent to consult or for checks on environmental statement (ES) prior to release.
The Netherlands	Yes	EIS is checked against guidelines and EIA Act by competent authority before release for public consultation.
Canada	Yes	Content prescribed in CEAA and some checks made by federal authorities to limit inadequacy of EA reports (especially EISs).
Australia	Yes	EIA reports are checked against project-specific guidelines and vetted by EPA before release.
New Zealand	No	Act provides strong guidance as to content but no checks on adequacy of EIA reports before release exist.

an EIA report and they are frequently utilised in the five jurisdictions for this purpose. However, the checks probably have more to do with the cooperation which scoping engenders between the proponent and environmental and decision-making authorities. Such checks sometimes take place informally in both the United Kingdom and New Zealand but the major impediment to the release of inadequate EIA reports in both these countries is the lack of experience of many local authorities. In California, where local government is also responsible for many EIAs, the system has been in place for nearly 25 years and the risk of court action provides an incentive to check the content of EIA reports before their release.

More formal checks on EIA quality in the United Kingdom and New Zealand may come either from the adoption of scoping or from the acquisition of experience. In developing countries participation by the national environmental agency (or agencies) in scoping and/or checking EIA reports probably provides the way forward.

Notes

1. See, for example: Clark et al. (1980); Ministry of Housing, Physical Planning and Environment (1984); Ortolano (1984); Shopley and Fuggle (1984); Bisset (1988); Lee (1989); and Glasson et al. (1994). A distinction is often drawn between 'methods' which, as Bisset (1988, p. 47) says, 'have been described alternatively as methodologies, technologies, approaches, manuals, guidelines and even procedures' and 'techniques'. Methods are frequently comprehensive or aggregative, and attempt to address several tasks within the EIA process. Techniques, on the other hand, usually focus on a single task, such as the forecasting of noise levels or the organising of a public discussion about impacts. In practice, however, the terms are often used interchangeably. Most EIA methods and techniques have evolved from work in other fields: land use planning, cost–benefit analysis; multiple-objective decision making; checklists, matrix and network analysis; and modelling and simulation (Hyman and Stiftel, 1988, Chapter 1). Their use normally requires the involvement of a range of professionals working as an interdisciplinary team. In general, methods or techniques need to be selected to deal with the tasks involved at each step in the EIA process. It is the choice of an appropriate set of techniques which is the hallmark of good EIA practice, rather than the use of a comprehensive 'method'. Each technique chosen should be appropriate to the EIA task considered, replicable by other users, consistent in use on different alternatives and cost-effective in its use of data, financial, time and personnel reserves.
2. Forecasting and prediction both involve an estimate of the impacts likely to arise from the action. Strictly speaking, a prediction is a forecast with a very high probability (usually certainty). Most estimates in EIA involve an element of uncertainty and should therefore be termed forecasts. In practice, the terms are used interchangeably.
3. It is of interest to note that the term 'EIS' is never used in CEAA, despite its being defined in EARP Guidelines. RAs, the CEAA Guide (FEARO, 1993a) emphasises, are to continue to ensure that EISs are prepared for panel reviews.
4. Addendum: The late 1994 version of the Guide (see Chapter 6, Note 2) contains reference guides on addressing cumulative environmental effects, the public registry and determining where a project is likely to cause significant adverse environmental effects.

CHAPTER 12

EIA report review

Introduction

If there is only one point in an EIA process where formal consultation and participation take place it is during the review of the EIA report. Indeed, in some jurisdictions 'public review' is virtually synonymous with public participation. This is not to say that all jurisdictions provide for public participation once the EIA report has been prepared (that in Hong Kong, for example, does not). However, nearly all do. The public review of EIA reports provides an invaluable check on their quality, especially where such checks have not been applied earlier in the EIA process. This chapter advances a set of evaluation criteria for the treatment of the review of EIA reports in EIA systems. These criteria are then employed in the analysis of EIA report review procedures in the United States, California, the United Kingdom, The Netherlands, Canada, the Commonwealth of Australia and New Zealand.

Review of EIA reports

The formal review of EIA reports is handled differently in different EIA systems. In the US model, for example, the draft environmental impact statement (EIS) is used as the basis for consultation and participation and is duly succeeded by a final EIS. The power to require a supplementary EIS also exists. The Environmental Protection Agency reviews all EISs and publishes its opinions about both the adequacy of the EIS and the environmental impact of the proposed action using a rather general set of criteria (below). This 'EIA report–review–further EIA report' pattern has been emulated in other EIA systems (e.g. in the Commonwealth of Australia) where the comments by consultees and the public are published as part of the review process.

In yet other systems, no formal provision exists for the proponent to respond to public comment in this way (e.g. in the United Kingdom).

Clearly, while the treatment of EIA review varies in different jurisdictions, the fundamental requirement of this stage in the EIA process is that those bodies with responsibilities and expertise, and the public, should be able to comment upon the EIA report and the action it describes. This stage exists in almost every EIA process. Such comments should, of course, be considered by the decision-making/environmental authorities before any decision on the action is made.

One of the most difficult areas in the review of EIA reports, as in the preparation of EIA reports, is ensuring objectivity since the organisation charged with responsibility for formal review (if any) may have a vested interest in the decision about the proposal. There are various methods of ensuring objectivity, including the use of review criteria, the accreditation of EIA report review consultants, the setting up of an independent review body, the publication of the results of the review and the involvement of consultees and the public. There is probably no substitute for utilising the services of skilled professionals in the review process, whether within the decision-making/environmental authorities, within retained consultancies, or within consultee organisations, including public interest groups.

The existence of criteria can provide a useful focus for the review of EIA reports. Action-specific scoping guidelines, where they are prepared, provide a valuable checklist for review. Another checklist is normally provided by the set of statutory requirements for EIA reports contained in legislation or regulations (or sometimes in formal guidance). Very few jurisdictions, however, have as yet published formal criteria to assist in the review of EIA reports. Government-commissioned advice containing review criteria has, however, been issued in New Zealand and has been promised in the United Kingdom (below).

Several sets of criteria intended for use in the review of EIA reports have been published independently (see, for example, Beanlands and Duinker, 1983; Elkin and Smith, 1988). Lee and Colley's (1992) review procedure requires at least two reviewers to utilise criteria relating to the extent to which specific tasks are completed and reported and builds up through a series of stages to an agreed overall evaluation of the EIA report as broadly adequate or inadequate. Such criteria can be used to meet various objectives. For example, Box 12.1 presents a set of criteria, modified from Lee and Colley (1992), to deal specifically with the treatment of landscape and countryside recreation issues in EIA reports in the United Kingdom. Where the treatment of a particular EIA task, or group of tasks, is clearly adjudged inadequate, further information can be requested.

It is by no means uncommon for consultants to be retained to undertake, or to assist in, the review of EIA reports. The competence of such reviewers is an important issue, especially as they tend to be engaged to review the EIA reports for the more complex and significant proposals. Generally, review consultants are drawn from the same group which prepare EIA reports and the need for some form of accreditation applies similarly.

The appointment of an independent panel selected from acknowledged experts in the field to review EIA reports has two advantages. First, it should provide a means of reducing any bias in the relevant authority's decision on the action. Second, it should ensure that the quality of EIA

Box 12.1 Criteria for the review of UK EIA report landscape treatment

1. **Description of the development, the local environment and the baseline conditions**

- The purpose(s) of the development should be described, together with its physical characteristics, scale and design, including quantities of materials needed during construction and operation.
- The land requirements of the development should be described together with the duration of each land use.
- The types and quantities of wastes and energy created should be given, together with the expected rate of production and the proposed disposal routes to the environment.
- The landscape likely to be affected and, where relevant, its recreational use should be described in terms of its physical characteristics and the way these interact to form identifiable landscape types.*
- The elements identified in the baseline landscape and, where relevant, recreation description should include reference, where applicable, to:
  - Countryside Commission designations;
  - local authority statutory and non-statutory landscape and recreation policies and any other designations, for example, by the NCC or English Heritage;
  - cultural associations that the affected area may have, for example, with authors, poets or painters;
  - subjective assessment made by other bodies, for example, scenic routes or tourist board guides;
  - perceptions of any other interested parties not included above, for example, user groups, visitors, pressure groups.

- The methods and assumptions (if any) used in describing and assessing baseline landscape and, where relevant, recreation conditions should be clear.
- The baseline conditions should be projected to predict a 'no development' alternative to development. (This should facilitate comparison of expected non-development conditions with recognisable phases of development.)

2. **Identification and evaluation of impacts**

- Potential impacts of the proposed development on baseline landscape and, where relevant, recreation conditions should be identified and described in appropriate detail.* Impacts, both positive and negative, should be defined to cover:
  - construction, operation and, where appropriate, decommissioning/restoration phases;
  - direct and indirect impacts; temporary, permanent and cumulative effects;
  - short- or long-term effects;
  - interrelationships between landscape and: human beings; flora and fauna; soil; water; air; climate; material assets; and cultural heritage; and, where relevant, between landscape and recreation.
- A systematic method should be employed to identify impacts to ensure comprehensiveness (for example, a checklist, matrix or flow diagram) and its use should be explained.
- An explanation of the methods used, and reasons for selecting the key impacts and the appropriate level of detail of assessment, should be provided. (Not all impacts need to be studied in equal depth.)

163

- The magnitude of impacts identified should be quantified where appropriate.
- An explanation of the forecasting methods employed should be provided.
- Significance should be attached to identified impacts on landscape and, where relevant, recreation and its derivation explained. Reference, where applicable, should be made to:
  - Countryside Commission designations or definitions*;
  - local authority statutory and non-statutory landscape and recreation policies, and any other designations, for example, SSSIs*;
  - cultural associations which the affected area may have;
  - subjective assessments by other bodies;
  - perceptions of any other interested parties not included above.

3. Alternatives, mitigation and monitoring

- Alternative sites/processes to those originally proposed should be considered in terms of their landscape and, where relevant, recreation impacts.
- Criteria for the rejection of alternatives should be justified and any assumptions made explicit. These reasons should reflect the role of landscape and recreation impacts in decision making.
- All significant adverse impacts should be considered for mitigation and evidence should be presented to show that proposed mitigation measures will be effective.
- The developer should demonstrate commitment to, and capability in, the carrying out of effective measures to mitigate any identified significant landscape and recreation impacts of the proposed development.
  - The mitigation of any signifi-

cant adverse impacts on landscape and recreation baseline conditions should be considered and detailed proposals put forward.*
- The impacts of any mitigation proposals should themselves be identified and their magnitude and significance assessed.
- Non-mitigation of any identified significant adverse impacts on landscape and recreation baseline conditions should be justified, with any assumptions made explicit.
- Where uncertainty over the effectiveness of a mitigation measure exists, then a monitoring programme should be established to enable subsequent adjustment as necessary.
- The ES should incorporate a description of measures for monitoring impacts to ensure that the accuracy of predictions and the effectiveness of mitigation measures is checked.

4. Communication of results

- Information should be arranged in sections to enable the reader to find and assimilate any landscape and recreation information easily. A list of contents or index should be included*.
- Any data, conclusions or quality standards drawn from sources external to the ES should be acknowledged at the appropriate point in the text and in a full reference list.
- Care should be taken in the presentation of information on specific topics to ensure that it is accessible to the non-specialist; technical terms and acronyms should be fully explained.
- Information should be presented without bias and receive emphasis appropriate in the context of the ES. Visual presentations should be neutral, i.e. avoiding false camera positions, distorted perspectives, etc.

> - The significant landscape and re-creation impacts should be communicated in a clear, concise and well-written non-technical sum-mary, which should include the main conclusions about alternatives, etc., and how they were reached.

Criteria marked with * can only be assessed as satisfactory if full illustrations, including maps, diagrams, sketches or photo montages, are employed within the text of the ES.

Source: Stiles *et al.* (1991, p. 47).

reports improves over time, since its opinions, whether adverse or positive, should be both public and influential. This is certainly the case with the review panels in Canada and The Netherlands, for example.

In order that a review of the EIA report may be seen to have taken place, the outcome should be made public. This is most clearly seen to be the case where a formal review is published (as in The Netherlands or the Commonwealth of Australia, for example). In addition, the comments arising from reviews of the EIA report by the consultees and by the public (correspondence, notes of telephone conversations, minutes of meetings, etc.) should be placed in the public domain (either by publication – with or without editing – or by allowing access to the decision-making authority's files).

One of the more significant checks on the preparation of inadequate EIA reports by proponents is the right to demand, and the duty to provide, further information (including any necessary corrections) following submission. Ideally, such information should be requested after the decision-making or environmental authority has reviewed the EIA report and after there has been a response from consultees and the public, to ensure that a full range of expertise and opinion is brought to bear on the adequacy of the information in the EIA report. The approaches to ensuring the provision of further information vary from a right to request further information but no power to enforce its provision or to 'stop the clock' while it is furnished (as in the United Kingdom) to the formal preparation of a final EIA report in response to comments on the draft report (e.g. in the United States and in Australia).

As at the other stages of the EIA process, the existence of published advice on the procedures employed in either formal or informal EIA review and on the methods which may be used in reviewing EIA reports is invaluable not only to those engaged in EIA review (the decision-making and environmental authorities, consultees and the public) but to those involved in EIA report preparation. Such guidance might include a checklist or a set of review criteria (above).

As mentioned above, if there is one step in the EIA process where public participation takes place it is following publication of the EIA report. It is preferable that this participation occurs prior to requesting further information from the proponent. Even if this does not happen, suitable provision for participation (adequate availability of copies of the EIA report over a realistic time period, etc.) is necessary (see Chapter 16).

Similarly, where further information is submitted by the proponent, this too should be open to public inspection and comment.

If a formal review of the quality of the EIA report is carried out and made public, especially if this is in the form of a recommendation to the decision-maker, equity suggests that the right of appeal against the findings of the review should exist. In many jurisdictions, where no formal review is published prior to the making of the decision, such a demand can be combined with an appeal against the decision itself.

It is, of course, important that the review stage of the EIA process is carried out effectively and efficiently. In other words, the EIA report should be thoroughly reviewed by an appropriate range of participants and further information should be provided without undue demands upon resources or time. Balances are needed to ensure that requests for further information are both coordinated (and preferably all made at once), reasonable and not deliberately used as a delaying tactic. These can be introduced by the use of a 'single request' provision, by the use of time limits and by administrative or legal appeal. The various criteria discussed above are summarised in Box 12.2. These evaluation criteria are now used to assist in the analysis of EIA report review procedures in each of the seven EIA systems.

United States of America

There are comprehensive arrangements under the National Environmental Policy Act for the review of draft EISs. The lead agency must circulate the draft EIS for review to prescribed agencies. Not only must

Box 12.2 Evaluation criteria for the review of EIA reports

Must EIA reports be publicly reviewed and the proponent respond to the points raised?

Must a review of the EIA report take place?

Do checks on the objectivity of the EIA report review exist?

Do review criteria to determine EIA report adequacy exist?

Does an independent review body with appropriate expertise exist?

Must the findings of the EIA report review be published?

Can the proponent be asked for more information following review?

Must a draft and final EIA report be prepared?

Does published guidance on EIA review procedures and methods exist?

Is consultation and participation required in EIA report review?

Is consultation and participation required where further information is submitted?

Is there a right of appeal against review decisions?

Does EIA report review function effectively and efficiently?

---

agencies with any jurisdiction over, or special expertise with regard to, the proposal comment but other federal, state, tribal and local agencies and the public must also be invited to comment. Generally, 45 days are allowed for comment on a draft EIS.

It is customary for consultees and the public to check the draft EIS against the scoping report (which is often summarised in the EIS). In addition, the Environmental Protection Agency (EPA) employs review criteria (below). EPA also conducts detailed reviews of final EISs, especially where significant issues are raised at the draft EIS stage. For projects that it rates as environmentally unsatisfactory, EPA may refer the issue to the Council on Environmental Quality (CEQ). In addition, EPA will informally review environmental assessments that it receives if so requested by a lead agency. It also informally reviews 'draft' draft and final EISs if requested.

As mentioned in Chapter 11, the comments of EPA and other agencies reviewing the draft EISs must be made available to the public in the final EIS together with responses to them. There is a 30 day waiting period between the publication by EPA of a notice in the *Federal Register* that the final EIS has been filed and the issuing of the record of discussion (Fig. 2.2). The final EIS must be circulated to all federal agencies with jurisdiction or expertise, to the project applicant (where there is one), to persons requesting to be notified and to persons who submitted comments on the draft EIS.

There is no formal general guidance on EIS review but EPA (1984) has issued policy guidance concerning the quality of the draft EIS and the acceptability of the proposed action (summarised in Box 12.3). There have been some criticisms of lack of consistency in review findings from the different EPA regional offices but EPA records an improvement in EIS quality between its published review of the draft and its unpublished review of the final document. Thus, during the period 1989–91 EPA recorded a rating of 3, environmental objections, or environmentally unsatisfactory to about 25 per cent of draft EISs (the action rating is based upon the 'worst' alternative).[1] During the same period, the percentage of final EISs so graded fell to less than 10 per cent (based upon the 'chosen' alternative). The percentage of EISs for which environmental concern was recorded also fell substantially between the draft and final documents.

Generally, although there have been some criticisms of the time taken for EIS review, the process seems to work reasonably efficiently and there is no doubt that it is effective in ensuring that the concerns of the various relevant agencies and, perhaps to a lesser extent, the public are met.

California

Once the 'draft' draft EIR has been approved by the lead agency and the draft EIR finalised, public notice of its availability must be given. This must contain a brief description of the project, of its location and of the project's

Box 12.3   US Environmental Protection Agency review criteria

---

CRITERIA FOR DRAFT EIS ADEQUACY

Category 1: Adequate

- The EIS's treatment of impacts and alternatives is adequate.

Category 2: Insufficient Information

- The draft EIS does not contain sufficient information to fully assess all reasonable alternatives.

Category 3: Inadequate

- The deficiencies in impact analysis or alternatives are of such magnitude that a revised draft EIS should be recirculated. The action is a potential candidate for CEQ referral.

CRITERIA FOR ENVIRONMENTAL EFFECTS OF PROPOSED FEDERAL ACTIONS

LO: Lack of Objection

- EPA review has not identified any potential impacts requiring changes to the proposal.

EC: Environmental Concerns

- EPA review has identified environmental impacts that should be avoided to fully protect the environment.

EO: Environmental Objections

- EPA review has identified significant impacts that must be avoided to fully protect the environment.

EU: Environmentally Unsatisfactory

- EPA review has identified adverse impacts of sufficient magnitude that the action should not proceed as proposed. The proposal will be referred to CEQ unless the unsatisfactory impacts are mitigated.

Source: Adapted from Environmental Protection Agency (1984).

significant effects, specify the review period, identify any public meetings or hearings on the project and state where the draft EIR is available for inspection.

The lead agency must send the draft EIR to various local and other agencies for comment. In addition, 10 copies are sent to the State Clearing-house, which sends the draft EIR to the state agencies that it considers to be relevant, receives their comments and transmits them to the lead agency. The State Clearinghouse Handbook indicates when copies should be sent to the Clearinghouse for review (e.g. residential projects of more than 500 units) (Bass and Herson, 1993b, p. 54). Where the Clearinghouse is involved (probably the majority of cases), 45 days are allowed for public review.

CEQA does not require a public hearing on the draft EIR but, in practice, many agencies conduct such hearings. After the review period for the draft EIR closes, the lead agency assembles all written comments and transcripts of comments made at public hearings. These are then used either in preparing the final EIR or, if significant new information becomes available, the lead agency must provide a second public review period and recirculate the draft EIR.

Responses must be provided for all comments unless a response is not appropriate, in which case an explanation must be provided as to why a response is not warranted. The final EIR should consist of the draft EIR, copies of comments received during the public review, a list of those commenting and responses to the comments. These may be incorporated either by revising the draft EIR or by adding new material, or both. In some instances, the public comments and responses are very lengthy. One final EIR, consisting of several volumes of comments, weighed over 15 kg. The final EIR must be made available for public comment before any decision is made. It is quite common for changes to be made to projects to accommodate comments made on the draft EIR.

The criteria used for judging the adequacy of EIRs consist of the contents of the scoping documents, whether particular topics have been addressed and problems satisfactorily resolved, and whether judicial standards have been met. The lead agency is likely to review the 'draft' draft EIR most comprehensively and it is best acquainted with the scope of the document. It normally checks off the contents of the draft EIR report and reviews each part of the document against the legal adequacy criteria.

The State Clearinghouse acts purely as a post box. At one time the Office of Planning and Research reviewed the draft EIRs for the more controversial projects but it no longer has the staff to do this. The various agencies consulted are normally interested only in their statutory area of competence and are concerned to see that appropriate mitigation of impacts has taken place. It is often only at this stage of the EIA process that many agencies respond to projects – they tend not to involve themselves in screening or scoping because of staff shortages. Environmental groups and the public at large respond only if the project is particularly controversial and use a variety of different criteria in their reviews.

Legal action against the lead agency may be taken on the grounds that the final EIR is inadequate. Legal challenge against the final EIR on the grounds that comments on the draft EIR have not been adequately addressed is also possible. The courts do not require that an agency achieve perfection but rather that an EIR should show that the agency has made an objective, good-faith attempt at full disclosure and covers all the contents required by the Guidelines and scoping. The scope of judicial review does not extend to the correction of an EIR's conclusions, but only to its sufficiency as an information document for decision-makers and the public (Guidelines, Section 15151; Bass and Herson, 1993b, p. 77).

United Kingdom

There is a requirement for the environmental statement (ES) to be made available for consultative and public review in the United Kingdom, but not for the preparation of any formal review report by the local planning authority (LPA) as a separate stage in the EIA process. The LPA review of the EIA report is normally in two stages: an early evaluation of the environmental statement (ES) to see whether more information should be requested and a fuller review once the results of consultation and participation have been received. This latter stage is normally completed immediately prior to decision making. The responses from consultees and public participants, together with the LPA's own review, are usually open to public inspection, thus ensuring that some element of objectivity is seen to apply.

LPAs review ESs without the benefit of formal review criteria or of any specialised review body. They are able to use Schedule 3 to the Planning Regulations and the list in the official *Guide to the Procedures* (Department of the Environment – (DOE) 1989, Appendix 4) as checklists. The LPA may commission consultants to review the ES but there is no provision to charge the costs to the developer. There is, as mentioned in Chapter 11, some voluntary accreditation of EIA consultants. The Institute of Environmental Assessment is not only involved in this process but provides an ES review service to LPAs utilising a modified version of the review criteria advanced by Lee and Colley (1992).

As a result of its own review and the responses from statutory consultees and the public, LPAs may request further information (for example, on how certain objections are to be overcome). However, DOE has indicated that this should be the exception, not the rule: 'the use of these powers should not normally be necessary, especially if the parties have worked together during the preparation of the environmental statement' (DOE, 1989, para. 39). The Planning Regulations impose a time limit upon LPAs to process the review phase of the EIA process, and there is no mechanism for 'stopping the clock' while further information (including any corrections) is provided. Such information must be made available to the consultees and the public for further comment.

Should proponents choose to do so, they can decline to provide the information requested and subsequently appeal to the Secretary of State for non-determination of the planning application. However, this is frequently a lengthy process. There is no provision for the proponent to prepare a formal response to comments, for example, in the form of a final EIA report.

In over half of 24 cases surveyed by Wood and Jones (1991), the LPA requested further information or evidence as part of the EA process. This was usually done on a purely informal basis with the Regulations being invoked in only two cases. This further information was generally provided by the developer. While consultees were often able to request and receive further information, voluntary groups (many of which were dissatisfied with the content of ESs) tended not to be given the information they requested (Wood and Jones, 1991).

There is currently less guidance about how LPAs should review ESs than there is about how developers should prepare them, but advice on the review of ESs is to be published (Her Majesty's Government, 1990). A planning application accompanied by an ES must be advertised and copies of the ES must be made available to the public (for inspection and for sale) and to a set of statutory consultees. If further information is required by the LPA, this too must be made available for consultation and participation (DOE, 1994b).

When a sample of 24 ESs were subjected to an independent review, nearly two-fifths were found to be broadly satisfactory. The other three-fifths appraised as unsatisfactory were judged not to fulfil the minimum requirements of para. 2 of Schedule 3 to the Planning EIA Regulations. Many developers appeared not to include in their ESs the 'further information' described in para. 3 of Schedule 3 (Wood and Jones, 1991). These findings accord with those of Lee and Colley (1991), on whose methodology the reviews were based.

The 24 LPAs concerned took a more sanguine view as a result of their own reviews, however. Over two-fifths of the 24 LPAs found the general content of the ESs to be satisfactory and about one-third categorised them as 'neither satisfactory nor unsatisfactory'. Three quarters of the LPAs felt that the requirements of para. 2 of Schedule 3 of the Regulations were covered.

There is some evidence that the quality of ESs may be improving. Lee and Brown (1992) reported that the quality of samples of the approximately 300 ESs now being prepared annually varied depending on the regulations under which they are prepared, the length of the ES, the size of the project, the experience of the developer and whether a consultant was used. About 40 per cent of 1990/91 ESs were still considered unsatisfactory, however. While subsequent work indicates that some further improvement has probably occurred, there remains considerable scope for an increase in ES quality in the United Kingdom.

The Netherlands

Once the competent authority has accepted the EIS, the public must be notified and the EIS is made public together with the draft decision on the proposal. The public review period of at least five weeks for the EIS coincides with the public review period for the application to which the EIS relates. There is provision for a public hearing if this is requested (Environmental Management Act, 1994, Section 7.24) and this usually happens. The various statutory consultees are also asked to comment on the EIS. The results of the consultation process are then passed to the EIA Commission which will have received the EIS at the time of its publication by the competent authority.

The EIA Commission checks the EIS against the legislation and the regulations and against the scoping guidelines to see whether it is complete. The Commission must also make a judgement about whether the EIS is adequate for decision-making purposes (for example, are the predictions likely to be accurate, are all the reasonable alternatives analysed, is the EIS sufficiently objective?). There are no specific review criteria but it is usual to use the published scoping guidelines and the requirements of the Act as the basis for its check on adequacy as well as on completeness. The EIA Commission must first identify omissions and mistakes in the EIS and then evaluate which of these are relevant to the decision. It must take the opinions of the statutory consultees and the public into account in framing its recommendations. It endeavours to group its comments, utilising the latest detailed information about, for example, land contamination, according to their significance to the competent authority's decision and to keep them brief. The EIA Commission, if it finds that substantial information is lacking, will recommend how to overcome this deficiency.

The review findings of the EIA Commission are published. These findings, which are confined to the contents of the EIS rather than to the advisability or otherwise of the proposal, are almost always accepted by the competent authority. If the recommendations are negative the competent authority may then require a supplementary EIS (Fig. 5.1). This procedure is not specified in the Act, but is now well established and laid down in government advice. It is recommended that this further information be treated in the same way as the original EIS. However, it is not formally subject to review by the public (though most competent authorities arrange for this to occur) but it is reviewed by the EIA Commission. There is now Ministry guidance on the treatment of supplementary EISs.

The EIA Commission undertook 68 reviews of EISs in 1993. In 20 per cent of these cases it found 'substantial deficiencies in the content of EISs' (EIA Commission, 1994, p. 5) necessitating the supply of supplementary information to the competent authority. In addition, supplementary information, mostly of a relatively minor nature, was requested in a further 30 per cent of cases. The principal areas of weakness in practice relate to treatment of alternatives, miscalculation of extent of impacts, overlooking of impacts, use of outdated models or inadequate baseline information (Box 12.4).

In general the EIA review process is working reasonably well. The level of expertise within the competent authorities is growing, the public is becoming increasingly involved and increasingly sophisticated: as a result, the quality of EISs is improving. While there is no right of appeal against the EIA Commission's review findings, they are widely regarded as authoritative and are very seldom challenged by proponents, by the competent authority or by environmental groups. The Evaluation Committee on EIA (1990) saw the role of the EIA Commission's published review of the EIS as critical to maintaining and improving the quality of EISs.

Canada

The situation relating to public review of EA reports in Canada varies with the type of report. In the *self-directed assessment* procedure there are differences between the screening and comprehensive study tracks.

Box 12.4 Shortcomings noted during reviews of EISs in the Netherlands

- Objective of the activity is described too narrowly.
Example: The EIS describes the transport problem concerning the movement of persons and goods between two places only in terms of
road transport neglecting the potentials for rail or other means of
transport.

- The description of the activity does not cover the entire activity.
Example: The EIS describes the proposed construction of an industrial
plant but omits information about construction of a pipeline and
other facilities to transport and handle raw materials and fin-
ished products to and from the plant.

- Selection of alternatives does not take into account environment aspects.
Example: The EIS on a car racing circuit in a coastal dune landscape only
considers alternatives meeting motorsport requirements, visitors' 'needs' and public safety regulations while overlooking
environmental considerations such as noise abatement, the protection of geomorphology and ecology of the coastal landscape.

- Key problems affected by the activity are not described.
Example: The EIS describes the proposed construction of a coal-fired
power plant using surface water as cooling medium. The EIS
does not describe that the surface water body is already used by
other industrial activities for cooling purposes to the limit of its
cooling capacity.

- Sensitive elements in the existing environment are overlooked.
Example: The EIS on a pipeline project does not describe that the proposed
alignment of the pipeline will dissect certain areas of ecological
value.

*- Environmental target values and standards are not properly described and
observed.*
Example: The EIS for an extension of an airport describes the impacts up to
the standard of 25 per cent of people seriously affected by aircraft
noise whereas the target value aims at 10 per cent of people
seriously affected.

*- No alternative is described complying with legal environmental regulations and
standards.*
Example: The EIS for a sanitary landfill indicates that the soil types in the
area are very diverse ranging from sand and clay to peat. The
alternatives do not take into account the large differences in
compaction and subsidence of these soil types with subsequent
failures of underlinings and drainage systems underlying the
landfill.

- Possible promising mitigating measures are not considered.
Example: The EIS for a sanitary landfill does not describe a system for
collecting the methane gas produced in the landfill. Methane is a
greenhouse gas contributing to global warming.

*- The alternative offering the best protection to the environment is not described or
insufficiently described.*
Example: The EIS on a bridge or tunnel connection across an estuary does
not take seriously the alternative whereby the connection is
carried out as a drilled tunnel. A drilled tunnel underneath the
bottom of the estuary has considerable less impact on the
environment than a bridge connection or a tunnel composed of
segments on the bottom of the estuary.

- Serious impacts on the environment are not or not correctly described.

Example: In the case of the EIS for a sanitary landfill in an area with very variable soil conditions the EIS does not describe the impact on the environment following failure of the underlying sealing and drainage systems.

- Insufficient or outdated prediction models are used.

Example: The EIS on an urban development scheme made use of a mobility prediction model using national averages whereas local input data is available enabling a more precise prediction.

- In comparing the alternatives incorrect conclusions are drawn.

Example: In the EIS on a regional management plan for the disposal of municipal sewage sludge various alternative methods for disposal are compared. One alternative concerns composting the sludge into a low grade compost product. The comparison of the alternatives in the EIS presents the composting method as an attractive form of disposal as it greatly reduces volume. However, the comparison does not take into account the limited potential of applying the low grade compost as a soil conditioner or improver due to high heavy metal contents in the sludge.

Source: Scholten and van Eck (1994, Annex 2).

Public participation is discretionary in screening, but, should the responsible authority (RA) determine that the public should be given the opportunity to comment on the screening report, it must provide an opportunity to do so. In practice this is likely to happen in only a few cases, where there is public demand or opposition. Some screening decisions under the Environmental Assessment and Review Process (EARP) related to such minor matters as 'school bus purchase' or 'security fence' (Federal Environmental Assessment Review Office – FEARO, 1992a.) There has, in the past, been little effective public review of screening reports, so the situation under the Canadian Environmental Assessment Act (CEAA) is an advance. The RA has the discretion to seek comments by the relevant expert federal departments on the 'screening' report, and these are likely to be regularly sought.

The Act provides for more central control in the review of comprehensive study reports than did EARP for their predecessors. While the review of comprehensive study reports by expert federal departments is nominally discretionary, the report must then be submitted to the Canadian Environmental Assessment Agency (CEA Agency) for review and public comment. The Agency must explain when and where the report can be obtained and give a deadline for public comment. It then files both the comments and the summary of these that it is obliged to prepare in the 'public registry'. While the Agency's review of the report is mainly procedural, it will have to ensure that the content of the report is scientifically and technically accurate. The Agency thus asks, for example:

- Was the comprehensive study undertaken in accordance with the procedural requirements of the CEAA?
- Were all the relevant expert federal authorities consulted and all concerns adequately resolved?
- Were there appropriate and sufficient opportunities for public involvement in the comprehensive study and are there any out-standing public concerns?

(FEARO, 1993a, p. 110)

While there was no formal requirement for public review of initial environmental evaluations (IEEs) under EARP, it was customary to make these documents available for comment, and for the comments to be addressed. The tightening in procedures, and particularly the involve-ment of the CEA Agency, should improve the quality of public review under CEAA. It may also result in the proponent being asked to produce further information.

Public review has always been an important element in the independ-ent panel review procedure. Over time, panels have organised the preparation of draft EISs and public review of the actual EIS (apart from expert federal department reviews) and, where necessary, have demanded that the proponent provide further information where public or government agency concerns justified this. The panel guidelines are used as review criteria. This step (which is not shown in Fig. 5.3) precedes the public hearings which panels hold into proposals. Since the EIS is the central document at these hearings, there has been adequate, if some-times lengthy, public review and comment on the EIS. This situation will continue under CEAA where similar procedures will be applied and panel reports will continue to be published. There is, necessarily, little EA report review in the mediation procedure, since no further documents are prepared and the negotiations (which often involve representatives of the public) take place in private.

There has been considerable Canadian interest in, and original research on, the quality of EISs (which are sometimes voluminous, running to thousands of pages) (see Beanlands and Duinker, 1983; Ross, 1987). There has also been some emphasis on the quality of IEEs and initial assessment reports (Elkin and Smith, 1988). The quality of EA reports depends heavily on the professionalism of the RA and has varied considerably in the past, though it has improved over time as greater experience has been gained. While the quality of IEEs and EISs bears comparison with the best international practice, it is much harder to generalise about the quality of initial assessment reports under EARP.

Commonwealth of Australia

The requirements relating to public and agency review of the environ-mental impact statements are specified in considerable detail in the Administrative Procedures (Commonwealth of Australia, 1987). As men-tioned in Chapter 11, the provision relating to the approval of advertisements announcing the availability of the draft EIS usually provides an invaluable check on the premature release of an inadequate draft EIS. The project-specific guidelines provide a useful checklist for reviewing the EIS but no other review criteria are employed and there is no published guidance on EIA report review. However, it is clear that the guidelines, and the adoption of scoping, have helped to improve the quality of draft EISs over time.

As in the United States, the proponent is required to produce a final EIS, following consultation and participation. In reality, this final document is usually a listing of the various comments received on the draft EIS, together with the proponent's response to them. This practice of preparing a supplement to the draft EIS, rather than a 'true' final EIS, has been criticised and can lead to over-long and confusing documentation. It is, however, usual for further modifications to the project design to be made at this stage by the proponent and this mechanism certainly provides accountability.

The proponent has to make copies of the EIS available to the Commonwealth Environment Protection Agency (EPA) and to the public 'by sale or otherwise'. The charge made for these environmental documents is usually about $A20, enough to deter casual requests but not interested parties. Further, the proponent must make a copy of the final EIS available free of charge to any person that has commented in writing on the draft EIS.

There is no provision in the Commonwealth for the preparation of both a draft and a final public environment report. In this sense the PER is similar to the UK environmental statement for which no draft is required. In practice, the difference between the information contained in Commonwealth PERs and EISs has proved to be small.

Notwithstanding the various checks and balances on the preparation and review of EIA reports, including the role of EPA as an independent (supposedly impartial) arbiter, criticisms have been levelled at the frequently lengthy EISs (which may well be accompanied by detailed appendices). Formby (1987) has stated that they are often of indifferent quality and limited effectiveness. It has been claimed that the documents are difficult for the public to understand, that cumulative impacts are neglected, that alternatives are not treated properly and that, in particular, the no-action alternative is frequently neglected (Chapter 8). While these criticisms are not universal, it appears that, while EPA's various checks on the quality of EISs are invaluable, the documents still contain bias and need to be seen as what they are: proponent statements.

New Zealand

The review stage in the New Zealand EIA process serves two purposes. The first is to review the information provided by the developer to determine its adequacy and to permit the decision on the action to proceed. The second is to act as a second screening stage, not unlike the Canadian EIA review process, in which case further information is usually requested from the developer. In effect, this provides for a two-tier EIA system in which the initial EIA report is used to determine whether the project is likely to have significant effects. If the developer can mitigate the effects of the project so that residuals are only minor, and to the satisfaction of those affected, the local authorities need not notify the proposal and the further EIA information requirements will not be triggered.

The first step is to review the submitted EIA report to determine whether more information is required (Fig. 5.5). If it is, the developer must provide it. The New Zealand Resource Management Act imposes time limits upon councils to process applications once they have sufficient information.

Having reviewed the initial EIA report and any further information, the local authority then determines whether the development is likely to have significant effects or not (in consultation with the people affected by the proposal). If the effects are judged to be significant, the application will be notified publicly and certain consultees will be informed. In this case (and there may be several hundreds of these each year) the local authority may request further information on alternatives and on the consultation undertaken by the applicant (Section 92(2)(a)) (if this has not already been provided by the developer (Chapter 11)), and/or commission an independent review of the EIA report (Fig. 5.5).

EIA materials are made available to the public for inspection. All submissions by organisations and by members of the public must be copied to the developer and the council has the power to ask the developer to explain how these are to be dealt with. The Act provides that, if submissions requesting a public hearing at which opinions and objections can be expressed are received, one should be held by the council before any decision is reached. (There are also provisions for pre-hearing meetings to resolve conflicts.) The Parliamentary Commissioner for the Environment retains the right to intervene in EIA cases, but does not possess the resources for such intervention to be other than exceptional.

All the costs incurred by the local authority (LA) in dealing with the application (including staff and hearing costs) can be recouped from the applicant. The ability in New Zealand to recruit consultants to review EIA reports without time or financial costs to the LA is clearly likely to help in ensuring that the review is adequate. This power is supported by a provision for further information to be required of the developer at any time prior to the hearing (Section 92(1)).

There is some advice on review procedures (Ministry for the Environment – MfE, 1992b). Morgan and Memon (1993, pp. 61, 62) have suggested that, rather than relying solely on the checklist provided by the Fourth Schedule to the Resource Management Act, local authorities should include the points listed in Box 12.5 in evaluating the proponent's EIA report. In practice, it appears that local authorities are suffering from a distinct lack of expertise in reviewing EIA reports and that, unsurprisingly, practice is very variable, with marked problems in meeting deadlines at regional council level (Morgan, 1993). Some district councils

Box 12.5   Checklist for evaluating New Zealand EIA reports

1. If a brief or framework for the EIA was agreed upon by the proponents and the consent granting authority, has the impact assessment team followed the agreed format in a satisfactory way?

2. Did the impact assessors consult the local communities and have they, in their study, shown evidence of having taken note of the community attitudes and feelings?

3. Is the nature of the proposed project clearly described? In particular, are the key processes likely to interact with the environment identified and explained?

4. Is there evidence of a rational, coordinated approach to reviewing the potential effects of the proposed project on the environment?

5. Are there obvious gaps in the coverage of the study? Are social impacts included? Are long term as well as short term effects considered? Are indirect effects as well as direct effects considered?

6. Have the assessors made predictions about possible impacts? Are the predictions based on sound methods and data? Do they indicate the probability of an impact occurring, and its likely severity or magnitude? Do they identify beneficial effects as well as adverse effects?

7. To what extent have cumulative impacts been addressed? [that is potential impacts from the proposed project, which might be small in themselves, but which might add to existing impacts from activities already operating in the local area and thereby bring about unacceptable environmental consequences].

8. Have the impact assessors examined environmental impacts that might arise as a result of abnormal operating conditions (such as the implications of accidental fires, or particular natural hazards)?

9. Have the impact assessors handled the information about predictions in a reasonably balanced and objective manner? Do they attempt to impose their own assessment of the social (or political) significance of the possible effects?

10. What steps have been taken to determine the views of the affected communities (at the local, regional or national scale) concerning the social significance of the predicted impacts?

11. What suggestions are made for mitigating adverse predicted effects? Have the predicted impacts been clearly stated, separate from the proposed mitigation measures? Have the consequences for the environment, of implementing the mitigation measures, been clearly stated?

12. What form of monitoring programme is proposed, and are key indicator variables identified for future monitoring?

13. Has the impact assessment team produced a summary document outlining the potential effects, both beneficial and adverse, on the environment? Can members of the local community understand the information? Is the material presented in a neutral way, without apparent favour to the proposal? Is technical information easily available to those people or organisations wishing to follow up on specific points?

Source: Morgan and Memon (1993, pp. 61, 62).

appear to be depending on the professionalism of the applicant's consultants for quality assurance, while others may be relying on public hearings for evaluation. There appears to be growing use of pre-hearing meetings and combined hearings between regional and district councils into all the consents required. Perhaps because of the risk of antagonising developers (who must pay for them), few independent reviews of EIA reports have been commissioned. There remains a clear need for the development of review methods and criteria, to be utilised on receipt of the EIA report, to improve practice (Dixon, 1993b).

Summary

Only the United Kingdom does not fully meet the EIA report review criterion, as shown in Table 12.1. Even in the United Kingdom it is not the public review of EIA reports which is missing but the duty on the proponent to respond formally to the points raised. In practice, the proponent usually provides further information if it is requested by the local planning authority (and such information must be made available

Table 12.1   The treatment of EIA report review in the EIA systems

Criterion 7: Must EIA reports be publicly reviewed and the proponent respond to the points raised?

Jurisdiction	Criterion met?	Comment
United States	Yes	Lead agency must respond to agency and public comments on published draft EIS in final EIS.
California	Yes	Lead agency must respond to all relevant comments on published draft EIR in final EIR.
United Kingdom	Review: Yes Response: No	LPA may request further information and proponents usually provide it. Proponents under no duty to respond to comments.
The Netherlands	Yes	EIA Commission reviews the EIS and, where necessary, supplementary information is requested by competent authority.
Canada	Yes	Discretionary public review of screening reports, public review of comprehensive study reports, and extensive public review, with proponent response, of EISs.
Australia	Yes	Proponent responds in final EIS to relevant points raised on published draft EIS.
New Zealand	Yes	Local authority power to commission independent review of public EIA report at developer's expense and to demand more information for notified projects.

for a further period of consultation and participation). The EIA systems in many developing countries fail to meet this criterion because EIA reports are frequently regarded as confidential and, in any event, there may be no real culture of participation (see Appendix).

The three longest-established EIA systems (those in the United States, California and Australia) all require the preparation of both draft and final EIA reports. In practice, the situation in three of the other four jurisdictions does not differ greatly since in The Netherlands, Canada and New Zealand, the proponent can be formally asked for further information. This supplementary information, which may consist of additional material, an elaboration of existing information or a response to comments, must be provided in all three countries. It is rare for supplementary information to be provided in developing countries. The form of this additional material may be disparate and may consist of several different documents. One advantage of final EIA reports is that they bring all the further information together within a single document.

Generally, most EIA reports in all the jurisdictions appear to require supplementing with additional data following formal or informal review. Apart from the obvious increase in quality of the information provided between the initial EIA report and the final documentation, there seems to be a general trend towards gradual improvement in the quality of EIA reports over time. There is, however, still much room for further increases in EIA report quality, perhaps especially in developing countries.

Note

1. These percentages are based upon undated diagrams for the years 1989–91 produced by the Office of Federal Activities, Environmental Protection Agency, Washington, DC, entitled 'DEIS rating distribution' and 'FEIS rating distribution'.

CHAPTER 13

Decision making

Introduction

Decision making takes place throughout the EIA process. Many decisions are made by the proponent (e.g. choices between various alternatives – Chapter 8). Others may be made jointly by the proponent and the decision-making and environmental authorities (e.g. screening and scoping decisions). However, the main decision in the EIA process, whether or not to allow the proposal to proceed (or, less frequently, which alternative to implement) is always taken in the public domain. While the decision-making body may have given previous indications of the likely outcome of this decision, it is normally taken by a government agency, following consultation and public participation. The typical decision taken at this stage in the EIA process is not usually a choice between alternatives, but a seemingly simpler choice between authorisation and refusal.

This chapter presents a discussion of the decision-making stage in the EIA process. It presents a set of criteria for the evaluation of decision making in EIA systems. These criteria are then used to assist in the review of decision-making procedures and practice in the United States, California, the United Kingdom, The Netherlands, Canada, the Commonwealth of Australia and New Zealand.

Making decisions about actions

There is considerable literature about decision-making methods and their use in EIA.[1] While the use of quantified decision-making methods has been roundly criticised on a number of occasions (see, for example, Bisset, 1978; Hollick 1981b), much effort continues to be devoted to them. One reason for this, as Bisset (1988) stated, is that such methods are often devised and used by engineers and others whose training emphasises the use of quantified methods. Another:

> is the desire of many decision makers to be faced with an easy decision, especially when comparing a complex variety of impacts from a number of alternatives.
>
> (Bisset, 1988, p. 60)

Much of the literature on quantified decision making (and especially about subjective weighting) has indeed focused on the choice between alternative proposals,[2] rather than on the more usual choice about whether to approve or refuse a proposal.[3]

In practice, decisions on proposals subject to EIA, whether they are yes/no decisions or involve choice between alternatives are frequently made incrementally and often in the cyclical manner characteristic of the EIA process (Lawrence, 1994). Thus, it may become apparent at the design stage of the process that certain impacts are likely to be unacceptable, leading to withdrawal or to redesign. Equally, the weight and force of objections raised to the proposal during review may lead to the proponent modifying the action further. Decision making is seldom straightforward (e.g. it may well involve more than one decision-making body) and is dependent not only upon the merits of the proposal but on political circumstances. The environmental impact of the proposal will usually be only one of the factors to be considered by the decision-makers.

The making of any decision will involve a large number of trade-offs in the information base: between simplification and the complexity of reality; between the urgency of the decision and the need for further information; between facts and values; between forecasts and evaluation; and between certainty and uncertainty. The people making a decision on a proposal involving EIA will frequently be elected central, state or local government politicians. They will seldom have time to read the EIA report and other EIA documentation and will therefore be dependent upon their officials for some form of summary evaluation of the earlier stages of the EIA process.

This evaluation, which will often be very brief, will typically summarise the objectives of the proposal, any alternative to it, its principal positive and negative impacts and their significance, the mitigation measures proposed, the principal representations about the proposal and how objections have been met. Compatibility with relevant policies is usually discussed but formal decision-making methods are seldom employed. The evaluation is bound to subsume some element of decision making and may well contain recommendations.

However, while the use of quantified decision-making methods may be helpful in reaching a consensus among similarly trained professionals, their use in public decision making is regarded with suspicion and is discouraged by politicians. The furthest most politically accountable decision-makers would wish formal evaluation to proceed is the preparation of a summary set of quantitative forecast impacts, together with a separate set of adjectival indications of their significance.

Other, non-environmental objectives and political factors may well out-weigh the findings of the 'technical' evaluation in the interactions between elected or appointed representatives from which the decision emerges. These are likely to be trade-offs between environmental and other factors. It is likely, given the positive benefits that most proposals confer (e.g. employment), that the decision-makers will seek to approve the action, unless there are politically overwhelming reasons to refuse it, but to negotiate increases in benefits and further mitigation of its negative impacts.

The original intention of EIA was that environmental considerations should be given greater weight in the design of proposals and in the decisions taken upon them. EIA was intended to constrain, but not to control discussions. Accordingly, most jurisdictions forbid the taking of a decision on the action until an EIA report has been prepared and subjected to review. This is a fundamental requirement of any EIA system.

If the decision on the action is to reflect meaningfully the EIA process, it should be possible to require modifications, to impose conditions or, in the last analysis, to refuse permission for the proposal to proceed. Clearly, the power of refusal is very important in ensuring that the aims of EIA are met. Refusal is the ultimate sanction on the proponent. This discretionary decision (i.e. where the decision-maker possesses the discretion to refuse, rather than the power only to set performance conditions which must be met) usually falls to the authority responsible for land use control (Wood, 1989b). Other bodies will obviously also be involved in decision making, but they do not usually possess the power of refusal (for example, air pollution control authorities must grant permits provided emission and/ or ambient air quality standards are met). Any conditions imposed upon the approval should be phrased to take account of: (a) the forecasts made in the EIA report (to help to ensure that the commitments made in that document are implemented) and (b) the uncertainty in the forecasts upon which the decision is based (to ensure that the conditions are realistic).

For the decision on the proposal to be seen to be fair it is obviously preferable that it should, in general, be made by a body other than the proponent. Further, any summary evaluation prepared for the decision-makers by their advisers (e.g. the Australian assessment report, the Canadian panel report, the Dutch EIA Commission report or the British report to the local authority planning committee) should be made public.

Without such a check, it would be too easy to meet procedural EIA requirements relating to preparation of the EIA report and review without meeting the substantive obligation to consider the outcome of these procedures in reaching a decision. An additional, and even more important check, is that the decision, the reasons for it and the conditions attached to it, should be published. It is obviously desirable, if the extent to which the EIA process is actually taken into account in decision making is to be clear, that their reasons for making the decision include an explanation of how the EIA report and review influenced the decision. This is provided for in many EIA systems (for example, in the United States and Dutch EIA procedures).

As at other stages of the EIA process, advice on the factors to be considered by decision-makers in reaching their decision is a valuable way of ensuring that procedures are complied with and that environmental factors are given appropriate weight. Such advice could provide guidance on whether or not certain impacts were likely to be environmentally acceptable. For example, a jurisdiction might well indicate that noise levels above a certain level would not be tolerated and that the proposal should either be modified or refused.

Some jurisdictions allow for consultation and participation once the evaluation has been prepared for the decision-makers but before any decision has been reached. This is clearly not possible where no separation of the steps in decision making takes place. However, the provision of a public right of appeal against the decision can increase public confidence in the EIA process. Such an appeal could either be administrative (e.g. to a tribunal, as in New Zealand) or to the courts, as in the United States.

The possibility of a proponent eluding these various checks exists (though the more there are, the more difficult evasion becomes). It is, therefore, important that practice in decision making reflects the results of the EIA report and review in practice, i.e. that it is effective. If practice does not appear to be influenced by the EIA process, the implication must be that additional checks and balances are required, even though this will probably conflict with the criterion of efficiency (i.e. that the decision should be taken without undue expenditure or delay). The various criteria for the evaluation of decision making are summarised in Box 13.1.

Box 13.1 Evaluation criteria for decision making

Must the findings of the EIA report and the review be a central determinant of the decision on the action?

Must the decision be postponed until the EIA report has been prepared and reviewed?

Can permission be refused, conditions be imposed or modifications be demanded at the decision stage?

Is the decision made by a body other than the proponent?

Is any summary evaluation prepared prior to decision making made public?

Are the decision, the reasons for it, and the conditions attached published?

Must these reasons include an explanation of how the EIA report and review influenced the decision?

Does published guidance on the factors to be considered in the decision exist?

Is consultation and participation required in decision making?

Is there a right of appeal against decisions?

Does decision making function effectively and efficiently?

These criteria are employed in the review of decision making in each of the seven EIA systems below.

United States of America

No federal decision on a proposed action can be made until 30 days after the Environmental Protection Agency (EPA) has published a notice that the final environmental impact statement (EIS) has been filed. This is an important provision but perhaps symptomatic of the National Environmental Policy Act (NEPA), which contains no absolute requirement to protect the environment. The lead agency has the power to refuse consent on the basis of the EIS (i.e. to choose the no action alternative) but no other agency can override it. It is likely only to refuse consent in responding to an application for a permit or for funding. NEPA does require that federal agencies disclose the environmental effects of their actions and identify alternatives and mitigation measures. Agencies may not select an alternative unless it has been discussed and evaluated in the final EIS, but they are not required to adopt the environmentally preferred alternative.

While the lead agency, in its role as proponent, will clearly wish to implement the proposed action, other agencies may, and do, voice objections. These may lead to the abandonment of the proposal or, in rare cases, to a referral to the Council on Environmental Quality (CEQ). The referral process is intended to resolve interagency disagreements. Although a CEQ decision is not binding, agencies usually abide by its recommendations and disagreements have generally been effectively resolved by the referral process (Rand and Tawater, 1986). Of the 25 referrals made between 1974 and 1993, 15 were originated by EPA.[4] The threat of referral is often sufficient to bring environmental concessions from the lead agency, since CEQ may submit its recommendation to the President for action, elevating the issue to a level that would embarrass a lead agency. This referral process, based upon the impacts of the action, is thus a substantive element in the EIA process that has been criticised as being essentially procedural.

At the time of its decision, following the preparation of the final EIS, the federal agency must prepare a record of decision (ROD). This is a written public record which was introduced only in 1978 with the intention of ensuring that the EIS actually influences agency decisions. It must contain:

- a statement explaining the decision
- an explanation of the alternatives considered and those that are environmentally preferable
- the social, economic and environmental factors considered by the agency in making its decision
- an explanation of the mitigation measures adopted and, if all practicable mitigation measures were not adopted, an explanation of why not

- a summary of the monitoring and enforcement programme which must be adopted to ensure that any mitigation measures are implemented (Regulations, Section 1505.2).

Some agencies publish their RODs in the *Federal Register* though this is not a formal requirement.

Agencies, in making their decisions, must balance the relevant environmental and other factors, as in other jurisdictions. In practice, the EIS frequently influences the decision, though often not as a direct result of the information it contains. As Hyman and Stiftel (1988, pp. 426, 427) put it:

> while decision makers generally have not turned to EISs in the expected ways – for specific data, evidence or policy implications – ... [EISs have] succeeded in jarring the consciousness of many decision makers and administrators.

NEPA, the CEQ regulations and the agency's own regulations provide the necessary guidance and the environmental review procedure leading to the decision is open to appeal in the courts. However, providing the appropriate procedures have been followed, to ensure that the agency has genuinely considered the environmental consequences of the action, including the making public of the final EIS and the ROD, the courts have not attempted to contradict agency decisions (Mandelker, 1993b).

California

The lead agency must certify the final environmental impact report (EIR) before making its decision. This certificate must state that the final EIR was prepared in compliance with the California Environmental Quality Act (CEQA) and that it was presented to the lead agency's decision-making body which reviewed and considered the final EIR before approving the project (Guidelines, Section 15090). The value of these certificates must be questionable: given the length and general lack of readability of EIRs it is very unlikely that they will be reviewed by the elected decision-makers themselves.

Once the final EIR has been certified, agencies may, under the provisions of CEQA:

- disapprove a project because it has significant environmental effects
- require changes in a project to reduce or avoid a significant environmental effect
- approve a project despite its significant effects.

To support its decision, an agency must prepare written findings of fact for each significant environmental impact identified in the EIR. In practice, tentative findings are normally presented to the decision-making body (commission, council, committee, etc.) with the official's report and recommendations. After any redrafting they are filed and are available to the public, but do not have to be circulated for public review.

For each significant impact, the decision-makers must reach one of the following conclusions or findings:

1. Changes have been made to avoid or substantially reduce the significant environmental impact.
2. Changes are within another agency's jurisdiction and have been or will be adopted by that agency.
3. Specific economic, social, or other considerations make mitigation measures or alternatives infeasible.

Each conclusion must be supported by substantial evidence and must contain an explanation that links the evidence to the finding (Bass and Herson, 1993b, pp. 84–6). The findings must be clear and well organised.

CEQA requires the decision-maker to balance the benefits of a proposed project against unavoidable environmental risks in determining whether to approve the project. If it approves a project with unavoidable significant environmental effects, an agency must prepare a written statement of overriding considerations explaining why it is willing to accept these impacts. This statement must be filed and be available to the public.

An agency must file a notice of determination after deciding to approve a project for which an EIR is prepared (this is equivalent to the federal record of decision). This must summarise project impacts and state where the final EIR may be inspected by the public. There are provisions regarding the circulation of the final EIR to the various relevant agencies. Each agency is required to make its own CEQA findings, use the final EIR to reach its own decision and, where necessary, adopt a statement of overriding considerations.

CEQA thus has the most comprehensive set of requirements relating to the integration of EIA in decision making of any of the EIA systems examined. Notwithstanding, there is a widespread feeling that decision-makers rely upon their staff to make sure that the lead agency complies with CEQA while making their decisions regardless. This is not to say that CEQA has no effect on decisions (many mitigation measures are incorporated and implemented as a result of negotiation by agency staff) but that it has little effect on elected decision-makers beyond their desire to avoid legal action. It is difficult to see what further procedural steps could be taken to enhance the effect of EIA on the grant or refusal of permission by decision-makers (as opposed to the mitigation of impacts) beyond prohibiting the granting of approvals if significant environmental impacts remain. Olshansky (1991) reported that 58 per cent of project approvals involving an EIR required statements of overriding considerations. Very few projects involve refusals: however, about one-third of projects are abandoned or postponed indefinitely because of delays caused by the CEQA process itself, or by actions brought once the decision has been announced (Olshansky, 1993).

United Kingdom

The local planning authority (LPA) is required to have regard to the 'environmental information' (the environmental statement – ES – and the various submissions by statutory consultees and the public) in making its decision, which cannot be reached before this environmental information has been considered. As with any other planning application, the LPA may refuse permission or grant it with or without conditions. In reaching this decision, the LPA attempts to weigh all the planning advantages and disadvantages: the environmental impacts of the proposal are only one factor in the decision.

Planning permission is the only genuinely discretionary consent in the United Kingdom: other consents relating to, for example, pollution control are virtually always granted, providing the appropriate conditions are met (Wood, 1989b). LPAs can determine their own applications (a very small proportion of ESs) but must use the same procedures as for ESs prepared by external proponents.

It is normal practice for planning officers to prepare a report (to the planning committee) on applications to be determined by the LPA. This report should summarise the salient parts of the EIA process, including the ES, the comments received during consultation and participation, and the officers' own evaluation. This evaluation report is a public document.

The planning decision, and the conditions attached to it, are published in the planning register that every LPA is required to maintain. If the application is refused, reasons must be given. If, however, it is approved, there is no requirement to provide reasons. While there is no formal specification of what areas the reasons for refusal should cover, it would be expected that the results of the EIA process would be relevant and should therefore be detailed. It is mandatory that the LPA states in writing that the environmental information has been taken into account in reaching the decision (DOE, 1994a).

Very little guidance on decision making exists in the United Kingdom. Basically, LPAs are required to make development control decisions in accord with the development plan 'unless material considerations indicate otherwise' (Department of the Environment – DOE, 1992, para. 1.3). In addition, the LPA is 'of course required to have regard to the environmental statement, as well as to other material considerations' (DOE, 1989, para. 41). Nevertheless, the adherence of new sources to pollution standards is a relevant matter for LPAs to consider in reaching decisions (Wood, 1989b; DOE, 1992), as are other criteria of environmental acceptability. Although failure to meet such criteria would be a valid reason for refusal, there is no prohibition on the grant of planning permission if the EIA process reveals that the proposal does not meet them.

There is no consultation and participation requirement during the decision-making process, though members of the public are often permitted to address the LPA elected representatives while the decision is being discussed. The public has no right of administrative appeal against the decision, unlike the developer. Planning appeals are normally heard at public inquiries conducted by the Planning Inspectorate. Inspectors frequently make recommendations about decisions to the Secretary of State for the Environment, usually based on similar criteria to those employed by the LPA. While such appeal decisions provide important precedents, they do not have legal force (there are restricted rights of further appeal to the courts).

Of a sample of 20 applications where a planning decision was made following submission of an ES, two-fifths were refused (and a further two were withdrawn). While the sample is small, and represents experience in the first 18 months of operation of the Regulations, the refusal rate is significantly greater than for planning applications generally (most of which relate to householder and other minor developments). It is also greater than might be expected in any EIA system in which developers first discuss projects with LPAs at very early stages of conceptualisation. (It is notable that this did not take place in about one-third of the 24 cases analysed (Wood and Jones, 1991).)

It may be that some LPAs regard environmental assessment (EA) as controversial and prefer to leave final decisions on applications accompanied by ESs to the Secretaries of State on appeal. However, there is little empirical evidence about the weight given to EA in decision making, though it is known that some decisions have had little regard to EA findings (see, for example, Lambert and Wood, 1990). There is no reliable evidence to date that public inquiry findings have been influenced by EA practice, or vice versa. It must be concluded that, while the environmental information is a material consideration in planning decisions, it is not yet a central determinant in many of them.

The Netherlands

The competent authority (which may be the same as the proponent) either publishes its draft decision with the EIS or following the review of the EIS in licence application cases. However, this draft decision is frequently modified as a result of the recommendations of the EIA Commission and the intervention of the public.

The competent authority is obliged by the Environmental Management Act 1994 (Section 7.37) to incorporate the findings of the EIS, of the EIA Commission's review and of the comments of the consultees and the public in its deliberations on the decision. This must include a discussion of the alternatives considered and a statement (if relevant) as to why the environmentally preferable alternative was not selected. The competent authority must explain fully the reasons for its decision in writing. These reasons must also indicate the weight which has been attached to environmental parameters in comparison to other factors (Ministry of Housing, Physical Planning and the Environment (VROM) and Ministry of Agriculture, Nature Management and Fisheries (LNV), 1991).

Irrespective of what sectoral legislation may say on the subject, all the environmental impacts must be considered when a decision requiring EIA is taken (Section 7.35). The competent authority taking a decision requiring EIA has the power to prescribe any condition, regulation or restriction necessary to preclude or limit the impacts on the environment. This includes a specification of how post-auditing is to be carried out (Chapter 14).

Furthermore, the competent authority may not take a decision when no EIS is provided (Section 7.28) or when the information in the EIS is out of date because the circumstances under which the statement was prepared have changed considerably (Section 7.27). These substantial and comprehensive decision-making requirements together place a considerable duty, which it is difficult to avoid, on the competent authority to take the EIS into account. A copy of the decision, and the reasons for it, must be sent to those commenting earlier on the EIS, including the statutory consultees and the EIA Commission (Section 7.38). The decision must also be made public through the media.

It is rare for applications to be refused. Of the 116 decisions made before the end of 1993 (EIA Commission, 1994) only three or four were refusals. However, a significant number of proposals are not pursued or are modified as a result of negotiation between the proponent and the competent authority, often on the basis of advice from the EIA Commission. It is customary for conditions to be applied to permissions.

There is a general third party right of appeal against decisions and this is exercised in 30–50 per cent of cases involving EIA. These appeals involve a hearing, which often only takes a few hours, and take place under the relevant sectoral legislation. A poor EIS may be a reason for instituting an appeal against a decision for which the EIS was drawn up (VROM and LNV, 1991). The number of court cases involving EIA has been very small (Chapter 6).

It is generally accepted that EIA is actually affecting decisions in The Netherlands, even if the reasons given for the decision are occasionally invented to justify it. Some decision-makers decline to read EISs or to consider their findings, so that environmental arguments are sometimes used to explain economic decisions. However, proponents are often more positive than competent authorities about the EIA process and they (and the authorities) are modifying proposals as a result of it. The Evaluation Committee on EIA (1990) found that the openness of decisions was leading to improvements in the way environmental considerations were taken into account in decisions.

Canada

It is a requirement of the Canadian Environmental Assessment Act (CEAA) that no responsible authority (RA) may 'exercise any power or perform any duty' in relation to a project until the EA is complete (Section 11(2)). In the *self-directed assessment* the RA can take action to allow the project to proceed if it is not likely to cause significant adverse environmental effects, after taking into account the implementation of

New Zealand

Environmental impact assessment should be central to decision making in New Zealand because the Resource Management Act 1991 requires both plans and consent decisions to avoid, remedy or mitigate any adverse effects of activities on the environment (Box 5.1). Section 104 of the Act specifically states that the consent authority must have regard to the provisions of relevant policies and plans and of any EIA report when considering an application for a resource consent. The EIA report (and the various submissions made) is intended to be central to the decisions since 'the consent authority shall have regard to any actual and potential effects of allowing the activity' (Section 104(1)).

It is customary for the local council members to make the decision, on the basis of an evaluation report and recommendations prepared by officers. The decision must be taken within 15 working days of the hearing, if one is held. The applicant, or any person who has made a written submission about the application, can appeal against the consent authority's decision to the Planning Tribunal. The Tribunal has not hesitated to comment on the deficiencies of the documentation furnished in the past and it is expected that it will continue to provide a check on the quality of EIA reports in the future.

The results of appeal decisions by the Planning Tribunal are likely to prove influential in the development of practice, as a result of specific guidance furnished by them. Rather as in the United Kingdom, there are powers for the Minister for the Environment to call in applications of national significance for decision in New Zealand. However, unlike in the United Kingdom, it is intended that these powers should be employed only exceptionally.

In practice, decision making does not appear to be meeting the aims of the Act. Dixon (1993b) pointed out that very few checks exist on whether the EIA is actually considered when the decision is made and that there is a considerable danger of the parochialism of past practices continuing under the far more sophisticated Resource Management Act. There is no need, in the New Zealand EIA system, to produce the equivalent of a US record of decision. Morgan (1993) stated that elected politicians in regional and territorial authorities tend not to be aware of (or sympathetic to) the basic principles of the new Act and thus not to appreciate that EIA is intended to influence decision making.

Summary

Appropriately, the decision about whether to give a 'yes' or a 'no' to the EIA systems according to whether or not they meet the decision-making criterion is a delicate one. All the seven systems could have been stated to meet it partially, though most EIA systems in developing countries would fail to meet it (Appendix).

The criterion states that the EIA report and the comments upon it must

appropriate mitigation measures. In other circumstances a public review (Fig. 5.3) becomes necessary. The various decisions that can be taken by a RA following consideration of a comprehensive study report are shown in Table 13.1. The situation for screening reports is similar, save that the Minister of the Environment has no direct role. The intervention of the Minister is an advance on the Environmental Assessment and Review Process (EARP) in which the Minister acts only as an expert. Public notice of the RA's decision, and any conditions or follow-up measures, must be provided but no reasons have to be stated.

For *public review* projects the panel or mediator must prepare a report and make it available to the public. For panel reviews, this must set out:

- the rationale, conclusions, and recommendations of the panel, including any mitigation measures and follow-up programme

Table 13.1 Decisions in Canadian comprehensive studies

Comprehensive study report: Conclusions	The Minister	The responsible authority (RA)
1. The project is not likely to cause significant adverse environmental effects, or it has the potential to cause significant adverse environmental effects that can be prevented or significantly reduced by mitigation measures.	Refers project back to RA.	May exercise any power or perform any duty or function that would allow project to proceed; must ensure implementation of mitigation measures.
2. The project is likely to cause significant adverse environmental effects and these effects cannot be justified in the circumstances.	Refers project back to RA.	May not exercise any power or perform any duty or function that would allow project to proceed.
3. The project is likely to cause significant adverse environmental effects and it is uncertain whether the effects can be justified.	Refers project to mediator or review panel.	May not exercise any power or perform any duty or function that would allow project to proceed.
4. It is uncertain whether the project is likely to cause significant adverse environmental effects.	Refers project to mediator or review panel.	May not exercise any power or perform any duty or function that would allow project to proceed.
5. There are public concerns about the project.	Refers project to mediator or review panel if public concerns warrant.	May not exercise any power or perform any duty or function that would allow project to proceed.

Source: Federal Environmental Assessment Review Office (1993a, p. 115).

- a summary of public comments.

(Federal Environmental Assessment Review Office
– FEARO, 1993a, p. 139)

The report is submitted to the Minister and to the RA. The RA must decide whether the project can proceed (with appropriate mitigation measures). If the RA concludes that 'the project is likely to cause significant adverse environmental effects that cannot be justified in the circumstances' (CEAA, Section 37(1)(b)) it must not permit the project to proceed. If the RA is the proponent (as is frequently the case) then its decision will determine the fate of the project. In some other cases, the proponent may be able to proceed without federal approval (FEARO, 1993a, p. 140) but, if federal land, money or permits are required, the project will be stopped.

The RA may draw different conclusions from the panel or mediator but must explain its reasons in the public notice of decision. (There is no appeal against this decision except through the courts.) Normally, disagreements about panel recommendations are resolved between ministries, often in Cabinet. This (often lengthy) process also applies to comprehensive study and mediation reports under CEAA. In the early years of EARP, panel recommendations were accepted by initiating departments. More recently, as EARP has been applied on a less voluntary basis, they have tended frequently to deviate from panel recommendations.

In one controversial case (the Oldman River Dam), the panel's main recommendation was to:

decommission the dam by opening the low level diversion tunnels to allow unimpeded flow of the river.

(FEARO, 1992b, p. i)

Unsurprisingly, this recommendation (which arose because of court retrospective imposition of EARP) was not accepted. While this case is unusual, decisions have often been made so late in the design/construction process (largely because of uncertainties over EARP application) that the 'no-go' option has been virtually precluded (Fenge and Smith, 1986). Weston (1991, p. i) reported, following a study of initial assessments, that 'often, a federal government department has decided to undertake environmental studies after deciding to undertake a project'. It is, therefore, perhaps unremarkable that the Executive Chairman of FEARO has stated that:

the next challenge is to ensure that all decisions are reached only *after* environmental factors are identified, considered and integrated into the very process used to reach decisions.

(Dorais, 1993, emphasis in original)

Since affecting decisions is the objective of EA (Chapter 1) this statement must be taken as damning self-indictment of the federal Canadian decision-making process. Under EARP, therefore, decision

making cannot be said always to have been either effective or efficient though there have been many successes involving panel reviews. It remains to be seen whether CEAA, which retains the same basic approach, will improve this situation by, *inter alia*, eliminating such exceptions from the federal EA process as the Oldman Dam.

Commonwealth of Australia

The Commonwealth Environment Protection Agency (EPA) must prepare an 'assessment report' for the Environment Minister (Commonwealth of Australia, 1987, Procedures, para. 9.1). This assessment report is intended to stand alone, to be read in isolation from the proponent's EIS, to provide an analytical summary overview of the environmental effects of the proposal based upon the proponent's documentation and the responses to it. These reports may be lengthy (that for the highly controversial Third Sydney Airport, for example, ran to some 202 pages). This report is made available for sale to the public and is widely distributed.

The Minister must make any comments, suggestions or 'recommendations' and/or suggest conditions relating to the proposal within 42 days of receiving the final EIS (28 days in the case of a public environment report – PER) (Procedures, para. 9.3.1). The Minister's recommendations are separate from, but normally coincide with, those in the assessment report and must be made public.

In practice, if there is any disagreement between the Environment Minister and the minister responsible for the action, liaison takes place, sometimes in Cabinet, to resolve any differences between them, before the assessment report and the Minister's recommendations are finalised. Such an arrangement is not dissimilar to that in The Netherlands. There is no requirement for the Action Minister's decision, and the reasons for it, to be publicised but this often occurs. Notwithstanding the imprecations to take the environmental impact assessment into account in reaching the decision (Procedures, para. 9.5) and the accountability ensured by the publication of the assessment report and the Environment Minister's recommendations, they need not determine the decision. They must, however, be 'taken into account' (Procedures, para. 9.5).

By the end of 1991, only three or four proposals which had been through the full Commonwealth EIS or PER procedure had been refused. Virtually every proposal is modified during the EIA process to reduce impacts and it could be claimed that some unsatisfactory proposals drop out of the procedure before the final decision is taken, or that conditions placed upon some consents are so onerous that the proponent declines to progress them. In the United States there is also a very low refusal rate and, in both countries, the main aim of EIA is the mitigation of impacts. Notwithstanding, a refusal rate of 2 per cent of an average of eight EISs/ PERs annually over 19 years is a remarkable testimony to the negotiating skills of the EPA staff.

be 'a' (not 'the') central determinant of the decision. EIA was never intended to provide the sole basis for decision making (Chapter 1). However, to meet the criterion, an EIA system needs to demonstrate not only that the decision should be influenced by the EIA (all seven do so) but that, in practice, the EIA report actually influences the decision and is not just 'boiler-plate' paper. Only one of the seven EIA systems (that in The Netherlands) meet this interpretation of the criterion (Table 13.2).

California has perhaps gone furthest in trying to ensure that decisions are influenced by EIA by requiring a statement of overriding considerations where significant impacts will result from the approval of an action. However, in practice it is still possible for decision-makers to ignore the EIA. The United States of America, the United Kingdom, Canada, Australia and New Zealand use differing mechanisms for ensuring that the EIA is considered. In practice, however, it is still common practice for decision-makers to circumvent these EIA mechanisms where this is convenient. In The Netherlands the recommendations of the highly influential EIA Commission are published and the competent authorities

Table 13.2 The treatment of decision making in the EIA systems

Criterion 8: Must the findings of the EIA report and the review be a central determinant of the decision on the action?

Jurisdiction	Criterion met?	Comment
United States	No	Explanation of decision and disclosure of environmental effects mandatory. In practice, EIS often influences decision.
California	No	Statement of overriding considerations must be written if project approved, as often happens, with unavoidable significant environmental impacts.
United Kingdom	No	Environmental information is a material consideration but not necessarily a central determinant. Practice varies.
The Netherlands	Yes	Explanation of way environmental impacts considered in decision is mandatory. In practice, EIA generally does influence decision.
Canada	No	Conclusions of self-directed assessment report determine decision: reasons must be given when responsible authority disagrees with recommendations of public review report.
Australia	No	EIS, suggestions and recommendations must be taken into account but need not determine decision.
New Zealand	No	Act makes EIA central to decision but, in practice, EIA is often not given appropriate weight.

are effectively obliged to accept them (as used to be the case in Canada). The trick in the seven jurisdictions, and in developing countries (see Appendix), is to ensure that EIA remains, or becomes, a central determinant of decisions in practice as well as in theory.

Notes

1. See, for example, Munn (1979), Bisset (1980, 1988), Hart *et al.* (1984) and Lee (1989).
2. As Lee (1989) pointed out, the use of scoring methods (for example, of intervals, of ranking, of a binary yes/no approach or of normalisation) in the comparison of alternative values of the same type of impact (e.g. noise) should be clearly distinguished from the use of weighting to compare different types of impact (e.g. noise and air pollution).
3. Checklists and matrices and other index approaches to the identification and communication of impacts have been modified to become decision-making methods by the integration of scales and weights as in the Leopold *et al.* (1971) matrix and in the 'environmental evaluation system' advanced by Dee *et al.* (1973). Here, and in simulation modelling, value judgements are implicit in the weights or scales used. There are various means of making these value judgements less opaque. One is to list the impact forecasts clearly and then to explain the basis on which scaling and weighting is undertaken. Another is to stop short of overall aggregation and to quantify only groups of impacts (e.g. different types of air pollutants). Another is the use of sensitivity analysis in which the weights are varied to determine their effects on the outcome. Yet another is the opening up of the attribution of weights to consultees to make the judgements more representative.
4. Environmental Protection Agency (1993), personal communication, Office of Federal Activities, EPA, July. See, for a listing of 24 of the 25 cases, Bass and Herson (1993a).

CHAPTER 14

Monitoring and auditing of actions

Introduction

There are numerous definitions of monitoring and auditing in EIA. Thus Tomlinson and Atkinson (1987a) identified a total of seven different types of EIA audit. The first distinction to make is that between the monitoring of individual actions and of the EIA system as a whole (the latter is dealt with in Chapter 17). The second distinction is between the three main types of action monitoring and auditing: implementation monitoring, impact monitoring and impact auditing. This chapter is concerned with these three types of monitoring and auditing. It presents a set of criteria for analysing the treatment of action monitoring in EIA systems. These criteria are then used to assist in the review of the monitoring of actions in the EIA systems in the United States, California, the United Kingdom, The Netherlands, Canada, the Commonwealth of Australia and New Zealand.

Monitoring and auditing of action impacts

As mentioned in Chapter 11, the EIA report is no more than a record, or 'photograph' of the forecast impacts of the action as it is designed at a particular point in time. Further design work will take place once approval of the action has been granted (again on the basis of information available at a particular point in time) and this may well lead to modifications. Further, since even the best design may need to be altered to meet unexpected problems encountered during construction, further modifications may well take place during the process of implementing the action. With the best will in the world, therefore, the implemented action may well differ from that envisaged when the EIA report was prepared. If the proponent's intentions fall short of this, a significant 'implementation gap' may occur.

The necessity of maintaining surveillance of, and· control over, the implementation of actions has tended to be a somewhat neglected area in EIA. Inadequate checks have been applied in many jurisdictions to ensure that developers negotiated any necessary changes with the appropriate authorities and thus that implementation did not result in significant unanticipated impacts (Sadler, 1988).

'Implementation monitoring' involves checking that the action (normally a project) has been implemented (constructed) in accordance with the approval, that mitigation measures (e.g. sound-proofing) correspond with those required and that conditions imposed upon the action (e.g. noise emission limits) have been met. Such checking may involve physical inspection (e.g. of building location, wall construction or waste storage/disposal) or measurement (e.g. of noise emissions) using various types of instrument, together with the application of professional judgement. This type of monitoring can be carried out either by the decision-making or environmental authorities or by the proponent (with appropriate checks and balances) or, as is frequently the case, may be divided between them.

Implementation monitoring frequently takes place under the provision of more than one set of legislative requirements (e.g. land use planning, building approval or pollution control procedures). Any necessary action to enforce compliance with the terms of the approval may be taken under powers other than the legal underpinning of the EIA system. Implementation monitoring is therefore essentially reactive, its principal purpose being to ensure that the action adheres to the conditions of its approval (Hollick, 1981a; Tomlinson and Atkinson, 1987a; Sadler, 1988).

'Impact monitoring' involves measurement of the environmental impacts (e.g. on ambient noise levels or upon a species of bird) that have occurred as a result of implementing the action. A variety of measurement techniques is likely to be needed, coupled with the exercise of expert opinion. This type of monitoring serves two purposes:

1. Where monitoring of the environment reveals unexpected or unacceptable impacts (e.g. elevated noise levels at night) further design changes (e.g. baffling) or management measures (e.g. ensuring closure of doors and other openings) may be necessary. The monitoring results may indicate that the approval conditions (e.g. on noise emissions) have been breached. Even where this is not the case, voluntary action by the proponent may take place or action may be required under the provisions of other legislation.
2. Impact monitoring can provide useful feedback for the assessment of other similar actions by helping to ensure that relevant areas of concern are identified. It can also assist in indicating where existing environmental knowledge is deficient and thus where further research may be needed to improve environmental management practice.

In most EIA systems impact monitoring is carried out by some combination of the developer and the environmental authorities, though this is increasingly becoming a responsibility of the proponent. As with implementation monitoring, impact monitoring may be covered by a variety of legislation and there may be little coordination between, or compilation of, measurements. In some EIA systems, the proponent is required to specify proposed implementation and impact monitoring proposals in the EIA report (e.g. in New Zealand where an environmental management programme may be required). Figures 1.1 and A.1 show the role of monitoring in the project planning cycle.

'Impact auditing' involves comparison between the results of implementation and impact monitoring and the forecasts and commitments made earlier in the EIA process (and especially in the EIA report). It is also frequently referred to as 'post-auditing'. The principal purpose of impact auditing is to enable the effectiveness of particular forecasting techniques to be tested and thus to improve future practice (e.g. by reducing the uncertainty in impact prediction (Bisset and Tomlinson, 1988)). A secondary purpose is in the management of the impacts of the action concerned (see below).

Since the conditions applied at the decision-making stage should reflect the outcome of the earlier stages of the EIA process (Chapter 13), implementation monitoring should be designed to permit auditing of the proponent's commitments in the EIA report and any subsequent documentation. The same is obviously true of impact monitoring.

Auditing may be carried out by the decision-making or environmental authorities, by the proponent (possibly as part of internal auditing procedures) or later by, say, research investigators. Few EIA systems require impact auditing, as opposed to implementation or impact monitoring, for particular actions. This is clearly a matter of concern (Sadler, 1988), and follows from the orientation of most EIA systems to project authorisation, rather than to the management of impacts from projects (Tomlinson and Atkinson, 1987b).

In granting approvals, it will not always be possible to impose conditions to cover every eventuality and environmental impacts may arise which are uncontrolled. If forecasts about these impacts have been included in the EIA report (for example, forecasts about the general appearance, landscaping and tidiness of a project), the proponent may be obliged to fulfil the undertakings made, for fear that a public commitment will be seen not to have been honoured. A requirement upon the proponent to produce impact auditing reports after a certain number of years (say, two and five years), subject to checks by the environmental authorities, would be invaluable in meeting both the purposes of such auditing.

First, impact auditing, in which the EIA report and subsequent documentation provide the basic point of reference, would generate a body of experience about the results of auditing in EIA systems. This would not only assist in the development of forecasting and monitoring methodology but would provide public reassurance about impact management. As Buckley (1991a, p. 21) has stated, 'impact audits provide a means for both industry and government to demonstrate their competence in environmental management to the public'. Second, the achievement of satisfactory auditing results could provide the basis of agreement between the environmental authorities and the proponent to terminate impact monitoring programmes (e.g. on certain air or water quality parameters) though implementation monitoring would need to continue.

By far the largest proportion of the literature on EIA monitoring and auditing concerns impact auditing.[1] The main conclusions may be summarised as follows:

- There appear to be no standardised audit methodologies.
- Monitoring needs to be considered and designed very early in the EIA process.
- Monitoring requires coordination, information management and resources.
- Post-EIA report design changes invalidate many forecasts.
- Many EIA reports contain very few forecasts.
- Many EIA report forecasts are vague and qualitative.
- Monitoring has often been inadequate for auditing purposes.
- Only a minority of EIA forecasts have proved accurate or almost accurate.
- Few EIA report forecasts have proved totally inaccurate.
- Few unanticipated impacts have been detected.
- There is little evidence of systematic bias in forecasts.

There is disagreement about the role of hypotheses in impact auditing. Many, especially those with an ecological background, have argued that EIA report forecasts should be framed as falsifiable hypotheses to facilitate auditing (see, for example, Caldwell et al., 1983; Beanlands and Duinker, 1984). Others have opined that, while precision is desirable where it is feasible and appropriate, the value of EIA report impact forecasts does not depend of their strict auditability (see, for example, Culhane et al., 1987). Given that the purpose of EIA is to ensure that appropriate mitigation measures are utilised to minimise the impacts of approved actions, the crucial question becomes 'did it result in appropriate management action being taken?' (Bailey and Hobbs, 1990, p. 165).

It is unsurprising, given the dichotomy between the procedural and the technical inherent in EIA, that these two views should co-exist. If nothing else, efficiency in EIA suggests that EIA report forecasting and impact monitoring and auditing should not involve the writing of PhD theses. As Culhane (1987, p. 236) stated: EIA reports are not intended 'simply to gratify theorists trained in the scientific method' but to force consultative and participative consideration of the consequences of proposals.

As at the other stages of the EIA process, the availability of clear guidance on the procedures and techniques of action implementation monitoring, impact monitoring and impact auditing will be helpful to proponents, the decision-making and environmental authorities, consultees and the public.

The publication of monitoring results is clearly a necessary check on the operation of monitoring procedures. Such information is frequently currently available from a variety of sources but is often difficult to obtain and is seldom collated. The availability of all the relevant monitoring data at a single location would be a significant advance on most current EIA

Box 14.1   Evaluation criteria for the monitoring and auditing of action impacts

---

Must monitoring of action impacts be undertaken and is it linked to the earlier stages of the EIA process?

Must monitoring of the implementation of the action take place?

Must the monitoring of action impacts take place?

Is such monitoring linked to the earlier stages of the EIA process?

Must action monitoring arrangements be specified in the EIA report?

Can the proponent be required to take ameliorative action if monitoring demonstrates the need for it?

Must the results of such monitoring be compared with the predictions in the EIA report?

Does published guidance on monitoring and auditing action implementation and impacts exist?

Must monitoring and auditing results be published?

Is there a public right of appeal if monitoring and auditing results are unsatisfactory?

Does action monitoring function effectively and efficiently?

---

procedures. As mentioned above, the publication of the results of audits (and the environmental authorities' response to them) would be a considerable step forward. Sadler (1988, p. 141) supported the use of publishing auditing results because public participation 'drives many innovations in EIA practice'. Public scrutiny would be even more effective if it were supported by a right of appeal, say against any proposal by the environmental authorities to agree to the cessation of impact monitoring, where the monitoring results to date were demonstrably unsatisfactory.

Finally, it is important that, as at the other stages of the EIA process, monitoring should function effectively (i.e. that it should provide relevant information about implementation and impacts, linked to the earlier stages of the EIA process) and efficiently (e.g. that needless monitoring is not undertaken). This and the other requirements for action impact monitoring and auditing discussed above are summarised in Box 14.1 in the form of evaluation criteria. These criteria are now used to assist in the analysis of the monitoring of actions in each of the seven EIA systems.

United States of America

The National Environmental Policy Act 1969 is silent on the issue of monitoring. However, as stated in Chapter 13, the Regulations (Section

1505.2 (c)) require that a 'monitoring and enforcement program shall be adopted and summarized [in the record of decision] where applicable for any mitigation'. The Council on Environmental Quality (CEQ) has made it clear that:

> the terms of a Record of Decision are enforceable by agencies and private parties. A Record of Decision can be used to compel compliance with or execution of the mitigation measures identified therein.
>
> (CEQ, 1981a, question 34d)

This is significant since, providing the lead agency commits itself to mitigation measures, it must 'include appropriate conditions in grants, permits or other approvals' and 'condition funding of actions on mitigation' (Regulations, Section 1505.3(a)(b)).

Monitoring is, however, essentially discretionary. As the Regulations (Section 1505.3) state, 'Agencies may provide for monitoring to assure that their decisions are carried out and should do so in important cases'. The Regulations require some implementation monitoring, since they specify that the lead agency must, upon request, inform other agencies on progress in carrying out certain mitigation measures adopted (Section 1505.3 (c)). Further, it must (again upon request) make available to the public the results of relevant monitoring. If the monitoring requirements are specified in the record of decision, the lead agency is obliged to implement them.

In practice, despite these somewhat ambiguous requirements, monitoring is generally perceived as a weak link in the US EIA system (Blumm, 1990; Reed and Cannon, 1993). For example, Blaug (1993) found that monitoring of conditions in FONSIs was only required by half the federal agencies and Clark (1993) has stated that, in practice, many agencies fail to monitor the environmental impacts arising from projects. In general, monitoring is not given high priority and some monitoring commitments are not honoured because of budgetary constraints or communication lapses. However, monitoring practice varies very substantially. Some agencies have instituted impact monitoring programmes (for example, for water pollution). In addition, a few agencies (for example, the Tennessee Valley Authority (Broili, 1993)) have undertaken auditing of selected projects.

Culhane *et al.*'s (1987) auditing work remains the only major US study comparing the accuracy of predictions with monitored results. They found that:

> Despite some general cynicism about the veracity of government promises, agency managers prove to be quite responsible in carrying out promised mitigations (p. 254).

Dickerson and Montgomery (1993), in surveying other post-auditing studies and their own work, concluded that the NEPA process is reasonably effective in producing useful predictions of impact but that there is scope for considerable improvement (Chapter 20).

California

The monitoring of environmental impacts in the Californian EIA system was strengthened in 1989. Prior to that, the California Environmental Quality Act (CEQA) lacked any requirement for agencies to monitor the results of their decisions to determine whether or not required mitigation measures were implemented. The need to monitor mitigation measures had long been referred to as the 'missing link' in the CEQA process. As Bass and Herson (1993b, p. 90) stated:

> As a result of this serious limitation in the law, the mitigation measure requirements for many projects had been ignored and significant environmental impacts had been allowed to occur, often in direct violation of project requirements and conditions.

CEQA now requires public agencies to adopt monitoring and reporting programmes each time they approve a project that contains mitigation measures to reduce or avoid significant environmental impacts. Agencies must thus prepare a programme at the same time as they make their findings after certifying an environmental impact report (EIR) or when they adopt a mitigated negative declaration. The reporting or monitoring programme must ensure compliance with mitigation measures during project implementation. The aims of this environmental monitoring are:

- to ensure implementation of mitigation measures during project implementation
- to provide feedback to agency staff and decision makers about the effectiveness of their actions
- to provide learning opportunities for improving mitigation measures on future projects
- to identify the need for enforcement action before irreversible environmental damage occurs.

(Bass and Herson, 1993b, p. 88)

Where practical, mitigation measures must be made express conditions of project approval and agencies are expected to identify mitigation performance objectives. They may levy fees to pay for the monitoring programme.

Various enforcement measures may be employed if monitoring reveals that mitigation has not been carried out properly: work cessation orders, denial of building occupancy permits, revocation of project approval, sanctions such as fines or jail, the posting of performance bonds and the recording of monitoring results. These results are available for inspection by the public.

No formal guidance on implementing the monitoring requirements has been published but the Office of Planning and Research (1989) has issued an advisory manual about these requirements. Agencies are required only to monitor the implementation of mitigation measures, not the success of those measures. (For example, if a mitigation measure recommends that trees be replanted, the agency's responsibility is only to monitor the planting and not to determine whether the trees survive.)

There is no requirement to monitor action impacts as such. Although the project-specific monitoring programme does not have to be included in the EIR or negative declaration, many agencies do incorporate monitoring into their documents so that members of the public and responsible agencies can review them before they are adopted.

The new requirements are undoubtedly a step forward in ensuring that mitigation measures are actually implemented and environmental groups hoped that monitoring and enforcement would provide CEQA with more teeth and reduce 'paper compliance'. Generally, it appears that some agencies are adopting jurisdiction-wide, comprehensive monitoring programmes and then, as individual projects are approved, detailed, project-specific monitoring plans. However, since the notion of monitoring actions once they have been approved appears to be alien to the Californian (and American) culture, universal success is far from assured. It appears that, in practice, the monitoring performance of agencies is inconsistent and formal monitoring programmes are the exception rather than the rule. Roques (1993, p. 11) reported that planners regard monitoring as 'very time consuming and costly ... [and] cite the absence of performance criteria and standards for monitoring as a major obstacle'. Weaknesses in monitoring were also reported by Johnston and McCartney (1991). There had, by mid-1994, been no legal actions to force the issue.

United Kingdom

Like the European Directive on EIA, the Planning Regulations and other regulations implementing its provisions in the United Kingdom are silent on the question of monitoring. This is not to say that monitoring, especially implementation monitoring, does not take place. It is customary for local planning authorities (LPAs) to impose planning conditions on permissions and for compliance with these to be checked to a greater or lesser extent. Such conditions may include emission standards (especially relating to noise), and may sometimes require the proponent to monitor these. It has been more usual for local authorities to undertake implementation monitoring, however. Similarly, the monitoring of conditions on air pollution, water pollution or solid waste disposal permits granted under the pollution control legislation tends to have been carried out by the environmental authorities.

LPAs and the pollution control authorities may also impose conditions requiring impact monitoring (especially in relation to ambient pollution levels) which may be carried out by the LPA/environmental authorities or by the proponent. However, impact monitoring is by no means a general requirement for projects approved under the Planning Regulations, being confined to major developments like power stations or waste disposal operations. There is no formal linkage between either implementation or impact monitoring and the earlier stages of the EIA process (save, of course, for the making of the decision). Audits are not a requirement, though voluntary proponent auditing is increasing in importance. No auditing results for projects approved under the Planning Regulations have yet been published.

Where the proponent is shown to be in breach of the conditions on the planning approval or a pollution control permit, enforcement action can be taken. However, monitoring is split between agencies administering separate legislation and enforcement of both planning and pollution control conditions has left much to be desired in the past.

The existing guidance on EIA in the United Kingdom, including the checklist of matters to be considered for inclusion in the environmental statement in the *Guide to the Procedures* (DOE, 1989, Appendix 4), makes no reference to the monitoring of implemented project impacts. The requirements relating to the publication of monitoring results vary in the United Kingdom. However, there is at present no regulatory mechanism for bringing together the monitoring results arising from different legislative requirements. Indeed, LPAs are discouraged from imposing conditions which may overlap with the requirements of other environmental control authorities (Wood, 1989b). There is no formal right of appeal if monitoring reveals unsatisfactory omissions or impacts but monitoring results can be used to bring pressure to bear on proponents and/or environmental authorities.

The Netherlands

The Dutch Environmental Management Act 1994 does not require any information on monitoring to be submitted as part of the EIS. However, the EIA Commission advises that the proponent covers monitoring and auditing in the EIS by specifying, in the scoping guideline recommendations, that impact measurement and possible corrective arrangements should be described. Most guidelines issued by the competent authorities make this advice a requirement. As mentioned in Chapter 13, the Act requires the competent authority to specify how it intends to carry out auditing in its decision (Section 7.34(2)).

The Act also contains five sections devoted to 'evaluation', i.e. to auditing. In summary, these are as follows:

1. The competent authority must monitor the consequences of the implemented action (Section 7.39).
2. The proponent must provide the competent authority with monitoring information (Section 7.40).
3. The competent authority must prepare a post-auditing report (or evaluation) comparing impacts with those predicted in the EIS, publish it and send it to the EIA Commission and the statutory consultees.
4. The competent authority must take action (e.g. by tightening licence conditions) if impacts are more severe than anticipated.
5. Detailed regulations relating to monitoring can be made (none has as yet been issued).

Two of the Dutch guidance documents relate to auditing and a third was to be published in 1994. There is, therefore more than adequate provision for monitoring and auditing in the Dutch EIA system. However, this provision is not proving to be effective.

No mention has been made of monitoring in about half of the decisions reached to date. There has been detailed reference to auditing only in about 25 per cent of the decisions made. So far, very few auditing reports have been published (only two by the end of 1993 (EIA Commission, 1994)). These have been more concerned with implementation monitoring than with impact auditing, notwithstanding the Dutch legal requirements.

While relatively few projects have been fully implemented, the main reason for the failure to prepare more auditing reports is the lack of attention paid to monitoring by many competent authorities. The authorities are not pressing proponents for the necessary information and not preparing the monitoring reports for publication and transmittal to the EIA Commission. There may be a general right of appeal against a failure to audit but this has not yet been tested. The Evaluation Committee on EIA (1990, p. 13) stressed the importance of post-project evaluation in relation both to minimising the project impacts and to improving EIA generally.

Canada

Monitoring (or 'follow-up') is emphasised strongly in the new Canadian EA system. Section 14(c) of the Canadian Environmental Assessment Act (CEAA) states that the EA process includes where applicable 'the design and implementation of a follow-up program' and Section 16(2)(c) requires that EA reports other than for 'screenings' must include a section on 'the need for, and the requirements of, any follow-up program in respect of the project'. 'Follow-up program' is defined in Section 2 of the Act being a programme for:

(a) verifying the accuracy of the environmental assessment of a project, and
(b) determining the effectiveness of any measures taken to mitigate the adverse environmental effects of the project.

However, CEAA contains no means of ensuring that monitoring is implemented, since it is intended only to be an aid to good decision making and is not intended to be a regulatory tool.

While the Guide to the Act states that 'the need for and requirements of a follow-up program need not be considered during preparation of a screening report' (Federal Environmental Assessment Review Office – FEARO, 1993a, p. 117), the responsible authority (RA) may require 'details on monitoring programs to evaluate the effectiveness of mitigation measures as well as to determine the accuracy of the EA' (p. 102).

Further, for screenings, the RA:

> must make a decision about whether a follow-up program is appropriate. If so, it must ensure that one is designed and implemented.
>
> (FEARO, 1993a, p. 117)

For projects subject to comprehensive study or to public review the RA must decide whether to implement a follow-up programme. In the case of panel and mediation reports, the Guide (p. 144) revealingly states:

> The RA is not obliged to follow the recommendation for a follow-up program, but if it does not, it must justify the decision publicly. RAs should keep in mind that the report from a mediator or panel is the product of an open, fair and rigorous review involving all key interests, and that the report's recommendations cannot be treated lightly.

It goes on to state that the initial question in determining the need for a follow-up programme is one of uncertainty or unfamiliarity. Box 14.2 shows a box from the Guide, which is to be supplemented by more detailed advice on follow-up in a future reference guide.

The implementation of appropriate mitigation measures is discussed in Chapter 16. Together with the follow-up programme, these measures provide for monitoring of both the accuracy of predictions of the environmental effects of the project and the effectiveness of mitigation measures applied to projects subject to comprehensive study or public review. Though CEAA contains no absolute means of ensuring either compliance or monitoring by RAs, other regulatory tools can be used to improve mitigation requirements. For example, the scope of a permit issued under the navigable waters legislation could be expanded to cover a wide range of mitigation measures, but enforcement of both compliance and monitoring remains problematic (Gibson, 1992, 1993).

The monitoring programme is clearly linked to the earlier stages of the

---

Box 14.2 Canadian guidance on monitoring predicted effects

When should a follow-up program be implemented?

The RA should develop a follow-up program for a project whenever circumstances warrant. Examples include situations when:

- the project involves a new or unproven technology;

- the project involves new or unproven mitigation measures;

- an otherwise familiar or routine project is proposed for a new or unfamiliar environmental setting;

- the analysis is based on a new technique or model, or there is other uncertainty about the conclusions; or

- project scheduling is subject to change such that significant environmental effects could result.

Source: Federal Environmental Assessment Review Office (1993a, p. 144).

EA process in Canada and, in particular, to the EA report which must specify monitoring arrangements. The Act provides no mechanism for ameliorative actions to be taken where monitoring reveals the need for them: this is left to the regulatory approvals process.

The Canadian Environmental Assessment Research Council commissioned various studies on EA monitoring. Munro *et al.* (1986) made numerous recommendations about monitoring and about EA generally. Krawetz *et al.* (1987) outlined a framework involving a monitoring plan, process management and a monitoring objective. Their case studies revealed that satisfying impact management and prediction objectives at the same time was very difficult. Practice on auditing and monitoring under the Environmental Assessment and Review Process (EARP) appears to have been very mixed: an unsurprising outcome since EARP included no formal means of requiring monitoring.

In practice, monitoring and auditing results have sometimes been made available under EARP though many of the monitoring commitments in EISs have amounted mostly to rhetoric. This situation is likely to improve under CEAA, which makes provision for public availability of the results of follow-up programmes (but not for public appeal if the results fail to meet expectations).

Commonwealth of Australia

Monitoring provisions exist in the Australian EIA system but they are essentially discretionary:

> For the purpose of achieving the object of the Act, the Department may at any time, whether before or after a proposed action has been completed, review and assess all or any of the environmental aspects of the proposed action, including, in particular, the effectiveness of any safeguards or standards for the protection of the environment adopted or applied in respect of the proposed action and the accuracy of any forecasts of the environmental effects of the proposed action, and the Department shall report to the Minister.
> (Commonwealth of Australia, 1987, Procedures, para. 10.11)

The Minister of the Department of the Environment, Sport and Territories must inform the relevant minister and the proponent of the monitoring results and may make recommendations and suggestions for improvements in environmental protection (Procedures, para. 10.2.1). The results and suggestions are normally to be made public.

EIA reports must contain a section on the monitoring arrangements for ensuring that mitigation is effective (Box 8.4) but there is no provision for monitoring results to be compared with predictions. However, monitoring of action implementation and impacts appears to be but little employed in practice. Organisational arrangements, lack of staff and, perhaps, lack of political determination, have frequently rendered the

Commonwealth EIA monitoring provisions inoperative. On-going control over implementation and performance and comparison with impacts predicted in EISs has frequently been lacking because of the absence of feedback.

Buckley (1989) found few verifiable predictions in an audit of several Commonwealth and state EISs. There was little obvious bias in the limited number of forecasts he was able to check: 57 per cent of impacts were less severe than predicted, 43 per cent more severe. These results do not differ greatly from those reported elsewhere (e.g. Bisset and Tomlinson, 1988; Culhane *et al.*, 1987). Various recommendations to overcome these shortcomings in the Commonwealth system have been suggested, most involving the use of enforceable and auditable conditions and an active monitoring programme (Anderson, 1990; Martyn *et al.*, 1990; Australian and New Zealand Environment and Conservation Council, 1991; Ecologically Sustainable Development Working Groups, 1991; Fowler, 1991).

New Zealand

The Resource Management Act 1991 makes both general and specific reference to monitoring. Every local authority in New Zealand is required to 'gather such information, and undertake or commission such research, as is necessary to carry out effectively its functions under this Act' (Section 35(1)). Each local authority must also monitor the exercise of the resource consents that have effect in its area (Section 35(2)(d)). The Fourth Schedule (Box 7.3) specifies that the EIA report should contain proposals for impact monitoring where the scale and significance of impacts require it. This provides the opportunity for a strong linkage between assessment and monitoring.

Despite the specification of monitoring requirements in the Act, it is unclear whether the aim of the monitoring provisions is enforcement or EIA system enhancement, for which predictions need to be checked against actual impacts to test their accuracy and the extent to which cumulative impacts are occurring. Because of this lack of specificity, and because monitoring falls to regional, city and district authorities, Montz and Dixon (1993) stated that there are likely to be variations in the extent to which the monitoring provisions are implemented.

No guidance on monitoring has been issued in New Zealand and the recently published practical guide (Morgan and Memon, 1993) makes very little reference to monitoring. Where monitoring does take place, the results are normally made public. However, as Dixon (1993b) indicated, local authorities have few funds available for monitoring and have not developed monitoring systems or appropriate methods. Morgan (1993) confirmed that, notwithstanding the legal requirement to monitor, practice in enforcement and monitoring is very poor, with reliance on public complaint to indicate problems. It appears that there is considerable scope for improvement in monitoring practice in New Zealand.

Summary

Table 14.1 demonstrates that none of the EIA systems fully meets the impact monitoring evaluation criterion. Monitoring is an acknowledged weakness of the US EIA system and there is no provision for monitoring in the UK EIA system (though monitoring can be accomplished under other legislative means). The New Zealand Resource Management Act imposes a general duty upon local authorities to monitor project impacts but this is infrequently undertaken and there is no linkage to the earlier stages of the EIA process in these essentially discretionary monitoring requirements. The same is largely true of the Commonwealth of Australia EIA system where the monitoring provisions are quite clearly discretionary rather than mandatory. Very few EIA systems in developing countries make any provision for monitoring (see Appendix).

The Netherlands EIA system contains several impact monitoring and auditing provisions but, in practice, these are often not implemented. The same is true in Canada, where the Canadian Environmental Assessment Act also contains extensive impact monitoring requirements. The Cali-

Table 14.1 The treatment of action monitoring and auditing in the EIA systems

Criterion 9: Must monitoring of action impacts be undertaken and is it linked to the earlier stages of the EIA process?

Jurisdiction	Criterion met?	Comment
United States	No	Monitoring essentially discretionary but some requirements where mitigation measures specified in record of decision. Practice often weak.
California	Partially	Monitoring and reporting programmes required where project involves mitigation measures. Practice varies.
United Kingdom	No	No provision for monitoring. Uncoordinated implementation monitoring takes place under planning and other legislation unrelated to earlier stages in EIA process.
The Netherlands	Partially	Specific requirements relating to monitoring and comparison with EIS. However, in practice these are not often observed.
Canada	Partially	Extensive provision for monitoring action effects in CEAA but no mechanism for ensuring full compliance.
Australia	No	Discretionary provisions exist but, in practice, are seldom employed.
New Zealand	No	Duty of local authorities to monitor impacts of projects in Act often not complied with.

fornian EIA system requires monitoring where mitigation measures are agreed. This provision was added to the Californian Environmental Quality Act to close an obvious loophole but its observance has been found to be wanting. There is no clear linkage back to the EIA report. Because of these weaknesses, these three EIA systems can be adjudged only partially to meet the evaluation criterion. It is quite clear that monitoring and auditing is one of the weakest areas of EIA activity and that EIA systems need either to be strengthened or better coordinated with other environmental monitoring programmes.

Note

1. See, for example, Holling (1978), Bisset (1980, 1981, 1984), Sewell and Korrick (1984), Caldwell *et al.* (1983), Beanlands and Duinker (1984), Culhane (1987), Culhane *et al.* (1987), McCallum (1987), Tomlinson and Atkinson (1987a, b), Bisset and Tomlinson (1988), Sadler (1988), Bailey and Hobbs (1990), Bailey *et al.* (1992), Buckley (1990, 1991a), Glasson *et al.* (1994).

CHAPTER 15

Mitigation of impacts

Introduction

If the consideration of alternatives lies at the heart of the environmental impact statement (Chapter 8) then the mitigation of environmental impacts is the principal aim of the EIA process. In practice the consideration of alternatives is intertwined with the consideration of mitigation measures. The main purpose of EIA is, in essence, to allow the proposed development to proceed, while reducing its impacts to an acceptable level. The secondary purpose of EIA is to prevent unsuitable development by demonstrating that certain impacts cannot be mitigated to the point of acceptability. This chapter explains why the mitigation of environmental impacts is important and advances several evaluation criteria to assist in the review of this element of EIA systems. These criteria are then employed in the analysis and comparison of the EIA systems in the United States, California, the United Kingdom, The Netherlands, Canada, the Commonwealth of Australia and New Zealand.

Mitigation of impacts within the EIA process

Mitigation, or amelioration, of the severity of impacts arising from an action can take a variety of forms. The Department of the Environment (1994b) classifies mitigation measures into avoidance (using an alternative approach to eliminate an impact), reduction (lessening the severity of an impact) and remedy (which may involve some enhancement or compensation).

Examples of mitigation measures may include process alterations to reduce emissions, altering pollution control equipment to render it more effective, adjusting the hours of operation of a plant, changing site layout to reduce visual, noise or air pollution impacts, the requirement of fencing and walls, the amendment of vehicular access arrangements, the provision of mounding and planting, the creation of replacement habitat, and many others (see, for example, Canter *et al.*, 1991; Glasson *et al.*, 1994). Indeed the whole range of land use planning, pollution and other controls and measures should be considered during the EIA process to ensure that suitable mitigation measures are adopted.

As Fig. 1.1 demonstrates, mitigation is iterative: different measures may be proposed at the various stages of the EIA process. For example, the results of the review of the EIA report, and of the consultees' and public's comments upon it, may yield proposals for mitigation additional to (or different from) those proposed in the EIA report itself.

The adoption of some mitigation measures may involve considerable costs, though other effective mitigated measures may cost very little (e.g. alteration of road access, alternative material storage arrangements). There will normally come a point at which the developer may withdraw a proposed development because the additional costs associated with mitigation measures are deemed to be too high (Wood, 1989b). On the other hand, the decision-making authority may ask whether mitigation of the impacts of a proposal is enough to achieve sustainability. In other words, it may be that the authorities will seek either no deterioration of the environment, or a net improvement of the environment through off-sets or compensation (as in US air pollution control), rather than merely a reduction of (or a remedy for) impacts. In these circumstances, it may be necessary to consider radical alternatives to the proposal.

Mitigation measures can themselves have impacts which need to be identified, predicted and evaluated. For example, the use of earth mounding to provide noise baffles or to screen development from public view can create unnatural landforms which are themselves visually obtrusive.

Not only may there be a comparison of the benefits of mitigation but there may also need to be a trade-off between the mitigation of different impacts. Where these relate to pollution, the best practicable environmental option needs to be sought (Royal Commission on Environmental Pollution, 1988). However, it will frequently be necessary to trade off pollution and other types of impact in addition. These trade-offs can be complex. It is for this reason, among others, that consultation and participation are so important in the EIA process. Consultees and the public can provide invaluable assistance not only in suggesting mitigation measures, but in determining which residual impacts are tolerable and which cannot be countenanced.

Thus, in the United States, and in many other jurisdictions where widespread opportunities for public participation exist (such as the Commonwealth of Australia), a very high proportion (over 95 per cent) of actions which reach the 'final' EIA report stage are approved, but nearly all of these are substantially modified to mitigate impacts during the successive stages of the EIA process. On the other hand, in the United Kingdom, where 40 per cent of a sample of 20 applications were refused (Chapter 13), the mitigation of impacts would appear not to be assuming the importance attached to it in more mature EIA systems, perhaps as a consequence of the relatively limited opportunities for formal public involvement in the EIA process (Chapter 16).

Mitigation measures, to be effective, must be implemented. Several EIA systems (including those in the United States and California) initially possessed no effective means of ensuring that the mitigated measures proposed in EIA reports were actually carried through (Hollick, 1981a). Many of these shortcomings have now been corrected (see, for example, in relation to California, Bass and Herson, 1993b, p. 90). Perhaps as a consequence of the former lacuna in California, it has been suggested that the criteria shown in Box 15.1 should be employed in drafting effective mitigation measures.

Clearly, financial compensation or remuneration in kind have a role to play in gaining public acceptance of unmitigated environmental impacts. Though such payments may be controversial, and though there may be considerable problems in determining who should receive payments, the use of compensatory measures can help to resolve disputes and achieve the aims of EIA. Several instances of the use of financial compensation exist in siting decisions involving EIA in the United States (see, for example, Frieden, 1979), New Zealand and Australia. In the United Kingdom, compensation is often achieved through the use of 'planning gain', which may involve the provision of community facilities.

As mentioned in Chapter 8, preliminary documentation produced during the early stages of the EIA process should show clear evidence of the mitigation/avoidance of environmental impacts in the initial action designs. Similarly, documentation prepared for screening and scoping purposes should also address the question of mitigation of impacts.

Clearly, details of mitigation should be set down in the EIA report. The EIA report should provide a record of all the mitigated measures and modifications suggested or accepted by the proponent. Lee and Colley

Box 15.1   Guidelines for drafting effective mitigation measures

WHY:	State the objective of the mitigation measure and why it is recommended
WHAT:	Explain the specifics of the mitigation measure and how it will be designed and implemented
	• Identify measurable performance standards by which the success of the mitigation can be determined
	• Provide for contingent mitigation if monitoring reveals that the success standards are not satisfied
WHO:	Identify the agency, organisation, or individual responsible for implementing the measure
WHERE:	Identify the specific location of the mitigation measure
WHEN:	Develop a schedule for implementation

Source: Bass and Herson (1993b, p. 75).

(1992, p. 48) have suggested that EIA reports should deal fully with the scope and effectiveness of mitigation measures:

All significant adverse impacts should be considered for mitigation. Evidence should be presented to show that proposed mitigation measures will be effective when implemented.

Mitigation should therefore continue to be considered during the review and revision of EIA reports, during decision making and during the monitoring stages of the EIA process. The use of carefully worded conditions to any approval is frequently used to codify the set of designs and mitigation measures approved at the decision-making stage. These conditions need to be monitored and enforced to ensure that mitigation remains effective (Chapter 14). However, some flexibility is needed to ensure that, on the one hand, unexpected impacts are mitigated and that, on the other hand, unnecessarily expensive mitigated measures can be amended. The use of a negotiated environmental management plan with appropriate monitoring and opportunities for modification of monitoring arrangements, as in Western Australia (Wood and Bailey, 1994), is obviously desirable.

As at other stages of the EIA process, the existence of published advice on mitigation and modification of actions to render them environmentally more acceptable will be helpful to developers, consultants, environmental and decision-making bodies and the public. In practice, it appears that relatively little specific advice on mitigation in the EIA process, and especially at stages other than EIA report preparation, has been published.

Finally, the mitigation of action impacts should take place effectively (i.e. mitigated measures should actually ameliorate impacts) and efficiently (i.e. they should not involve the expenditure of unnecessary time, manpower or financial resources). Clearly, the earlier in the EIA process that the mitigation proposals are made, the more effective and efficient they are likely to be, since they will be progressively refined during the consideration of the proposal. This and the other criteria for reviewing the treatment of mitigation measures in EIA systems are summarised in Box 15.2. These criteria are now employed to help to analyse the procedures for mitigation in each of the seven EIA systems.

United States of America

Although the treatment of alternatives is the 'heart of the environmental impact statement' (EIS) in the United States (Chapter 8), mitigation of environmental impacts is probably the most important outcome of the American EIA process. (The abandonment of unsatisfactory proposals is rare.) Mitigation certainly pervades the US EIA system and most proposals are heavily modified by the end of the EIA process.

Mitigation is, of course, the central determinant of one outcome of screening, the mitigated finding of no significant impact (FONSI). Follow-

Box 15.2   Evaluation criteria for the mitigation of impacts

Must the mitigation of action impacts be considered at the various stages of the EIA process?

Must clear evidence of the mitigation/avoidance of environmental impacts be apparent in the action designs in preliminary EIA documentation?

Must details of mitigation and its implementation be set down in the EIA report?

Must evidence of the consideration of mitigation be presented during screening, during scoping, during EIA report review and revision, during decision making and during monitoring?

Does published guidance on mitigation and modification exist?

Does the mitigation of action impacts take place effectively and efficiently?

---

ing the preparation of an environmental assessment (EA) a mitigated FONSI may be prepared where the potentially significant environmental effects of a proposal can be rendered acceptable by the adoption of appropriate mitigation measures. The courts have ruled that mitigated FONSIs are legally adequate where, *inter alia*, the agency can convincingly show that its mitigation measures will reduce all the significant environmental impacts identified in the EA to less-than-significant levels (Bass and Herson, 1993a, p. 40). In practice, there is heavy reliance on mitigation measures to justify EAs and FONSIs, showing that environmental considerations are increasingly being integrated early in the decision-making process (Blaug, 1993).

Mitigation measures must be considered during scoping and summarised in the scoping report. Both draft and final EISs must contain a discussion of appropriate mitigation measures. The Council on Environmental Quality (CEQ) Regulations (Section 1508.20) specify that mitigation must involve avoiding, minimising, rectifying, reducing or compensating for significant environmental effects. The inference is that mitigation measures which are not specific and tangible (for example, proposals to consult, to conduct further studies or to monitor) will not generally solve the environmental problems identified.

All relevant, reasonable mitigation measures that could improve the action must be identified in the EIS, even if they are outside the jurisdiction of the lead agency (CEQ, 1981a, questions 19(a),(b)). However, the lead agency is not obliged to commit itself to implementing the mitigation measures identified in the EIS unless its own EIA regulations require their adoption.

Those mitigation measures adopted by the lead agency must be specified in the record of decision (ROD), together with a monitoring and enforcement programme for each measure (Chapter 13). As stated in Chapter 14, a ROD can be used to compel compliance with, or execution of, the mitigation measures contained in it. While these powers are valuable, the lead agency can easily circumvent them by failing to adopt relevant mitigation measures. Indeed the implementation of mitigation measures, like monitoring, is a weakness of the US EIA system.

The Environmental Protection Agency (EPA) conducted a study in 1987 which showed that, using the definition of adequacy in the Regulations (above), the effectiveness of the mitigation measures in about 20 per cent of the EISs that it reviewed was questionable (EPA, quoted in Bass and Herson, 1993a, p. 75). Even where mitigation measures are specified in EISs and RODs, implementation may be unsatisfactory because of budgetary constraints or failure to inform relevant personnel or to incorporate measures in construction contracts. However, the effective implementation of mitigation measures is gradually improving as public concern and knowledge grow.

California

Mitigation is, effectively, the main aim of the California Environmental Quality Act (CEQA) since few projects are actually turned down. Mitigation is important at each stage in the EIA process and should be considered by the applicant from the outset. It is a formal requirement that mitigation measures be discussed for any significant environmental effects identified in the initial study (Guidelines, Section 15063). If the initial study results in a negative declaration, this must contain 'mitigation measures, if any, included in the project to avoid potentially significant effects' (Guidelines, Section 15071 (e)). The concept of mitigation is, of course, central to the mitigated negative declaration. In this event, the proponent agrees to implement mitigation measures to avoid or substantially reduce potentially significant impacts. Monitoring programmes for mitigation measures are often incorporated into the mitigated negative declaration (Chapter 14).

In practice, since applicants are frequently anxious to avoid having to prepare an environmental impact report (EIR), they will incorporate or agree to meaningful mitigation measures in order to obtain a mitigated negative declaration. Olshansky (1992) reported that at least 16 per cent of negative declarations were mitigated. In the past the record of implementing mitigation measures in California has not been distinguished but CEQA was strengthened in 1989 in a partially successful attempt to close a loophole which permitted non-compliance (Chapter 14).

For every significant impact identified in an EIR, the lead agency must identify mitigation measures. The Guidelines provide for five categories of mitigation measures, virtually identical to those specified in the CEQ Regulations:

- avoiding the impact altogether by not taking a certain action or parts of an action
- minimising impacts by limiting the degree or magnitude of the action and its implementation
- rectifying the impact by repairing, rehabilitating, or restoring the affected environment

- reducing or eliminating the impact over time by preservation and maintenance operations during the life of the action
- compensating for the impact by replacing or providing substitute resources or environments (Section 15370).

The EIR is also required to discuss whether the effect has been avoided or substantially lessened and any significant side effects of implementing a mitigation measure. The Guidelines (Section 15126) state that the discussion should distinguish between measures that are proposed by project proponents and measures that are adopted as conditions of project approval.

When drafting mitigation measures, agencies should include only those that are feasible, i.e. capable of being accomplished in a successful manner within a reasonable period of time, taking into consideration economic, environmental, legal, social and technological factors. The final determination of the feasibility of a mitigation measure is made by the elected decision-makers when they agree findings. As explained in Chapter 13, the findings must contain a conclusion about mitigation measures for each significant impact. Finally, a monitoring and reporting programme must be adopted as part of the agency's findings.

In practice, there is considerable jockeying in the determination of mitigation measures. Proponents seldom reveal their intended position on mitigation early on since concessions may well be required at the draft EIR stage, at the final EIR stage, at the decision-making stage and possibly following monitoring. Equally, lead agencies often seek to achieve the best mitigation measures possible at each stage in the EIA process (though this is less true of favoured private projects or the agencies' own projects) but frequently fail to monitor their implementation (Johnston and McCartney, 1991).

Many argue that CEQA has been utilised with considerable success in winning significant mitigation concessions especially at the site-specific level. For example, CEQA has been employed as a means of helping to gain net compensation for infill on the shores of San Francisco Bay and for other environmental compensation measures. Others argue that the mitigation benefits achieved by CEQA have been picayune, especially in relation to cumulative impacts (of, for example, urban sprawl). Regrettably, there have been no research studies to help determine which argument should prevail.

United Kingdom

The UK Planning Regulations require that the environmental statement contains, where significant effects are anticipated, 'a description of the measures envisaged in order to avoid, reduce or remedy those effects' (Schedule 3, para. 2). The checklist in the Guide (Department of the Environment – DOE, 1989) covers site planning, technical, aesthetic and ecological measures and an assessment of their likely effectiveness (Box 15.3). There are no formal requirements for the treatment of mitigation

Box 15.3   UK guidance on mitigation measures

Where significant adverse effects are identified, a description of the measures to be taken to avoid, reduce or remedy those effects, e.g.:

a. site planning;

b. technical measures, e.g.:

    i)   process selection;
    ii)   recycling;
    iii)   pollution control and treatment;
    iv)   containment (e.g. bunding of storage vessels).

c. aesthetic and ecological measures, e.g.:

    i)   mounding;
    ii)   design, colour, etc.;
    iii)   landscaping;
    iv)   tree plantings;
    v)   measures to preserve particular habitats or create alternative habitats;
    vi)   recording of archaeological sites;
    vii) measures to safeguard historic building or sites.

Assessment of the likely effectiveness of mitigating measures.

Source: Department of the Environment (1989, Appendix 4, Section 4).

measures at other stages in the EIA process, though it is customary for local planning authorities to impose conditions designed to mitigate impacts, or to require a legal planning agreement for this purpose, when granting planning permission. Apart from the Guide (DOE, 1989) there is no other published guidance on mitigation and modification.

In a sample of 24 environmental statements, local planning authorities felt that the descriptions of measures to mitigate adverse environmental effects were broadly satisfactory in nearly two-thirds of the statements. In nearly two-thirds of the 24 cases studied, the proposal was modified as a result of suggestions made during the EA process. More than two-thirds of these alterations occurred following the submission of the planning application and ES but earlier changes were also felt to be significant. Developers and consultants believed that perhaps half of these modifications arose directly from the EA process, but that the other half would probably have been made in the course of normal discussions about the proposed development. Two-thirds of the changes made to the proposals in response to suggestions resulted from those put forward by the LPA, rather than by the statutory consultees and the public. However, LPA consultations with statutory bodies and the public were felt to have been significant in influencing more than half of these LPA suggestions (Wood and Jones, 1991).

It is apparent that while the scope for mitigation of project impacts to take place throughout the UK EIA process exists and that, in practice, this

219

occurs in many cases, there is considerable room for an increase in the effectiveness with which mitigation measures are employed, especially early in the EIA process. If mitigation were given greater weight in the UK EIA process, it would be expected that more planning applications would be withdrawn voluntarily and that a greater proportion of appropriately mitigated actions would be approved. The incorporation of mitigation measures into planning obligations and planning conditions would also be expected.

The Netherlands

Mitigation is not referred to by name in the Dutch Environmental Management Act 1994. It is not a requirement that mitigation be mentioned in the notice of intention and, frequently, mitigation measures are not specifically included in the scoping guidelines. EISs frequently (but not invariably) contain specific reference to mitigation measures. Perhaps surprisingly, the suite of EIA guidance documents does not include a volume on mitigation.

Mitigation is largely subsumed in the treatment of alternatives in the Dutch EIA system. The environmentally preferable alternative is required not only to incorporate all feasible mitigation measures but, following recent revisions to the Act, may also be required to include a description of compensation measures. For example, the compensation of habitat loss by the planting of trees is often regarded as an alternative to the proposed action rather than as mitigation. The Dutch would claim that strong mitigation measures thus provide the starting point for consideration of a project rather than being elicited from the proponent later. Nevertheless, little specific provision for mitigation measures to be considered (as demanded by the European Directive on EIA) is made in the Dutch legislation and there is no requirement to select the environmentally preferable alternative. It is generally left to the competent authority to implement the mitigation of the impacts of the proposed action by imposing conditions on its approvals.

There is no doubt that mitigation of environmental impacts is taking place as a result of the EIA process and the involvement of the EIA Commission. Though not universal, changes to improve the proposal's environmental compatibility usually take place by iteration in the EIA process. Changes in location, in design and in technical controls are all common. There is, perhaps, a view that many of these changes, while significant, are relatively minor in nature.

Canada

No Canadian EA report, from the screening report through to the EIS and panel report, can be written without a treatment of:

measures that are technically and economically feasible and that would mitigate any significant adverse environmental effects of the project (Canadian Environmental Assessment Act (CEAA) Section 16(1)(d)).

The responsible body is enjoined in Section 20(a) of the Act to 'ensure that any mitigation measures that the responsible authority considers appropriate are implemented'. The Act defines mitigation (Section 2(1)) as:

the elimination, reduction or control of the adverse environmental effects of the project, and includes restitution for any damage to the environment caused by such effects through replacement, restoration, compensation or any other means.

As with monitoring (Chapter 14), however, mechanisms for ensuring implementation are weak.

The Guide to the Act provides brief coverage of mitigation measures, emphasising that they should be considered first during the design phase of the project and be gradually refined as the assessment progresses. The use of specialists and public consultation is recommended where circumstances demand. The Guide suggests that mitigation measures can be implemented through, for example, the issuance of conditional approvals, hold-back provisions in funding arrangements, contractual arrangements and the placing of bonds by proponents (Federal Environmental Assessment Review Office, 1993a, p. 118).

To a very real extent, mitigation of adverse environmental effects, rather than affecting decision making, is the principal objective of EA in Canada. Under the Environmental Assessment and Review Process very few projects have been subject to panel review. In theory, this means that environmental effects have been mitigated to the point where they are not significant and that considerable expertise must have been developed in achieving this. In practice, while expertise has developed, it has been the express aim of virtually every department to avoid a panel review because of the expense, time and potential embarrassment involved. Hence the conclusion that effects are no longer significant following mitigation must be doubtful in many cases but has seldom been tested (in the courts or elsewhere). Where panels have been used, mitigation measures have been an important element of their reports.

This situation should improve with greater public access to the documents in the EA process under CEAA and the reference of comprehensive study reports to the Minister of the Environment. The political position of a relatively weak department telling powerful departments that, in effect, their mitigation measures leave significant residual adverse effects is a difficult one. While it is hoped that CEAA will lead to improvements in mitigation techniques and, especially, to improved implementation of these, there is no great optimism in Canada that this will happen quickly, despite a considerable research programme funded by the Canadian Environmental Assessment Research Council (CEARC) (see, for example, CEARC, 1988a). Gibson (1993), while noting that few laws provide a suitable mechanism for enforcing mitigation measures for projects requiring federal permits, was guardedly optimistic about the emphasis on mitigation in CEAA.

Commonwealth of Australia

As indicated in Chapter 13, the main aim of the Australian EIA system is the mitigation of impacts, and numerous references to mitigation measures are made in the Administrative Procedures (Commonwealth of Australia, 1987). Thus, the notice of intention must include a description of the mitigation measures to be taken to reduce the potential environmental impacts of the proposal (Chapter 9). The EIA report must describe any applicable safeguards or standards, the means of implementing these and appropriate monitoring measures. It is usual for further mitigation measures to be introduced in the final EIS, following consideration of the draft EIS.

The assessment report produced by the Commonwealth Environment Protection Agency always contains suggestions for the mitigation of environmental impacts. The Environment Minister, in making recommendations to the Action Minister, must include 'suggestions or recommendations concerning conditions to which the proposed action should be subject, that the Minister thinks necessary or desirable for the protection of the environment' (Procedures, para. 9.3.1).

In practice, the most important mitigation measures tend to be proposed in the final EIS, as a way of overcoming objections raised to the draft EIS. Further mitigation may well be suggested in the assessment report or in the Minister's recommendations but these are usually of less significance. It is generally accepted that mitigation is a genuine consideration in the Australian EIA system and virtually all proposals are extensively modified (Chapter 13). The main area of weakness in the treatment of mitigation appears to relate to implementation and action monitoring (Chapter 14).

New Zealand

The duty to mitigate the environmental impacts of activities is one of the main aims of the Resource Management Act 1991 (Box 5.1). The Act requires an application to contain an assessment of effects 'and the ways in which any adverse effects may be mitigated' (Section 88(4)(b)). The Fourth Schedule (Box 7.3) specifies that 'a description of the mitigation measures' to prevent or reduce impacts should be included in the EIA report.

There are no other specific requirements relating to mitigation in the Act beyond the general criteria contained in Section 108 relating to conditions, and no guidance relating to amelioration of impacts has been issued. This is somewhat surprising since the Environmental Protection and Enhancement Procedures (EPEP) lay great stress on mitigation measures and considerable expertise exists in New Zealand (though

EPEP mitigation recommendations have often been ignored). The principal method of ensuring that mitigation is actually working in practice (i.e. that local authority conditions on any permission specify mitigation measures derived from the EIA process) is public scrutiny of the implementation of conditions attached to resource consents and permits and the right to appeal to the Planning Tribunal.

In practice, the inexperience of developers and local authorities means that the treatment of mitigation in EIA varies very considerably in New Zealand. There are, undoubtedly, examples of good practice but too often mitigation measures have been belatedly conceived, ill-considered, or unimplemented.

Summary

Unsurprisingly, Table 15.1 shows that each of the seven EIA systems meets the mitigation criterion. 'Unsurprisingly' because mitigation is the principal aim of the EIA process. However, the length to which mitigation is taken in the EIA systems varies and the emphasis on the

Table 15.1   The treatment of mitigation in the EIA systems

Criterion 10: Must the mitigation of action impacts be considered at the various stages of the EIA process?

Jurisdiction	Criterion met?	Comment
United States	Yes	Formal requirement to incorporate mitigation measures in record of decision. Effectiveness of implementation varies, but is improving.
California	Yes	Mitigation and its implementation are central to EIA process. Practice varies.
United Kingdom	Yes	ES must cover mitigation and LPAs impose conditions upon permissions to mitigate impacts. Practice varies at various stages in EIA process.
The Netherlands	Yes	Mitigation is subsumed in treatment of alternatives but is not separately required. Practice often satisfactory.
Canada	Yes	Mitigation and its implementation are central considerations in EA process. Practice often satisfactory.
Australia	Yes	Mitigation measures are explicitly provided for at various stages in EIA process. Practice varies.
New Zealand	Yes	Mitigation of environmental impacts is one of the main purposes of Act. Practice varies at various stages in EIA process.

223

implementation of mitigation measures also differs both between and within jurisdictions. Mitigation of environmental impacts is frequently a prime consideration in the EIA systems in many developing countries but the implementation of proposed mitigation measures is often weak (see Appendix).

It is probable that, as concern over the sustainability of development grows, more emphasis will be placed on the avoidance of impacts by the consideration of alternative approaches, as in the Dutch EIA system. It is likely that increasing attention will be paid to the notion of 'no net deterioration' or 'net amelioration' of the environment. This will affect not only the choice of alternatives and mitigation measures (e.g. the establishment of a new and larger recreational open space as compensation for the loss of, or damage to, the original space) but also the implementation of these measures.

It is apparent that there is considerable scope for improving the implementation of mitigation measures in all seven jurisdictions, and in developing countries. This is bound up with impact and implementation monitoring (Chapter 14), where practice also needs to be improved. It is probable that the greatest single contribution to increasing sustainability which EIA could deliver (beyond the termination of environmentally unsatisfactory actions) is a major improvement in the nature of mitigation measures and their implementation.

CHAPTER 16

Consultation and participation

Introduction

As stated in Chapter 1, consultation and participation are integral to environmental impact assessment: EIA is not EIA without consultation and public participation. Almost all EIA systems provide for consultation and participation after publication of the EIA report, prior to the decision on the action (see Chapter 12). However, many jurisdictions either require or encourage consultation and participation at earlier stages of the EIA process, for example during scoping. Indeed, there appears to be a growing consensus that increased consultation and participation, using one or more of the large number of means of participation which exist, can produce significant benefits for the proponents of actions and for those affected.

This chapter presents an examination of the role of consultation and participation in the EIA process. It then advances a set of evaluation criteria for the treatment of consultation and participation within EIA. These criteria are then employed in the analysis of procedures for consultation and participation in the United States, California, the United Kingdom, The Netherlands, Canada, the Commonwealth of Australia and New Zealand.

Consultation and participation within the EIA process

The New Zealand Ministry for the Environment (1992b, p. 16) has identified the aims of early consultation and participation in the EIA process (identification and mitigation of impacts; prevention of environmentally unacceptable development) succinctly:

> [An] application preceded by a programme of community involvement where concerns are identified and addressed may result in few or no

225

objections. It may also result in substantial project modifications or abandonment if no other accommodation can be reached.

These aims have been echoed in the United States (Chapter 1) and in Australia (Australia and New Zealand Environment and Conservation Council – ANZECC, 1991). The principles for public involvement in EIA have been nicely summarised by ANZECC (Box 16.1). As Sheate (1991) has stated:

> By involving the public as early as possible issues may be identified which 'experts' might not have considered important, but which could prove to have a degree of importance out of all proportion to the magnitude of the impact.

There are several different types of public participation. These can be distinguished by the nature of the relationship between the public and the decision-making body or the proponent (Arnstein, 1969; Sewell and Coppock, 1976; Canter, 1977; Wathern, 1988a). These relationships range from the provision of information, through a range of types of consultation to direct public control. All three types of relationships may be identified in EIA systems at different times and in different circumstances. For example, Australian Aboriginal and New Zealand Maori people may sometimes control whether a particular mining project affecting lands over which they have acknowledged rights can proceed or not.

In addition to the various means of eliciting responses from consultees

Box 16.1   EIA public participation principles

(a) Participate in the evaluation of proposals through offering advice, expressing opinions, providing local knowledge, proposing alternatives and commenting on how a proposal might be changed to better protect the environment.

(b) Become involved in the early stages of the process as that is the most effective and efficient time to raise concerns. Participate in associated and earlier policy, planning and programme activities as appropriate, since these influence the development and evaluation of proposals.

(c) Become informed and involved in the administration and outcomes of the environmental impact assessment process, including:

- assessment reports of the assessing authority
- policies determined, approvals given and conditions set
- monitoring and compliance audit activities
- environmental advice and reasons for acceptance or rejection by decision-makers.

(d) Take a responsible approach to opportunities for public participation in the EIA process, including the seeking out of objective information about issues of concern.

Source: Australian and New Zealand Environment and Conservation Council (1991, p. 7).

and the public during the EIA process (below) the use of mediation or environmental dispute resolution in certain circumstances has been suggested, particularly in the United States. Mediation involves the assistance of a mediator in negotiations between the parties in a dispute over a new development and requires a willingness to compromise and utilise environmental mitigation. While it is not easy to state precisely when and if mediation will help negotiations towards completion, there appear to be four prerequisites to its success: a stalemate, or the recognition that stalemate is inevitable; voluntary participation; some room for flexibility; and a means of implementing agreements (Talbot, 1983; Bingham, 1986). However, these prerequisites appear to apply in only a small minority of siting decisions involving EIA.

Jeffery (1987) saw little use for mediation and negotiation in EIA, except during scoping. In particular, he did not consider mediation to be suitable for the use of public hearings in arriving at decisions. In many jurisdictions, such as the United Kingdom, there is often only one major siting decision and stalemates seldom apply. In these circumstances, the local planning authority often acts as mediator and negotiations are used to reduce the adverse effects of proposed developments. The same role is often played by inspectors at public inquiries. Buckley (1991b) was also wary of the use of mediation in environmental dispute resolutions:

[Environmental dispute resolution approaches] will not be able to substitute for planning and impact assessment legislation which provides for formal public information and participation at several stages, and third party recourse to the courts if planning agencies fail to discharge their responsibilities adequately.

Nevertheless, Canada has instituted a formal mediation procedure as one track in its new environmental assessment system, to be utilised when circumstances dictate (Sadler, 1994b).

Figure 1.1 shows that consultation and participation can be employed at every stage in the EIA process. While the involvement of agencies and the public in the very early consideration of alternatives and of preliminary design of the proposed action is not usually feasible, consultation and participation in screening can normally be organised without great difficulty and is usual in, for example, Western Australia (Wood and Bailey, 1994). Consultation and participation in scoping are commonplace in many EIA systems, and are a requirement in, for example, the United States and The Netherlands (Chapter 10). Similarly, the involvement of consultees and the public in EIA report preparation (as in Victoria (Wood, 1993c)) should lead to improved quality, or at least to improved acceptability (Chapter 11). As mentioned above, almost all jurisdictions provide for consultation and participation during the review process. In a fully participative EIA system, these rights should also extend to the review of further information submitted by the proponent, to any evaluation report relating to the action prepared and to the making of the decision (Chapter 13). Finally, the ability to comment upon monitoring results is a necessary check on the operation of monitoring procedures (Chapter 14).

Consultation and participation can only be effective if copies of EIA documents are made public at each stage of the EIA process (e.g. at the scoping stage as well as on completion of the EIA report). Such documents need to be made readily available at a number of locations convenient to those most likely to be affected by the proposal concerned. The documents also need to be accessible in the sense of being clear and comprehensible. This is especially true of the non-technical summary of the EIA report (Sheate, 1991).

Equity demands that copies of EIA documents can be obtained and/or purchased at a reasonable price for detailed perusal. In the United States such documents are generally free of charge whereas a substantial charge may sometimes be made in other jurisdictions. In the Commonwealth of Australia it is usual for a nominal charge for EIA reports to be made (say $A15) to discourage requests for large numbers of such documents from, for example, schools. The cost of EIA reports in the United Kingdom varies from free to over £100.

Proponents have, in the past, frequently invoked confidentiality and secrecy as reasons for not making information about a proposed action available to the public. Many EIA systems expressly permit restrictions on the availability of information where the case for withholding it can be demonstrated. In general, these restrictions are seldom invoked in most jurisdictions though instances of national security or commercial sensitivity do sometimes arise.

There exists a large number of consultation and participation methods, each with its own advantages and disadvantages (see, for example, Social Impact Unit, 1991; Ministry for the Environment, 1992b). Examples include:

- questionnaires and surveys
- advertisements
- leafleting
- use of media
- displays
- exhibitions
- telephone 'hot lines'
- open houses
- personal contact
- community liaison staff
- community advisory committees
- group presentations
- workshops
- public meetings
- public inquiries.

The choice of method should clearly be appropriate to the stage of the EIA process at which it is employed. For example, while a public inquiry may well be appropriate immediately prior to decision making, it is very unlikely to be appropriate at the screening stage (Glasson *et al.*, 1994). In practice, most jurisdictions leave considerable discretion to decision-takers and proponents in their choice of consultation and participation methods.

Intervenor funding is often difficult to arrange and control, but access to such funding can beneficially affect the outcome of the EIA process by making participation more effective. Without financial assistance, local groups may feel at a great disadvantage relative to the proponent at all stages in the EIA process. However, Canadian experience, among others, has shown that when funding has been made available to help participants prepare for public hearings they have frequently made well conceived and constructive contributions which have led to greater consensus about the environmental consequences of the proposed action (Lynn and Wathern, 1991).

As well as the public, various consultees will have valuable contributions to make at the different stages in the EIA process. While it is usual to consult the bodies which are thought likely to provide useful information on an *ad hoc* basis, there are advantages in specifying a list of consultees who must be consulted at the various stages of the EIA process by the proponent and/or the decision-maker.

It is clearly equitable that neighbouring authorities, states and countries be consulted where proposals are made which could affect their environments. Many EIA systems make specific provision for such consultation and the 1991 Espoo Convention on Transboundary Impacts has led to an increase in these. Environmental disputes may, of course, still arise but are considerably reduced by transboundary consultation.

As at other stages of the EIA process, the availability of clear guidance on the procedures and techniques for consultation and participation will be helpful to proponents, the decision-making and environmental authorities, consultees and the public.

The publication of the results of consultation and participation is clearly a necessary check on their use in the EIA process. There should be a right to inspect both public and consultee submissions and the use made of them by the responsible agencies. It is clear that the role of public involvement in the success of the US National Environmental Policy Act in influencing decisions on actions owes much to two factors: the first is the right to participate and to gain access to relevant documentation, including that relating to participation; the second is the public right of appeal to the courts over EIA decisions (Chapter 2). Other jurisdictions provide opportunities for appeal against the various decisions made during the EIA process. For example, an administrative appeal against the decision to permit an action can be made to an arm of government in Western Australia (Bailey and English, 1991) or to the courts in California.

Clearly, while such appeal rights should make the EIA process more effective by influencing decisions taken at different stages in the process they need to be tempered by the need to make it efficient. A balance has to be struck between the positive benefits of consultation and participation in ameliorating the impacts of actions and in reaching consensus on environmental outcomes and the financial and time costs involved. Since the expenditure of time, rather than money, often appears to be the principal criticism of those EIA systems with extensive consultation and participation requirements, such a balance could imply limiting the

Box 16.2   Evaluation criteria for consultation and participation

---

Must consultation and participation take place prior to, and following, EIA report publication?

Must consultation and participation take place prior to scoping, during scoping, during EIA report preparation, during review and following revision, during decision making and during monitoring?

Are copies of EIA documents made public at each stage of the EIA process?

Can copies of EIA documents be obtained/purchased at a reasonable price?

Do confidentiality/secrecy restrictions inhibit consultation and participation?

Are consultation and participation methods appropriate to the stage of the EIA process at which they are employed?

Is funding of public participants provided for?

Are obligatory consultees specified at various stages in the EIA process?

Must adjoining authorities/states/countries be consulted?

Does published guidance on consultation and participation exist?

Must the results of consultation and participation be published?

Do rights of appeal exist at the various stages of the EIA process?

Does consultation and participation function efficiently and effectively?

---

amount of time taken to complete each stage in the EIA process while providing adequate information and the opportunity for appeals to be dealt with effectively.

The various criteria for the evaluation of consultation and participation provisions within the EIA process are listed in Box 16.2. These criteria are used to assist in the analysis of consultation and participation procedures in each of the seven EIA systems which now follow.

United States of America

Consultation and participation have been the driving force in the evolution of EIA in the United States. The National Environmental Policy Act 1969 (NEPA) requires that relevant federal agencies be consulted during the preparation of the environmental impact statement (EIS) and that the public be involved. The Council on Environmental Quality (CEQ) Regulations specify that agencies must:

(a) Make diligent efforts to involve the public in preparing and implementing their NEPA procedures.
(b) Provide public notice of NEPA-related hearings, public meetings, and the availability of environmental documents so as to inform

those persons or agencies who may be interested or affected (Section 1506.6).

The Regulations make provision for agency consultation and public participation at the following stages of the EIA process:

- in screening (preparation of, and comment upon, the environmental assessment, and comment upon the finding of no significant impact)
- in publication of notice of intent
- in scoping
- in preparation of, and comment upon, the draft EIS
- in preparation of the final EIS
- on the record of decision
- on monitoring results following implementation.

Participation takes a variety of forms, from the making public of documents, to the circulation of documents, to meetings and hearings. EIA documents must be made available to the public either free of charge or for not more than the cost of photocopying. In cases where there is substantial controversy or interest, or where another agency with jurisdiction over the action requests it, a public hearing on the draft EIS must be held. In any event, the lead agency must consult other agencies with jurisdiction or special expertise.

NEPA has no enforcement mechanism specified in statute and CEQ and the Environmental Protection Agency have no enforcement authority. This is why public access to the NEPA process is so crucial. There are opportunities for recourse to the courts at various stages in the EIA process and interest groups, private citizens, state and local agencies and businesses have taken advantage of these by filing thousands of lawsuits, at great expense. The courts have always required good-faith efforts to comply with NEPA's full disclosure provisions. As Bass and Herson (1993a, pp. 101, 102) have pointed out, the Supreme Court has consistently supported the federal agency in each of the 12 cases it has heard. It has also held that NEPA is essentially procedural (notwithstanding the CEQ referral process, Chapter 14) and that agencies have no substantive duty under NEPA to protect the environment.

In practice, there are significant weaknesses in participation in the preparation of environmental assessments. Blaug (1993) reported that there was public involvement in less than half of the environmental assessment cases examined, despite the requirements of the Regulations.

If a charge is made for EISs, they might cost $US25–75 but they are made widely accessible for inspection. There is virtually no funding for participants in the EIA process. There appear to have been very few difficulties relating to commercial confidentiality but a limited number of EISs, prepared for actions involving national security (Fogleman, 1990, pp. 7, 8), have not been made available to the public. The response of the public is variable: some actions result in vociferous objections and well-attended meetings, but others elicit no public interest at all. This may depend partially on the nature of the action, and partially on the participation methods employed. On the other hand, agency consultation results in full and informed comments on the draft EIS from the relevant federal, state, tribal and local agencies, which must be made public.

California

One of the California Environmental Quality Act's (CEQA's) six objectives (Chapter 1) is to encourage public participation in the making of decisions from which significant environmental impacts might ensue. CEQA sets down several procedural requirements (see below) to ensure that this objective is met.

As in the case of the federal National Environmental Policy Act, it is the opening up of the government decision-making process to public scrutiny and to the courts which has been the most successful aspect of CEQA. The environmental review process introduced by CEQA has greatly expanded the opportunities for the public to participate in decision making. The EIR has become a tool by which the public can gain access to information and influence the outcome of a broad variety of project applications. The California Supreme Court has stated that members of the public hold a 'privileged position' in the CEQA process (Bass and Herson, 1993b, p. 12). The vigilance of private citizens and environmental groups has been instrumental in ensuring that agencies comply with CEQA and that its provisions are taken seriously.

While there is no formal provision for public participation in the initial study, the negative declaration must be made public and no action on the project may be taken for at least 21 days. Further, the lead agency must consider the comments received prior to project approval. Where an EIR is prepared, the public may comment at several stages in the EIA process:

- on the notice of preparation
- at public scoping meetings (if held)
- on the draft EIR
- at public hearings on the draft EIR (if held)
- on the final EIR
- at decision meetings of the lead agency
- on mitigation monitoring results following implementation.

In practice, most agency and public comments tend to be made on the draft EIR rather than earlier in the EIA process.

If the public feels that its comments have not been adequately taken into account in the EIA process, recourse to the courts can be taken once the agency's decision has been announced. Access to the courts is cheap in California. Further, if the courts find in favour of the plaintiffs, they may be awarded their legal costs. It is, perhaps, therefore unsurprising that so many court cases (perhaps 100 per year) are initiated. Olshansky (1991) has reported that about 1 in every 350 CEQA documents is challenged in the courts. Another survey demonstrated that fewer than 1 per cent of project applications subject to CEQA were the object of litigation and that most cases (over 75 per cent) were decided in favour of the lead agency. Despite these findings, the desire to avoid litigation is the driving force of CEQA: 'the perceived threat of litigation influences the CEQA process, and is charged with contributing to unnecessary documentation and obfuscation of material presented in EIRs' (Kaplan-Wildmann and McBride, 1992).

EIA documents have to be made accessible to the public and may be purchased for the cost of photocopying. There are no difficulties in relation to commercial confidentiality or to security issues in California (CEQA specifically excludes trade secrets from disclosure requirements). Obligatory consultees are specified but, with the exception of the area around Lake Tahoe, there is no requirement for other local governments, states or the Mexicans to be consulted. Other than the Guidelines, there is no published guidance on consultation and participation. There is no funding of public participants. Overall, public participation can work quite well, particularly during scoping. Participation is regarded as being very effective in the operation of CEQA but also as being responsible for many of the inefficiencies associated with it.

United Kingdom

The use of consultation and participation is officially encouraged throughout the environmental assessment process in the United Kingdom. Consultation can, and sometimes does, take place at the screening stage and, if a formal opinion is requested of the local planning authority (LPA), the material provided by the developer is made public and representations can be made about the need for environmental assessment (EA) (Chapter 9). Consultation often takes place at the scoping stage (Chapter 10) and the provision of information by the statutory consultees frequently occurs during the preparation of the environmental statement (ES) (Chapter 11). However, it is only once the ES has been submitted that the LPA *must* consult. Prior to this, public participation takes place only in a minority of cases. The LPA is required to forward, or arrange for the forwarding of, copies of the ES to the statutory consultees and to take their comments, together with those of the public, into account before reaching a decision.

Consultees must be allowed 14 days to comment, and the public has 21 days. Where additional information is provided by the developer following a request by the LPA, this too must be circulated to consultees and the public. As mentioned in Chapter 13, there is no formal consultation and participation requirement during the LPA decision-making process, though lobbying and, sometimes, the right to address LPA decision-makers while the decision is being discussed, are permitted. There is, of course, a right of the public and consultees to be heard at public inquiries. Similarly, there exists no public right to participation in the monitoring of implemented projects.

As mentioned in previous chapters, local planning authorities are required to keep planning registers and there is a public right to inspect many of the documents prepared during the EIA process, by visiting the local authority offices. Advertisements and site notices must be placed where environmental assessment is required. The environmental statement must not only be made readily accessible to the public, but available for purchase at a 'reasonable' charge. On the whole, purchase prices of ESs in the United Kingdom are indeed reasonable (many being free of charge) but a minority are expensive (Jones *et al.*, 1991) and some have been priced in excess of £100. Issues of confidentiality and secrecy have seldom arisen in relation to the EIA process in the United Kingdom.

No requirements as to consultation and participation methods are laid down in the Regulations, beyond those relating to the availability of environmental statements. However, it is suggested that:

> The authority and the developer may wish to consider the need for further publicity at this stage, for example, publication of further details of the project in a local newspaper, or an exhibition.
>
> (DOE, 1989, para. 37)

Similarly, there is no provision for the funding of public participants in the EIA process. Apart from the usual statutory consultees for planning applications, the Countryside Commission, English Nature and, for certain developments, Her Majesty's Inspectorate of Pollution must be consulted where an ES is received in England. As required by the European Directive, adjoining member states must be consulted by the British Government where a project is likely to have significant effects on their environment though this has seldom occurred. Consultation of neighbouring local authorities is at the discretion of the LPA.

There is no published guidance on consultation and participation beyond that in the Circular (DOE, 1988b) and the Guide (DOE, 1989). While there is no separate publication of the results of consultation and participation, it is normally possible to inspect both the responses and their analysis at the offices of the LPA. Although the developer has rights of appeal against the LPA's screening decisions (Chapter 9) and against its decision on the planning application (Chapter 13) no similar right of appeal by statutory consultees or by the public exists at these or any other stages in the EIA process.

In one-third of a sample of 24 cases, the distribution of the ES to the statutory consultees was undertaken, following prior arrangement, by the LPA. In just under a further third of the cases the developer or consultant sent a copy of the ES directly to the statutory consultees, and the LPA was informed of this action in all but one case. The consultation process, by both the LPA and the developers and consultants, included a variety of bodies additional to (and often instead of) the statutory consultees, such as the National Rivers Authority, the Health and Safety Executive and the water companies. In one-quarter of the cases the Countryside Commission was not consulted at all and did not receive a copy of the ES. Some breaches in the statutory procedure therefore took place. In over two-thirds of cases the consultations with the statutory consultees and the public based upon the ES were thought to have been helpful to the LPA (Wood and Jones, 1991).

To summarise, consultation and participation is not a requirement of the UK EIA system prior to the submission of the ES, though it frequently occurs informally as well as formally subsequent to submission. Practice varies substantially and there is clearly scope for consultation and participation to become more effective, especially in relation to the largely marginal role played by the general public in the EA process.

The Netherlands

The Dutch Environmental Management Act 1994 specifies two occasions in the EIA process on which the statutory consultees and the public must be given the opportunity to comment. The first is public participation in regard to the establishment of the scoping guidelines (Section 7.14(3)). The second is when the EIS is reviewed (Section 7.23(1)). There is usually a brief public hearing at the review stage. The notification of intent, the decision and the auditing report by the EIA Commission are also published. Confidentiality restrictions have rarely been invoked in The Netherlands EIA process.

While copies of EISs are available for purchase, they are often expensive (more than NLG60) though some are free of charge. Copies of the EIS must be widely available for inspection in the evenings as well as during the day. Proponents sometimes prepare leaflets and publicise the availability of EISs imaginatively but some EISs are somewhat lengthy and technical and thus inappropriate for public consumption, notwithstanding the availability of the summary. Copies of some of the documentation (e.g. the scoping guidelines) are made available at no charge to those participating in the process. The various comments of statutory consultees and the public must be made available for public inspection.

The Dutch EIA system did not contain any provision for the consultation of neighbouring member states. However, that shortcoming was rectified in the revisions to the Act in 1994. Notwithstanding, notifications of intent, scoping guidelines and EISs were sent to neighbouring countries for about 10 proposals thought likely to have transfrontier impacts prior to that date.

Some large environmental pressure groups like Stichting Natuur en Mileau receive public funding for intervenor activities. These groups choose how to spend these funds and involve themselves in selected projects requiring EIA. Several leaflets on public participation exist in The Netherlands but there is no formal guidance as to methods and procedures. The right to participation is supported by the possibility of administrative appeal against the decision (often used) and of appeal to the courts (rarely used). This right of appeal is important in ensuring that the EIA procedures are carefully followed.

Generally, consultation and public participation in The Netherlands work reasonably effectively. Many members of the public object to the proposal at both the scoping and EIS review stages but others make useful comments on such matters as alternatives, vulnerable people or receptors, potential damage to people and difficulties in predicting impacts (Commission of the European Communities, 1993c). There have been instances where an additional alternative has been put forward as a result of public participation, and public comments have often helped to refine proposals. More important to the decision, however, is the informal consultation between the proponent, the competent authority and the EIA Commission.

The Evaluation Committee on EIA (1990) stressed the importance of public participation in EIA ensuring that the main participants, and especially the competent authorities, discharged their responsibilities properly. The open nature of the Dutch EIA process, with consequent minimisation of the possibility of abuse, was seen as one of its great strengths.

Canada

One of the four purposes of the Canadian Environmental Assessment Act (CEAA) is 'to ensure that there be an opportunity for public participation in the environmental assessment process' (Section 4(d)). Every EA report must contain consideration of any comments received from the public. The provisions for public participation vary according to the type of assessment being undertaken. In screenings, notice and participation are at the discretion of the responsible authority (RA) which can provide for the screening report to be made available for comment. It must, in any event, be filed in the public registry for the project (see below) providing an opportunity for public scrutiny and post-decision protest and lobbying. In comprehensive studies, the public may, at the discretion of the RA, be involved in the preparation of the report and must be provided with the opportunity to comment on the report once it has been submitted to the Canadian Environmental Assessment Agency (CEA Agency).

The level of public concern is one reason for referring a project to a panel review or to mediation under CEAA (Section 25(b)). The Minister of the Environment must give public notice of the availability of reports of panels or mediators and of how copies may be obtained. In cases where a panel is convened, the public:

- provide input and comments throughout the panel process
- participate in public hearings convened by the panel.

> (Federal Environmental Assessment Review Office
> – FEARO, 1993a, p. 133)

This type of public involvement, at the scoping stage, in the preparation of the EIS (where appropriate), in review of the EIS, in hearings and in commenting on the panel's report is well established in Canada. It has evolved over time, the principal contributor being Justice Berger in the MacKenzie Valley Pipeline Inquiry (Sewell, 1981) who is credited with:

- broadening the interpretation of impacts to social perspectives
- utilising various hearing formats, including local hearings
- intervenor funding
- accepting individual concerns as valid information

236

- using the media to publicise the process (and popularise EA).
(Smith, 1993)

One of the most significant aspects of the Canadian situation is the public participation funding programme for panel review cases (Lynn and Wathern, 1991). Section 58(1)(i) of the Act states that the Minister may 'establish a participant funding program to facilitate the participation of the public in mediations and assessments by review panel'. This type of funding was first formally established in 1990 under the Environmental Assessment and Review Process (EARP), after several years of practice. An annual budget of over $CD2 million has been allotted to FEARO and criteria worked out for disbursement.

The money has been used by Inuit, Indian and other peoples to employ researchers to prepare and argue a case. Participant funding is widely regarded as being a great success and is credited with increasing the quality of information, of debate and, subsequently, of decisions. Generally, public participation, often supported by expert department information provision and comment, has worked well in panel review cases, but practice has been deficient in self-assessment EAs.

A reference guide to public participation is to be published, partly based upon the previous research and other work undertaken (FEARO, 1988a; Weston, 1991). Bush (1990) found that post-approval public participation can be useful to all stakeholders and should be part of the overall impact management process, as an extension of earlier public involvement.

The second significant provision of CEAA relates to the setting up and maintenance of the public registry for each project subject to EA by the RA. The public registry must include:

- any report relating to the assessment
- any supporting documentation
- any comments filed by the public in relation to the assessment
- any public notice of decision
- any advice or information provided by expert departments
- any terms of reference for public review
- any records produced as a result of the implementation of any follow-up program
- any documents requiring mitigation measures to be implemented.
(CEAA Section 55(3) and (4))

This ambitious public registry will consist of a computerised index to all EA cases, a listing of all the documents for each EA and the documents themselves. Documents are to be accessible electronically or can be copied at reasonable prices. In effect, this public registry system will extend the very widespread availability of information for panel review projects under EARP. Most documents, including many EISs, have been available free of charge in the past.

Commonwealth of Australia

Despite the number of different stages in the Australian EIA process and the necessity to produce documentation at each of these, public participation provisions are often permissive rather than mandatory. There are provisions for the confidentiality of EIA documents (Commonwealth of Australia, 1987, para. 6.2.2) but these have hardly ever been invoked. In practice, participation is more pervasive than the requirements specify but tends to rely heavily on written representations rather than allowing sufficiently for oral communication (Ecologically Sustainable Development Working Groups, 1991). Consultation of government departments and relevant agencies takes place throughout the EIA process.

Notices of intention are normally made public, though there is no requirement to do so (Chapter 9). Again, there is no requirement for public participation during the preparation of scoping guidelines but public involvement is, in practice, common (Chapter 10). There is no formal requirement for public participation in the preparation of the draft EIS and, unsurprisingly, this tends not to occur (Chapter 11). There is, of course, consultation and participation once the draft EIS has been produced and the proponent must respond to these comments in the final EIS. The Commonwealth Environment Protection Agency's assessment report must be made available to the public before the final decision is made. As mentioned in Chapter 13, there is no requirement for the Action Minister's decision, and the reasons for it, to be publicised, but this often occurs.

Some funding of public interventions has taken place but this has been very limited. As mentioned in Chapter 6, the number of legal actions mounted against the EIA procedure used in particular cases has been very low, in contrast to the US situation. In practice, public participation in Australian EIA has been very influential but is not so wide or effective as in the United States. There have been criticisms of inadequate public participation at the earlier and later stages of the EIA process (Formby, 1987; Anderson, 1990). There have also been demands for better funding to ensure fuller public participation at several stages in the EIA process, especially during scoping and monitoring (Fowler, 1991).

New Zealand

The Resource Management Act 1991 contains several provisions relating to public participation. The Fourth Schedule (Box 7.3) specifies that the proponent should include a list of affected or interested persons, the consultation undertaken and any response to the views of those consulted. As mentioned in Chapter 10, where an application is found to have major effects and is notified by the local council, in consultation with the people affected by the proposal, the developer can be asked to provide an explanation of 'the consultation undertaken by the applicant'. Consultation and negotiation with affected parties is not new in New

Zealand but the Act makes it necessary for the developer to include a description of discussions and their outcome in the EIA report, since the local council can insist on this. Consultation and participation prior to EIA report submission is thus not mandatory but is very strongly advised.

Local authorities are required to make EIA documentation available to the general public where projects are notified (though there are no provisions relating to purchase of EIA reports). Certain consultees must be informed of the existence of the EIA report. There exists the right for the applicant or an objector to insist that a public hearing be held (Section 100).

There is no funding of public participants in the EIA process. There is, however, a valuable official guidance document on public participation (Ministry for the Environment, 1992b). As Dixon (1993b) has pointed out, the role of the public is crucial in ensuring that the EIA system functions effectively. Morgan (1993) supported this view and suggested that non-technical summaries need to be made available and that the accessibility of EIA documentation ought to be improved.

Morgan (1993) discerned real shortcomings in the involvement of the public. He reported that councils rarely consult the public early enough in the process and often fail to encourage applicants to consult the public. Their publicity for applications often neglects to mention that environmental information is available for examination. Many EIAs were found to be overly technical and Morgan felt that guidance on EIA needed to be specifically targeted at the public. It is apparent that, in practice, the public participation provisions of the Act are not being implemented enthusiastically. It is expected that, with experience and with the precedents created by the Planning Tribunal, practice will improve.

Summary

All seven of the EIA systems meet the requirement that there must be consultation and participation following the release of the EIA report but four do not make consultation and participation prior to the EIA report mandatory (Table 16.1). While some consultation takes place in many developing country EIA systems, public participation is often lacking after, and especially before, EIA report release (see Appendix).

Table 16.1 masks some significant variations in consultation and participation between and within countries. The weakest requirements for pre-EIA report participation are those in the United Kingdom which, with New Zealand, is the only EIA system not formally to require scoping. While some consultation takes place in the United Kingdom on an informal basis prior to the ES, the involvement of the public is relatively rare. In New Zealand, the local authorities have the power to demand that consultation and participation take place prior to submission of EIA reports for notified projects while in Australia the existence of a scoping stage effectively ensures public participation, even though this is not mandatory.

239

Table 16.1   The treatment of consultation and participation in the EIA systems

Criterion 11: Must consultation and participation take place prior to, and following, EIA report publication?

Jurisdiction	Criterion met?	Comment
United States	Yes	Consultation and participation take place at several stages in EIS preparation.
California	Yes	Public participation and consultation take place at various stages in preparing EIRs.
United Kingdom	Partially	Some voluntary consultation and participation takes place following ES release.
The Netherlands	Yes	Formal requirements for consultation and public participation in both scoping and review.
Canada	Partially	Participation and consultation mandatory throughout panel reviews, required following comprehensive studies and discretionary in screenings.
Australia	Partially	No formal requirement for public participation prior to EIA report, but generally occurs. Agency consultation takes place throughout EIA process.
New Zealand	Partially	Duty to consult following EIA report publication, local council strongly recommended to require developer to consult earlier.

In Canada, although full participation takes place during panel reviews, there is no mandatory requirement for involvement in the preparation of comprehensive studies (which will, in effect, be the EIA report in the Canadian EA system). Participation in the preparation of Canadian screening reports in entirely discretionary.

In the United States and in California there is full provision for early participation and consultation in the preparation of the main EIA report (the EIS and the EIR respectively) but provisions relating to public involvement in the preparation of environmental assessments in the United States are often not observed, and do not exist in relation to Californian initial studies. Public participation and consultation are most strongly embedded in the EIA system in The Netherlands. Generally, however, participation provisions could be strengthened in many EIA systems and especially in those in developing countries.

CHAPTER 17

Monitoring of EIA systems

Introduction

In addition to the monitoring and auditing of impact actions (Chapter 14) it is increasingly being recognised that some form of EIA system monitoring is needed (Canadian Environmental Assessment Research Council, 1988a). The principal purposes of EIA system monitoring are the diffusion of EIA practice and the amendment of the EIA system to incorporate feedback from experience and remedy any weaknesses identified. However, the degree of detail of the records kept on the numbers of EIA reports produced, types of projects assessed, decisions reached, numbers of implemented projects, availability of documents, etc., may vary. This chapter outlines the issues to be covered in the monitoring of EIA systems and puts forward a set of evaluation criteria for EIA system monitoring. These criteria are then utilised in the comparative review of the EIA systems in the United States, California, the United Kingdom, The Netherlands, Canada, the Commonwealth of Australia and New Zealand.

EIA system monitoring

There are numerous elements of any EIA system which can be monitored with a view to diffusion of best EIA practice and amendment of the system to incorporate feedback from experience. However, the diffusion of best practice does not depend only upon EIA system monitoring but also upon such matters as the provision of published guidance and of training and the undertaking of research.

In any EIA system, a definitive record of the number of EIA reports undertaken should be maintained and made public. This record should relate both to total numbers of EIA reports and to EIA reports for different types of action. Clearly, sufficient details relating to the precise title of each document, its length, its date, its price, where it may be inspected or obtained and any other relevant matters should be made available. Such records are available in many EIA systems.

Where formal reviews of EIA reports are undertaken (e.g. the formal assessment reports drawn up in the Commonwealth of Australia EIA system) a record of these and the results obtained should be kept and made public. In addition, the existence of other EIA documents such as scoping reports should be recorded. Similar details to those listed for EIA reports should be maintained for this other EIA documentation generated within the EIA system. Practice in the maintenance of such records tends to vary from one EIA system to another.

Ideally, all EIA reports and other EIA documents should be publicly available at one or more central locations during reasonable hours. Collections of EIA reports provide an invaluable source of information to those engaged in preparing such documents, to those responsible for reviewing them, to those likely to be consulted, to the public and to those undertaking research. In practice, such documents may be consulted in many EIA systems, with varying degrees of difficulty.

There are considerable difficulties in obtaining accurate information about the financial costs involved in undertaking EIA. However, while the costs of EIA report preparation may be difficult to distinguish from other activities associated with the action (Chapter 18) some information about expenditure incurred in preparing and processing EIA documents in every EIA system should be obtained, perhaps on a sample basis, and centrally recorded. Details about numbers of staff involved in EIA, as well as about consultancy costs and any fee payments should be maintained and made public. As with other aspects of EIA system monitoring, such information is easier to collect if a single agency is responsible (as, for example, in The Netherlands) and practice varies accordingly.

Similar information should be collected and maintained in relation to the time required to undertake EIA. Data obtained should include the amount of time needed to process each EIA report once it has been received. As elsewhere in the EIA process, there are numerous measurement difficulties (Hart, 1984) but reasonably reliable records, possibly on a sample basis, can be kept if the will to do so exists.

Experience of specific EIAs may reveal that changes in practice or procedure within the EIA system need to be made. Provision for effective feedback should therefore exist. This may take the form of practice advice notes, circulars, regulations, training, amendment of project-specific or generic guidelines or other means. Clearly such feedback tends to be most effective where only a limited number of responsible authorities are involved in the EIA process and practice varies between jurisdictions.

More generally, reviews of any EIA system should be carried out from time to time and any necessary changes to the system implemented. The better the EIA system monitoring information obtained, the easier such a review will be. As with other elements of the EIA process, the role of consultation and participation in reviews of the EIA system is important and should be adequately provided for. Most jurisdictions have carried out reviews of their EIA systems from time to time and most would probably wish to implement further modifications at any specific time.

Box 17.1   Evaluation criteria for EIA system monitoring

Must the EIA system be monitored and, if necessary, be amended to incorporate feedback from experience?

Is a record of EIA reports for various types of action kept and made public?

Are records of other EIA documents kept and made public?

Are EIA reports and other EIA documents publicly available at one or more locations?

Are records of the financial costs of EIA kept and made public?

Is information on the time required for EIA collected and made public?

Are the lessons from specific EIAs fed back into the system?

Have reviews of the EIA system been carried out and changes made?

Is consultation and participation required in EIA system review?

Does the monitoring of the EIA system function efficiently and effectively?

Finally, and again in common with other elements of the EIA system, the monitoring of the EIA system should be effective (i.e. lead to achievement of its goals) and efficient (i.e. not consume disproportionate financial, managerial or time resources). This and the other criteria discussed above are summarised in Box 17.1. The various evaluation criteria are used in the comparative review of EIA system monitoring which now follows.

United States of America

The Council on Environmental Quality (CEQ) was created by the National Environmental Policy Act (NEPA) and given the responsibility for environmental policy development and the duty to review and appraise federal agency compliance with NEPA. Part of this CEQ oversight involves the preparation of annual reports which summarise the trends in the implementation of NEPA (numbers of statements, numbers of court cases, significant developments, etc.). The annual reports of CEQ provide an invaluable picture of the operation of NEPA over the years (and of environmental trends in the United States generally).

Since the Environmental Protection Agency (EPA) must notify each draft and final environmental impact statement (EIS) in the *Federal Register*, listings of all EISs can readily be obtained from the EPA, if necessary broken down by type of action and/or by agency and geographical location. Copies of EISs must be filed by EPA. EPA only

243

maintains a library of EISs in hard copy form for two years. However, EISs are available for inspection or loan at the library of Northwestern University, Illinois. Each year EISs are copied on to microfiche by a commercial agency (Cambridge Scientific Abstracts) from whom copies may be purchased. Monthly and annual catalogues are prepared and copies of both the catalogues and of the microfiches may be perused at the EPA Library in Washington, DC.

There is no central record of other NEPA documents (e.g. environmental assessments, findings of no significant impact, records of decision) but each of the agencies maintains at least some statistics on these documents. Copies of other NEPA documents are not kept centrally or filed on a long-term basis by the relevant agencies. No regular records of the financial costs or time requirements of EIA are maintained.

A number of federal reviews of the operation of the whole EIA system have been carried out, most notably those leading to the 1978 Regulations and that initiated during the early Reagan years (Chapter 2). There have also been federally funded reviews of parts of the EIA system, for example, on agency compliance (Environmental Law Institute, 1981), on the scientific quality of EISs (Caldwell *et al.*, 1983), on auditing (Culhane *et al.*, 1987) and on environmental assessments (Blaug, 1993). In addition, federal agencies have conducted reviews of their own NEPA procedures and fed lessons from specific EIAs (perhaps as a result of court cases) back into their systems. These reviews have generally involved extensive consultation. Further, there have been numerous academic reviews of the US EIA system, especially as it has come of age (see, for example, Blumm, 1990; Hildebrand and Cannon, 1993).

California

There is no provision for all the environmental impact reports (EIRs) or other documents prepared under the provisions of the California Environmental Quality Act (CEQA) to be recorded centrally. However, the State Clearinghouse maintains a record of all the EIRs and other CEQA documents which are returned to it and enters them into its database. For a charge, it is possible to obtain a record of all the EIRs for a particular type of project which the Clearinghouse received within a particular period of time. The Clearinghouse does not keep copies of the EIRs or other documents it receives. Most lead agencies maintain copies of the EIRs for which they are responsible and the California Department of Water Resources has a good (but incomplete) collection of EIRs. All the EIA documents kept by public agencies are open to public inspection.

No records of the financial costs of EIA or of the time required for EIA are kept. There is, of course, some knowledge of cost and time requirements but this has been obtained mostly by experience and by anecdotal evidence. Olshansky (1993) suggested that the mean cost of a draft EIR was about $US38,000 in 1990, excluding administration costs. Some of these costs would have been incurred in any event. As Bass and Herson

(1993b, p. 95) have indicated, some 'highly complicated and controversial EIRs prepared for very large projects' have cost a great deal more.

As a result of legal actions and their implications for future inter-pretations of CEQA, lessons from particular cases involving EIA are fed back into the system by means of amendments to CEQA. This legal lesson-learning is, however, very different from modifications resulting from the practice of EIA. Indeed, it appears that huge quantities of environmental data are being generated which are not used again and that few lessons from specific EIAs are being learned or fed back into the system.

Amending CEQA appears to be a Californian pastime, though there have been very few systematic reviews of the EIA system. In 1993 almost 100 proposals were advanced to amend the Act. These were eventually whittled down to some dozen front-runners prior to the compromises which led to the latest round of CEQA amendments. The California Bar Association set up a working party on the topic, as did the California Chapter of the American Planning Association jointly with the Associa-tion of Environmental Professionals. While this level of review activity is unusual, it has been an exception for a legislative session to pass without CEQA being reviewed and modified.

The Guidelines contain a requirement (Section 15007) that they be reviewed and amended from time to time. Unfortunately, the Office of Planning and Research has not been allocated sufficient staff to carry out the reviews so the Guidelines have not been kept up to date. Amend-ments to the Act are thus put forward as a response to interest group pressure for greater certainty. The various reviews and amendments are all subject to consultation and participation.

The monitoring of the Californian EIA system is *ad hoc*: it is neither effective nor efficient. It is remarkable that the best statistics about the operation of the EIA system have been generated by an out-of-state academic (Olshansky, 1992) and by practitioners (Kaplan-Wildmann and McBride, 1992) 20 years after CEQA was enacted. Further, it is ironic that the Act itself is amended almost annually when the Guidelines (which can be altered much more efficiently) have not been amended to match the Act. This is probably a fair reflection of Californian environmental politics: symbolic rhetoric and law making provide publicity and are cheap; state implementation lacks a high profile and costs money.

United Kingdom

There is, at present, no single official comprehensive listing of all the environmental statements (ESs) which have been published in the United Kingdom. The Planning Regulations for England and Wales and for Scotland require local planning authorities to send three copies of any ES to the appropriate government department when it is received. On the basis of this information, the Department of the Environment (DOE) regularly prepares lists of ESs prepared under the Planning Regulations which are published in the *Journal of Planning and Environment Law*. Information about the name of the local planning authority (LPA), about the nature of the development and about the category of project within Schedule 1 or 2 to the Regulations is provided. Because of non-compliance, or late compliance, with the requirement for LPAs to send copies of ESs to central government, these lists tend to be incomplete or not up to date. Nevertheless, the monitoring situation with regard to planning ESs is much better than for projects approved under other regulations (Jones *et al.*, 1991). Unofficial lists of ESs have been prepared for 1988–90 by Jones *et al.* (1991) and for 1991 by Jones (1993). The Institute of Environmental Assessment (1993) and Frost *et al.* (1994) have also prepared lists of ESs.

The Circular (DOE, 1988b) also requests LPAs to notify the Secretary of State about various decisions relating to EA prior to submission of the ES. Summaries of opinions, notifications and directions are also published in the *Journal of Planning and Environment Law*. The published information is generally regarded as valuable by LPAs and others, although it is not without weaknesses (Wood and Jones, 1991). There is no monitoring of public inquiry decisions on planning appeals or call-in cases involving EA. This is unhelpful, since such decisions provide important precedents, even though they do not have legal force.

There is no single repository of ESs for all types of projects for the whole United Kingdom. English ESs prepared under the Planning Regulations may be inspected at the Library of the Department of the Environment in London but the collection is far from complete as ESs may still be in use within DOE. Other collections such as those at the EIA Centre, Manchester University, at the Institute of Environmental Assessment and at Oxford Brookes University, are also incomplete. No record of the monetary costs and time required for EIA are kept though some information on these topics was collected for a sample of ESs by Wood and Jones (1991) (see Chapter 18).

Inevitably, as experience has been gained with EIA, practice has improved (see, for example, Lee and Brown, 1992) and minor modifications have been made to the operation of the EIA system by DOE. These have mostly related to discussions regarding screening. DOE also commissioned a monitoring review of the operation of the EIA system (Wood and Jones, 1991). As a result of experience and the recommendations arising from the review, guidance additional to that provided by the Circular (DOE, 1988b) and the Guide (DOE, 1989) was commissioned. Changes to the Planning Regulations were made which extended the scope of EA to various other projects and changed a number of procedures relating, for example, to consultation and participation where further documentation is required by the LPA. The proposed changes arising from the review were circulated for comment, but there was no participation in the review itself (though some consultation took place).

In brief, while the only formal requirement for EIA system monitoring involves the provision of copies of ESs for central government, lists of ESs exist, some monitoring of the system takes place, a review of the EIA system has been undertaken and amendments have been made.

The Netherlands

The Netherlands EIA system is subject to several monitoring provisions. Perhaps the most important is the requirement of Section 7.2 of the Environmental Management Act 1994 for the EIA system to be reviewed every five years. The previous Act required a first review after three years. To undertake that review an evaluation committee was set up to advise the Ministry of Housing, Physical Planning and the Environment (VROM) and the Ministry of Agriculture, Nature Management and Fisheries (LNV) on how the EIA system was working (Chapter 5). It commissioned three background studies and conducted interviews with government officials, developers, consultees and interest groups. Many of its recommendations (Evaluation Committee on EIA, 1990; VROM, 1991) have now been implemented. A similar mechanism, involving a call for the general submission of evidence, is likely to be employed in 1997.

Perhaps the second most important system monitoring provision in the Act is the requirement for the EIA Commission to make an annual report on its work (Section 2.18). These annual reports provide a summary of EIA activity during the year and a complete list of all the proposals which have been referred to the EIA Commission together with a bar chart showing their progress through the EIA system. The reports also contain discussions about aspects of EIA practice. A copy of every set of guidelines, every EIS, every EIA Commission review, every decision and each auditing report is available in the library of the EIA Commission in Utrecht, principally (but not exclusively) for internal use. In addition, a further copy of many of these documents is available at VROM in The Hague.

No records of financial costs or time are kept. However, it is believed that the total costs of EIA are generally limited to 0.001–0.01 per cent of the cost of large projects but, perhaps, constitute 1 per cent or more of the cost of smaller projects (EIAs usually cost NLG50,000–500,000) (Commission of the European Communities, 1993c). In addition to the costs of the EIA Commission (about NLG4.5 million annually) and the costs of ten people working on EIA in VROM, LNV employs one person and most of the twelve provinces employ one or more persons on EIA. Each provincial and municipal competent authority used to receive a sum of NLG50,000 when it had to deal with a development requiring EIA but this support has now been reduced substantially. The total government annual budget on EIA is about NLG6 million, including the research it funds, but excluding its own staff. This is a very substantial sum.

The Act provides for a minimum of 21 weeks to elapse between the submission of the notification of intent and the decision, though EIA cases sometimes take several years to decide. The principal cause of delay is the preparation of the EIS though it sometimes takes a lengthy period of time to make the decision.

Despite the amount of information available about the EIA system in The Netherlands, the results of individual EIAs are not sufficiently fed back into the system and VROM is examining ways of trying to improve this situation. The EIA Commission is able to utilise previous experience in drawing up scoping guideline recommendations and, to a lesser extent, in reviewing EISs but many developers and consultants are failing to utilise fully the available information. Competent authorities are not utilising sufficiently the experience of similar projects gained elsewhere. There is, therefore, some scope for improving practice in the well-monitored Dutch EIA system.

Canada

The Canadian Environmental Assessment Act (CEAA) contains a provision for a five-year review of the Act to be submitted to Parliament:

> Five years after the coming into force of this section, a comprehensive review of the provisions and operation of this Act shall be undertaken by the Minister (Section 72(1)).

The Auditor-General will review the Act's efficiency and effectiveness after three years.

Numerous reviews of the Canadian Environmental Assessment and Review Process (EARP) and the quality of EA have taken place in the past (including several damning internal evaluations) (see, for example, Beanlands and Duinker, 1983; MacLaren and Whitney, 1985; Canadian Environmental Assessment Research Council, 1988a; Federal Environmental Assessment Review Office – FEARO, 1988b; Sadler, 1989; Smith, 1993, p. 130). These reviews, which incorporated extensive public involvement, led to the production of the EARP Guidelines in 1984 and to the Act in 1992.

The EARP Guidelines were adjudged to have resulted in a system which was open to panel bias (which has, in practice, not been a problem), had variable public participation, had secretive screening, occurred late in the planning process, lacked a legal mandate, precluded the no-go option, was subject to uncertain compliance (and to uncertainty generally) and duplicated other controls (Fenge and Smith, 1986). A Canadian Environmental Assessment Research Council funded study found the initial assessment phase to be seriously flawed under EARP (Weston, 1991). It was widely believed that 'EARP should incorporate procedural checks and balances to ensure greater accountability, transparency and openness, and to minimise duplication and reduce inefficiency' (FEARO, 1988b). It is to be hoped that the reviews of CEAA will be less damning.

An annual report has to be made to Parliament about the activities of the Canadian Environmental Assessment Agency (CEA Agency) and the administration and implementation of CEAA. The requirement to maintain the public registry (Chapter 15) provides the basis for statistical summaries of operations, since the various individual project summaries can be aggregated to provide an up-to-date national picture.

Previous experience with monitoring the system has been mixed. FEARO has maintained a list of the projects subject to initial assessment notified by the initiating departments and has published these as a quarterly bulletin. The latest edition available in mid-1994 was a year old (FEARO, 1992a) and there were no plans to produce any more editions.

It is strongly believed that many initial assessments have not been recorded. This situation may well continue to apply under CEAA despite the requirement to set up a public registry for EAs conducted under the Act (Chapter 16). It is, however, likely that record keeping will improve because the electronic index system will make it easier to access information and to make checks on the responsible authorities. Gaining access to screening reports should be much easier for the public under CEAA than under EARP.

No record has been kept of initial environmental evaluations under EARP and no copies of these documents are available centrally. This situation will change, since a copy of all comprehensive study reports must be sent to the CEA Agency where they will be available for public inspection. The situation relating to these quasi-EISs will parallel that for EISs under EARP.

A register of all the panel reports is kept and published by FEARO and the reports, the EISs and all the public comments are available for inspection at one of FEARO's two offices (in Vancouver and Hull/ Ottawa). This situation will continue under CEAA, allowing the diffusion of best practice. Overall, therefore, EA system monitoring should be much more effective under CEAA than EARP.

There are no records of the financial costs of EA, nor of its time requirements, except for anecdotal evidence. There can be no doubt that lessons from specific EAs have been fed back into the EA system as a result of administrative experience with EARP, court cases and the development of technical expertise (see, for example, Munro et al., 1986). Future revisions of CEAA and its regulations seem certain as experience with the legislation is gained.

Commonwealth of Australia

Australia's Commonwealth EIA system has been in place since 1974. Since that time there has only been one major challenge in the courts and that was dismissed (Chapter 6). There have been several detailed reviews and critiques either of the EIA system or dealing with the EIA system. Recent reports (many of which have involved consultation and participation) have been sponsored by environmental organisations (Fowler, 1991; Martyn et al., 1990), by business (Bureau of Industry Economics, 1990; Kinhill and Phillips Fox, 1991) and by government (Australian and New Zealand Environment and Conservation Council, 1991; Ecologically Sustainable Development Working Groups – ESDWG, 1991; ESDWG Chairs, 1992). However, little fundamental change has occurred over the years beyond the introduction of scoping and of public environment reports.

The Environment Minister announced in 1993 that a comprehensive review of the Commonwealth EIA process was to be undertaken by the Australian National Audit Office. This review was intended to maximise both the effectiveness and the efficiency of the EIA system. It was to examine, in particular, the issues of participation, transparency, certainty, accountability, integrity, cost-effectiveness, flexibility and practicality (Commonwealth Environment Protection Agency – EPA, 1993; Anderson, 1994). It is possible that this review will result in significant changes to the Commonwealth EIA system.

Neither the Environment Protection (Impact of Proposals) Act 1974 nor the Administrative Procedures (Commonwealth of Australia, 1987) require any review or monitoring of the EIA system. In practice, a limited amount of monitoring occurs. A record of all the EISs and public environment reports produced is kept by EPA. However, no record is kept of other EIA documents, apart from assessment reports and there is no central collection of EIA reports for reference purposes. Neither is any record kept of the costs or time requirements of the EIA system, though anecdotal evidence of, and opinions about, costs and delays exist.

Various methods of strengthening the monitoring of the Commonwealth EIA system have been proposed, especially the provision of better environmental data, including a national inventory and depository of Commonwealth and state EIA documents (Anderson, 1990; Kinhill and Phillips Fox, 1991). There is also scope for the collection of better information about the time and resource demands of the EIA system. Indeed, EPA could assume many of the functions assumed by the Council on Environmental Quality in the United States.

New Zealand

The Ministry for the Environment is required to monitor the effect and implementation of the Resource Management Act 1991 (Section 24(f)). There is, however, no duty to review the operation of the Act in a specified number of years. Nor does the Act contain any specific monitoring requirements, such as a duty to collect documentation at a central point of reference, or to record EIA reports. Morgan (1993) has stressed the need for a register of EIA documents to assist in diffusion of best practice, scoping and other EIA activities.[1]

This lack of provision for recording and collection is perhaps surprising since a copy of each environmental impact report and the corresponding audit prepared under the Environmental Protection and Enhancement Procedures is housed in the library of the Ministry for the Environment (MfE). However, the Ministry for the Environment continues to play an informal but important role in the dissemination of information about EIA. This type of information exchange is common in New Zealand where professional networks tend to be small.

The Parliamentary Commissioner for the Environment also helps to ensure that lessons learned from specific EIAs are fed back into the system since its reports are published and disseminated. It is likely, however, that the major element in EIA system development will be the decisions of the Planning Tribunal which are widely read by environmental professionals. Some information on the EIA system is available within universities and consultancies but no data on the costs and duration of EIAs appear to be collected.

Some monitoring of the EIA system is taking place independently. In particular, Morgan's (1993) study of the operation of Resource Management Act EIA procedures throws light on current practice. Unsurprisingly, he reported that procedures for major developments are operating as before with the retention of consultants and the preparation of professional EIA reports. However, smaller proposals are creating difficulties since large numbers of very short EIA reports are being prepared by inexperienced applicants and reviewed by inexperienced and overwhelmed council officials. Many of these EIA reports will probably not be required once screening criteria are set down in operative regional and (especially) district plans.

Summary

The EIA systems in the United States, The Netherlands and Canada all meet the EIA system monitoring criterion (Table 17.1). The EIA systems in California, the United Kingdom, the Commonwealth of Australia and New Zealand are adjudged not to meet it. There is, however, far more documentation and monitoring information available relating to these EIA systems than there is about the EIA systems in most developing countries (Appendix).

It is probably no coincidence that the EIA systems which are monitored all possess a single body with overall responsibility for EIA and have a legal duty to review or oversee the EIA system. Of the four, only the US EIA system does not possess a legally imposed quinquennial review requirement though it does demand an annual report on the operation of the system.

It is the major task of the Council on Environmental Quality to oversee the US EIA system but it has recently had to make do with far fewer staff resources than it needs to undertake its oversight role effectively. Fortunately, the Environmental Protection Agency has the resources necessary to undertake the EIA system monitoring function in the United States. Both the Canadian Federal Environmental Assessment Review Office (Canadian Environmental Assessment Agency) and the Dutch EIA Commission (which are solely concerned with EIA) possess adequate staff resources to undertake EIA system monitoring.

This is also true of the Commonwealth Environment Protection Agency in Australia, which is not under a legal obligation to review the EIA system. The Californian, United Kingdom and New Zealand requirements do not include a formal duty to review or monitor the EIA systems and the jurisdictions do not allocate sufficient resources to the central bodies responsible for EIA to permit effective system monitoring to take place. The Canadian initiative with a public registry could provide a model for EIA system monitoring worthy of consideration elsewhere. The same absence of a duty and adequate resources effectively prevent EIA

Table 17.1 The treatment of system monitoring in the EIA systems

Criterion 12: Must the EIA system be monitored and, if necessary, be amended to incorporate feedback from experience?

Jurisdiction	Criterion met?	Comment
United States	Yes	Council on Environmental Quality charged with general oversight of EIA implementation. Numerous reviews undertaken and amendments made.
California	No	Little system monitoring but frequent amendments made to CEQA.
United Kingdom	No	No formal general requirement to monitor but some records published. EIA system review undertaken, and changes made to improve operation.
The Netherlands	Yes	EIA Commission prepares annual report and a comprehensive quinquennial EIA system review is undertaken.
Canada	Yes	CEAA contains five-year review requirement and public registry facilitates monitoring of EA system.
Australia	No	No formal requirement for monitoring or periodic review but reviews undertaken. Records of EIA reports maintained.
New Zealand	No	Duty to monitor operation of Act as a whole but not to collect data, review or amend EIA system.

system monitoring and review in most developing countries. The first steps in improving EIA system monitoring are clearly to make it a legal requirement and to provide the resources available to undertake it.

Note

1. The majority of environmental impact reports and 'environmental impact assessments' released between 1973 and the end of 1990 are listed by Morgan and Memon (1993), according to type of activity.

CHAPTER 18

Costs and benefits of EIA systems

Introduction

As mentioned in Chapter 1, there has been, as yet, no reliable quantification of the effectiveness of EIA, and it may be that it can only be measured subjectively and qualitatively by examining the attitudes and opinions of those involved. While the existence of firm justification remains scarce, the continued diffusion of EIA requirements around the world demonstrates the prevailing belief that EIA is an effective and efficient environmental management tool. This chapter discusses the various ways of attempting to evaluate the costs and benefits of EIA systems, relying mainly upon the opinions of the participants in the EIA process. A set of evaluation criteria for the costs and benefits of EIA systems is put forward. These criteria are then employed in the analysis and comparison of the EIA systems in the United States, California, the United Kingdom, The Netherlands, Canada, the Commonwealth of Australia and New Zealand.

Costs and benefits of EIA systems

For a variety of reasons, the costs of EIA systems are difficult to distinguish from other costs incurred in obtaining approvals. The chief reason, however, is the integration of EIA into decision-making processes. As Hart (1984, p. 340) has stated:

> The costs associated with [EIA] activities become harder to identify as environmental considerations are better integrated into planning and decision making. Thus, a 'successful' EIS program is defined, in part, by its inability to be evaluated accurately in terms of economic efficiency.

Hart (1984, pp. 348–9) distinguishes four principal elements of the cost of the EIA process:

- document preparation, review, circulation, and administration of the law costs
- delay (inflation and foregone opportunity) costs
- uncertainty costs (due to risk of failure)
- mitigation costs (which increase or may be decreased).

While the costs of programme administration and document preparation, circulation and review are not easy to calculate, they are less difficult than those associated with delay, uncertainty and mitigation.

Apart from the costs involved in preparing the EIA documentation, the proponent may have to pay a fee to the decision-making authority for EIA report review (as in the United States) or may have to pay the authority's review consultants (as in New Zealand). Decision-making authorities may have to maintain a specialist EIA staff unit, or to shift time and resources from other activities. Similarly, consultee organisations and the public will have to expend resources if they are to participate effectively. While the additional costs attributable to EIA are not known, it is widely held that the 'costs of environmental review are generally insignificant when compared to other accepted planning, design and regulatory costs' (Hart, 1984, p. 340). Costs of EIA as a proportion of total project costs generally appear to range from about 0.1 to 1 per cent, with 0.5 per cent being a commonly quoted figure (see, for example, Hollick, 1986). In some instances, as in a study of the costs of the EIA of wastewater treatment facilities carried out by the US Environmental Protection Agency, it has been claimed that the benefits of EIA included 'cost-savings that were the result of project changes prompted by the EIS process' (Council on Environmental Quality, 1990, p. 31).

Closely allied to the question of cost is that of delay. Most EIA systems specify the times within which the various stages of the EIA process should be completed. There may be lengthy periods for public participation at several stages in the EIA process (as in the United States) as well as a specified period during which the decision, based upon the EIA report, should be made (as in the United Kingdom). The time taken by the decision-making authorities frequently exceeds that specified (although this is frequently because of inadequacies in the information provided by proponents) and this is a source of major complaint by proponents. Delays have been a constant source of complaint since the US National Environmental Policy Act came into effect. However, many developers have built considerable lead-times into their project planning procedures to accommodate the EIA process. Others submit 'draft' EIA reports for informal scrutiny and amendment before formal submission to try to avoid the problems of delay caused by complying with later requests for further information. While proponents have complained of delay in EIA procedures, they have frequently not implemented approvals promptly once they have been granted. Sewell and Korrick (1984) found that 15 per cent of a sample of approved 100 projects were abandoned and 32 per cent were still on-going six years later. They confirmed the findings of an earlier US study of delays by suggesting that EIA imposed little or no significant time penalty on proponents.

Nevertheless, despite the absence of reliable information about expenditures, delays remain a recurring source of anxiety in many jurisdictions (for example, in the Commonwealth of Australia). These concerns are frequently coupled with complaints about 'moving goal-posts': unreasonable requests for further information or for changes to the design of the action. One method of overcoming these problems is the agreement of an action-specific timetable between the proponent and the decision-making authority which allows for the submission of further information only under specified circumstances.

EIA is intended to improve the quality of decisions having environmental implications by amending the behaviour of proponents, consultants, consultees, the public and the decision-making authorities. Examples of such changes in behaviour include an increase in public participation in decision making and increased coordination between the authorities responsible for environmental protection. It is generally accepted that such changes take time but that they have taken place in the more mature EIA systems (see, for example, in relation to the United States: Wandesforde-Smith and Kerbavaz, 1988; Caldwell, 1989c; Council on Environmental Quality – CEQ, 1990).

The crucial question in relation to the efficacy of EIA is whether the quality of decisions has actually increased and whether they have become more acceptable as a consequence of its use. Examples of the effectiveness of EIA would include the increased use of modification or mitigation (Chapter 16), the use of more stringent conditions upon permissions and the refusal of proposals which might previously have been approved. Once again, it is very difficult to obtain concrete evidence of such changes, though the US Environmental Protection Agency study mentioned above, as well as demonstrating cost savings and improved opportunities for public participation in the decision-making process, showed that EIA 'was effective in (1) causing major changes in projects, (2) providing more protection for the environment' (CEQ, 1990, p. 31). It is, therefore, generally necessary to rely on participants' opinions about the effectiveness of EIA systems in improving the environmental quality of decision making. Such opinions will, no doubt, rely on anecdotal evidence from particular examples of the use of EIA, some of which have found their way into the literature (see, for example, CEQ, 1990).

Since it should be the aim of any EIA system not only to minimise costs but to maximise environmental benefits, a final question about the effectiveness of EIA must be posed: is there any evidence that EIA has led to any improvement in the quality of the environment generally? Since EIA is only one of an array of environmental management measures, it is extremely difficult to distinguish its effect from those of other anticipatory controls, pollution controls, environmental standards, environmental designations, etc. It is, therefore, doubtful whether evidence of general environmental improvement attributable to EIA (as opposed to anecdotal evidence of the effectiveness of EIA in improving, or preventing the deterioration of, a particular local environment) can

Box 18.1   Evaluation criteria for the costs and benefits of EIA systems

---

Are the financial costs and time requirements of the EIA system acceptable to those involved and are they believed to be outweighed by discernible environmental benefits?

Do the financial costs of the EIA process to proponents, consultees, the public and the decision-making authorities exceed those which would have been incurred in any event?

Do the times required to complete the various stages of the EIA process exceed those specified?

Do the participants in the EIA process believe that it has altered the behaviour of proponents, consultants, consultees, the public and the decision-making authorities?

Do the participants in the EIA process believe that the environmental quality and acceptability of decisions are improved by it?

Does empirical evidence exist that the EIA process has significantly altered the outcome of decisions?

---

ever be adduced. This does not mean that this ultimate criterion should not be advanced as a yardstick, simply that it is likely to remain theoretical. It has therefore not been included with the other criteria summarised in Box 18.1. These criteria are now employed to analyse the costs and benefits of each of the seven EIA systems.

United States of America

The costs of EIA in the United States are substantial, and most exceed those that would have been incurred had the National Environmental Policy Act 1969 (NEPA) never been passed. Despite the fact that some environmental impact statements (EISs) cost hundreds of thousands (and occasionally millions) of dollars, the cost of EIA is generally seen as 'part of the cost of doing business'. This, perhaps, is why authoritative EIA costs are so elusive: they may be inextricably tied to other related costs.

While the Council on Environmental Quality (CEQ) has only a very small staff, total employment in EIA must run to several thousands in the United States. The Environmental Protection Agency alone employs several hundred personnel to meet its EIA commitments.

Generally, EIAs take longer to complete than the times specified in the Regulations. However, many EIAs take no more than 12–18 months, though occasional cases may take 30 months from initiation to the record of decision. The participants in the EIA process have adapted to these times, but proponents still resent unexpected delays. The participants in the EIA process firmly believe that their own behaviour and that of others has been affected by the EIA process. There have been numerous independent confirmations of these changes of behaviour, most notably in the study by Taylor (1984). The participants in the EIA process also believe that both the environmental quality and the acceptability of decisions have been improved by NEPA, though few are entirely satisfied with the process or the product. In particular, it is widely felt that projects are now much better designed and impacts better mitigated than was previously the case (Dickerson and Montgomery, 1993) and that certain projects have been abandoned as a result of EIA. The various participants in the EIA process have always supported it when its future was threatened (e.g. during the Reagan regime – Chapter 2).

Despite this almost universal view, it is not possible to produce unambiguous evidence that the EIA process has significantly altered the outcome of decisions. EIA has been in being for a quarter of a century and parallel changes in environmental management have made it impossible to unravel the effect of EIA from all the other factors determining the outcome of decisions. In the last analysis, as explained in Chapter 1, the opinions of the participants in the EIA process are probably the only measure of success or failure.

Bear (1988) found very little empirical evidence of the effect of EIA on decision making. She cited reduced EIA litigation as one indicator of success, but this is, at best, ambiguous. Clark (1993, p. 4) was remarkably frank about CEQ's view of the success and failures of EIA:

> Certainly, many environmental impact statements (EISs) are too long, take too long to prepare, cost too much, and many times do little to protect the environment. Some EISs are prepared to justify decisions already made, many agencies fail to monitor during and after the project, some agencies do not provide adequate public involvement, and few agencies assess the cumulative effects of an action.

Others have criticised weaknesses in the treatment of cumulative impacts, biological diversity and global climate change in EIA (e.g. Hildebrand and Cannon, 1993).

Yost (1990) believed that the EIA provisions in NEPA are successful because they are essentially procedural, but that the Supreme Court has undone the promise of NEPA. Caldwell (1989b, p. 19) took a similar view:

> The purpose of NEPA was not ... the preparation of environmental impact statements ... [but] to adopt a national policy for the environment ...

Caldwell (1989b) argued that, while environmentally destructive projects are still being proposed, NEPA 'has worked to reduce the extent of officially sponsored environmental damage' (p. 22). This view is almost unanimous: the benefits of EIA in the United States are generally perceived considerably to outweigh its costs.

California

Environmental impact assessment is big business in California. As mentioned in Chapter 2, several thousand professionals make a living from EIA in the state. There appears to be no doubt that there are considerable additional costs associated with the operation of the California Environmental Quality Act (CEQA), though it is extremely difficult to distinguish EIA costs from other project appraisal costs (Chapter 17).

Generally, the time taken to undertake an EIA in California exceeds the six months specified, especially for large private projects. It takes over the permitted 12 months to reach a decision on many private projects requiring an environmental impact report (EIR). This is widely seen as too lengthy (especially by developers) but as a necessary, if unwelcome, cost of EIA. This may be a consequence of the fact that many EIA professionals believe that legal defendability, rather than the disclosure of environmental information, often drives the preparation of the EIR (Kaplan-Wildmann and McBride, 1992). This may also account for the over-long and descriptive nature of many EIRs.

There seems to be a general consensus that EIA in California has affected the behaviour of most of the actors in the process. Generally, the EIA procedures appear to be followed as a result of court challenges by environmental groups. The approach of these groups, of EIA professionals in most agencies, and of many developers has been modified as a result of the CEQA provisions but there appears to be little evidence of a change in attitude to the environment among local political decision-makers. These crucial players seemingly still take decisions in which the environment plays little part. It is perhaps as well that California has such active environmental groups to insist on the enforcement of CEQA.

Many projects have been modified and some stopped by legal actions under CEQA. Californian EIA has increased the acceptability of decisions by making them more transparent and participative. Olshansky (1991) found that planners in California believe that CEQA has resulted in a more thorough analysis of environmental impacts, leads to the adoption of mitigation measures, informs the public about environmental issues and, overall, protects the environment. Many EIA professionals in California believe that CEQA affects more decisions than NEPA. While there is no evidence that EIA has led to an improvement in the overall quality of the environment, it has clearly resulted in many project-level environmental successes. As elsewhere, it is impossible to distinguish the environmental benefits of EIA from other factors, especially as EIA is not formally integrated with other environmental quality laws, such as those on air and water quality.

Overall, there appears to be little doubt that the Californian system, like the federal one, has proved effective in preventing some developments from taking place which would have caused adverse environmental consequences and in modifying the nature of other projects to make them more environmentally acceptable. (Recent court decisions have restricted CEQA's ambit and decreased somewhat the chance of using the Act to prevent or postpone development.)

These benefits, most would argue, outweigh the costs of CEQA, if only marginally. The situation was probably justifiably exaggerated by the Planning and Conservation League (and others) (1993) who were fighting (successfully) to prevent any dilution of the CEQA requirements by its opponents:

> CEQA costs are almost always very small in comparison to project costs. Complying with CEQA often results in unknown environmental costs being revealed. ... Not undertaking CEQA review poses risk of much greater long-run cost to the state and its citizens.

United Kingdom

Under the Planning Regulations, most of the cost of environmental assessment (EA) is borne by the developer and by local planning authorities (LPAs). Of a sample of 24 cases, just under half of the developers prepared the environmental statement (ES) themselves, although some used consultants for specific aspects (e.g. noise, ecology). Consultants were used in the remainder of cases and most appeared to charge £10,000–£50,000 for their services. Nearly two-thirds of developers felt that consultants would have been employed in the absence of EA, but not necessarily to the same extent. Overall, nearly two-thirds of developers felt that EA had caused a slight increase in their costs of obtaining planning permission. One-fifth of LPAs used consultants to evaluate the ES, their costs ranging from less than £1000 to over £20,000 in addition to staff time (Wood and Jones, 1991; see also Coles, 1991).

The Regulations extend the time allowed to the LPA to reach a decision from 8 weeks to 16 weeks. Of the 20 applications in the sample where a decision is known to have been reached, one-half were decided in less than the required 4 months, one-fifth took 4–6 months, one-tenth 6–8 months and one-fifth 8–10 months. Just over half of the planning officers questioned were of the opinion that EA had made no difference to the time taken to decide planning applications. However, two-thirds of the developers and half of the consultants thought that EA had caused a delay to the decision time (Wood and Jones, 1991).

There is little evidence, to date, that EIA has significantly altered the outcome of decisions. However, in general, it is believed that the benefits of EIA in the United Kingdom outweigh its costs. If EIA has not yet greatly increased public participation it does appear to have led to an increase in coordination between the relevant agencies. Most of the participants in the 24 cases studied, particularly LPAs and developers, expressed the opinion that EA was a worthwhile and helpful procedure which made a positive contribution to informed decision making in the planning system. Further, several of these participants remarked that practice in EA was improving (Wood and Jones, 1991).

The Netherlands

There is no doubt that the substantial sums of money spent by the government, by proponents and by third parties in the EIA process considerably exceed those which would have been expended on proposals in any event. However, extra costs tend to impinge most on small projects: the additional costs of EIA for large projects are small and there are no proponent fees for EIA. There have been no significant complaints about the cost of EIA in The Netherlands. In some instances it is felt that EIA may have saved money by demonstrating that certain lower cost alternatives or mitigation measures may be environmentally preferable.

In general, the time limits imposed on the EIA Commission are met. There have been cases where guideline recommendations have been late but significant slippage is rare. The main delays in the EIA process result from the activities of the developer (especially if supplementary information is required) and the competent authority. There has been a reduction in the number of objections to projects, thus saving time on the appeal process. Indeed, some EISs have been submitted voluntarily as a means of overcoming objections to projects. This is seen as an indication of the success of the EIA process. The Evaluation Committee on EIA felt that EIA often resulted in a streamlining of existing procedures.

It is widely believed that the EIA system has changed the behaviour of the participants but this is very difficult to measure. There is no empirical evidence to prove that EIA has altered the outcome of decisions, especially if this is measured in terms of the number of cancellations of projects. However, some projects have been cancelled and in other cases a less damaging alternative has been chosen. In almost every case, further consideration has been given to the environment than would have been given without EIA.

The outcome of the EIA process is therefore often different from the initial proposal. As the Evaluation Committee on EIA (1990, p. 6) put it, EIA contributes a great deal of environmental information to the decision-making process and:

> The procedure obliges the initiator of a proposed activity and the competent authority to give more intensive consideration to the effects on the environment at an earlier stage.

Some observers feel than an element of EIA fatigue may be setting in among competent authorities and that changes to proposals may be becoming more peripheral. As elsewhere, the most important factor is the attitude of the participants. Where developers and competent authorities take a positive approach to EIA real benefits often ensue.

Canada

There is no doubt that the financial costs of the EA process in Canada are very considerable. While some EA costs would have been incurred in any event and while costs for larger projects are thought always to be far less than 1 per cent of total costs, they have sometimes run into tens of millions of Canadian dollars. A proportion of this cost has been attributable to the public hearings and to participant funding but the major component has been the cost of preparing EA reports, which has frequently exceeded $CD1 million.

Large public and private teams are dedicated to the EIA industry in Canada: government officials, consultants, developers, researchers and academics are involved. The level of resource devoted to EIA per head in Canada is probably the highest in the world: the Federal Environmental Assessment Review Office (which has become the Canadian Environmental Assessment Agency) employs approximately 90 staff (see Chapter 6), Environment Canada over 20, provinces such as Ontario over 60 and Crown corporations such as Ontario Hydro over 40.

No time limits are specified under the Environment Assessment and Review Process (EARP) or under the Canadian Environmental Assessment Act (CEAA). Typical average times for scoping hearings for panel review projects are about 2 weeks and for the hearings on the project about 2–6 weeks. The typical panel review under EARP takes 12–18 months (with EIS preparation taking most of the time). The briefest panel review has taken less than a year, but the longest (into existing low level military flying in northern Canada) has taken over 7 years as a result of proponent procrastination. Canadian EA is sometimes very expensive and time consuming. Delays are widely seen as the most serious problem associated with EARP. On the other hand, initial assessments under EARP have generally been very quick and inexpensive (but often inconsistent and inadequate).

There is virtual unanimity that EA has altered the behaviour of the participants in the EA process. However, the performance of proponents has not always changed, especially in the initial assessment phase. Weston (1991, p. i) reported that:

Notwithstanding its original intent, with its emphasis on self-assessment, EARP has not significantly altered the basis upon which federal government departments make decisions.

In the panel review process, however, proponents and federal departments have had to think much more about alternatives and about mitigation measures than was previously the case.

The expertise of consultees, especially of federal departments like Environment Canada, has increased substantially. The same is true of environmental consulting firms. The behaviour of the public has certainly changed. The public has become increasingly professional and adept in using participant funding to mount a strong case to argue for the adoption of an alternative or a particular mitigation measure. As a result, the acceptability of decisions which implement panel recommendations has increased.

Gibson (1993, p. 23) believed that EIA has:

evolved from regulatory origins in narrowly focused largely technical impact assessment requirements imposed on already selected projects into an increasingly broad, open, and anticipatory approach to the integration of environmental considerations throughout the planning of a wide variety of undertakings.

The Canadian federal EA system probably falls mid-way along this continuum. Anecdotal evidence of the benefits of federal EA abound in Canada but almost all relate to the formal panel review process and not to self-assessment. The quality of the constructed project is generally believed to be better as a result of changes occasioned by the EA process. While a few projects have been abandoned or found to be unnecessary as a result of EA, its main benefit has been in mitigating the environmental effects of proposals.

One of the main problems of EIA in Canada has been overlap between the requirements of the federal and provincial EA systems and the confusion, conflict, duplication and delay this engenders. CEAA attempts to reduce this inefficiency by providing for inter-jurisdictional agreements leading to, for example, joint panel reviews. The first environmental assessment harmonisation agreement, between Canada and Alberta, was signed in 1993/94. Overall, it is clear that there is considerable scope for CEAA to improve the efficiency and effectiveness of EA in Canada. As Gibson (1993, p. 21) has stated: 'the greater clarity and predictability of the new federal requirements can be expected to reduce such problems'.

Commonwealth of Australia

There can be no doubt that the financial costs of EIA exceed those which would have been incurred in any event. Significantly, however, there is no debate about the costs of the EIA process in a system where an EIA for a highly significant project can cost more than $A1 million and where 30 governmental staff are employed on EIA at Commonwealth level alone. As one authoritative review put it:

> The direct compliance cost of the assessment process is not a significant problem for large companies, especially if environmental assessment is integrated with feasibility studies. These costs represent only a small proportion of total project costs.
>
> (Bureau of Industry Economics, 1990, p. v)

The duration of the EIA process is the main cause of complaint. There are several stages where delays can arise, even though no time periods for them are specified (for example, scoping, informal review, evaluating responses to the draft EIS, etc.). The one time period specified in the Administrative Procedures: 28 days for the preparation of the Minister's recommendations following receipt of the final EIS (Commonwealth of Australia, 1987, para. 9.4) is sometimes exceeded.

These delays are exacerbated by uncertainty about whether or not an EIS or public environment report will be required. Delays can be expensive if proponents fail to plan adequately for EIA or if belated demands are made by government (Bureau of Industry Economics, 1990).

There have also been substantial criticisms of the overlap between

Commonwealth and state EIS systems but these have diminished as a result of the activities of the Australia and New Zealand Environment and Conservation Council (1991). These have resulted in a draft National Agreement on EIA which was designed to eliminate duplication between jurisdictions and which was expected to be ratified during 1994/95.

As mentioned in preceding chapters, as well as complaints about lack of certainty there have also been criticisms of the EIA system in relation to coverage, screening, scoping, review and monitoring. Despite these weaknesses, there is little debate about the overall value of the Commonwealth EIA system, which is perceived to have improved the prior evaluation of project environmental impacts, principally because it has raised the level of awareness and knowledge about the environment among agencies and the public alike (Australian and New Zealand Environment and Conservation Council, 1991). This increased awareness is widely believed to have altered the behaviour of the main proponents in the EIA process, though not, perhaps, of the political decision-makers, and to have led to environmentally better and more acceptable actions.

As elsewhere, however, empirical evidence of decision modification is almost impossible to obtain. It is necessary, as usual, to rely upon opinions. The Ecologically Sustainable Development Working Group Chairs (1992, p. 213) have summarised the Commonwealth and state EIA situation succinctly:

> Recommendations for improving specific aspects of the EIA regime ... included the need for greater public involvement, extension to some areas not already covered (for example, mineral exploration activities), the adoption of more effective conflict resolution procedures and the need for a more structured follow up of conditions of approval through formalised environmental monitoring and auditing arrangements where these do not already exist. Nevertheless, ... given a more effective better resourced approach to evaluating EIAs, the present system does not require fundamental structural change.

New Zealand

It is fortunate, given the level of inexperience in the operation of New Zealand's new EIA system, that there is virtual unanimity among environmentalists, business, government, local councils and academics about the merits of EIA (Wood, 1993a). Morgan (1993) reported similarly positive attitudes towards environmental assessment (EA), despite problems with the numerous requirements of the Resource Management Act 1991 (RMA):

> interviewees were almost unanimous in their support for EA and the principles in the RMA on which EA is built. It is seen as an essential part of an integrated approach to planning natural resources and

263

managing the environment in New Zealand. Consequently, most councils seem to be tackling the EA issue in a positive, supportive manner, rather than resisting it.

It is apparent that most participants in the EIA process believe that the benefits of EIA outweigh the costs. While there is little empirical evidence available about the financial costs of EIA, experience with the Environmental Protection and Enhancement Procedures indicates that they are significant. Equally, Morgan (1993) has reported that regional councils have had considerable difficulties in meeting the detailed time constraints imposed by the Act. It is perhaps too early to comment definitively on behavioural changes brought about by the new all-pervasive EIA system but it is apparent that the attitude of many local council officials (Dixon, 1993b) and elected members (Morgan, 1993) to EIA shows considerable scope for further development. It is widely believed in New Zealand that the Mark I EIA system did alter the outcome of many decisions and the hope now is that the much more widely applicable Mark II EIA system will do the same.

Table 18.1 The costs and benefits of the EIA systems

Criterion 13: Are the financial costs and time requirements of the EIA systems acceptable to those involved and are they believed to be outweighed by discernible environmental benefits?

Jurisdiction	Criterion met?	Comment
United States	Yes	Virtually unanimous view by proponents, consultees and the public that benefits of EIA exceed its substantial time and other costs.
California	Yes	Costs and time requirements high but outweighed (for most participants) by improved project mitigation measures.
United Kingdom	Yes	Consensus (but not unanimity) as to utility of EA in improving project mitigation measures.
The Netherlands	Yes	Virtually unanimous belief that benefits of EIA outweigh its financial and time costs.
Canada	Yes	Costs and (especially) time requirements of panel reviews under previous EA system often high but significant mitigation has occurred: benefits of initial assessment less clear. Benefits of new EA system should be greater.
Australia	Yes	Complaints about uncertainty and delays generally outweighed by belief that EIA delivers real environmental benefits.
New Zealand	Yes	Virtual unanimity of view that benefits of EIA system outweigh costs but considerable unfamiliarity remains.

Summary

It is, perhaps, a testament to the inherent effectiveness and efficiency of EIA that, despite the marked differences in each of the seven EIA systems, there should be a unanimity of view that the benefits of all the EIA systems outweigh their costs (Table 18.1). It is nevertheless quite clear, from the criticisms of the seven EIA systems that have been listed in this chapter and earlier in the book, that the effectiveness of EIA can be substantially improved (Chapter 20). It is therefore unsurprising that because many developing country EIA systems are, as yet, less effective than those of the seven EIA systems, there should be no general agreement that the costs of EIA are exceeded by its benefits in these countries, despite these time and financial costs being relatively low (see Appendix).

It is noticeable that complaints about delays in project approvals as a result of EIA were most vociferous in the United States, California, Canada and Australia. These are, of course, the jurisdictions with the most formalised EIA systems and those which require proponents to undertake the greatest number of EIA steps. However, while complaints about delay are probably least in the UK and New Zealand EIA systems, they do occur in these jurisdictions, and in The Netherlands.

It is clear that delay is the major criticism in most EIA systems. Delay is the main reason why some members of the development industry in, particularly, California and the Commonwealth of Australia do not share the general view about the net benefits of EIA. The most significant advance towards the unanimous acceptability of EIA could be made by reducing the delays engendered by it and by explaining fully that these delays are generally attributable much more to proponents than they are to EIA agencies. Needless to say, such efficiency gains must not be made at the expense of EIA effectiveness.

CHAPTER 19

Strategic environmental assessment

Introduction

It is apparent from Chapter 18 that the benefits of project EIA are generally believed to considerably outweigh its costs. The widespread acceptance of the utility of EIA in improving the quality of decisions about proposed projects has led to active consideration of, and some practice in, strategic environmental assessment (SEA) or the environmental assessment of policies, plans and programmes. This is a consequence of the growing belief that project EIA may occur too late in the planning process to ensure that all the alternatives and impacts relevant to sustainability goals are adequately considered.

This chapter explains the nature of, and the need for, SEA. It then reviews the current provisions for, and practice in, SEA in the United States, California, the United Kingdom, The Netherlands, Canada, the Commonwealth of Australia and New Zealand.

Environmental assessment of policies, plans and programmes

Generally, there exists a tiered forward planning process which starts with the formulation of a policy at the upper level, is followed by a plan at the second stage, and by a programme at the end. A policy thus may be considered as the inspiration and guidance for action, a plan as a set of coordinated and timed objectives for implementing the policy, and a programme as a set of projects in a particular area. The tiered system can apply at the national level and also may apply at regional and local levels (Fig. 19.1). It can apply to sectoral actions and to physical planning actions (Wathern, 1988c; Wood, 1988a; Lee and Walsh, 1992; Therivel *et al.*, 1992; Wood and Djeddour, 1992; Glasson *et al.*, 1994; Sheate, 1994).

Partly because the advantages of project EIA are so widely recognised,

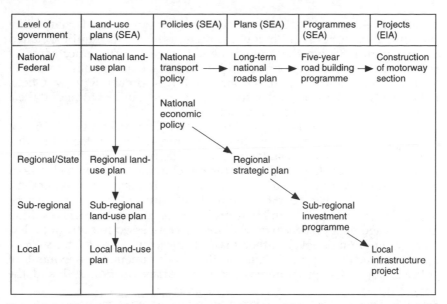

Level of government	Land-use plans (SEA)	Policies (SEA)	Plans (SEA)	Programmes (SEA)	Projects (EIA)
National/ Federal	National land-use plan	National transport policy ⟶ National economic policy	Long-term national ⟶ roads plan	Five-year road building ⟶ programme	Construction of motorway section
Regional/State	Regional land-use plan		Regional strategic plan		
Sub-regional	Sub-regional land-use plan			Sub-regional investment programme	
Local	Local land-use plan				Local infrastructure project

Fig. 19.1 Chronological sequence of actions within a comprehensive EIA system.
Sources: Adapted from Lee and Wood (1978a); Lee and Walsh (1992).

the desirability of taking the environment into account earlier in the planning process has been generally accepted throughout the world.[1] Thus, the World Conservation Strategy pinpointed the need to integrate environmental considerations with development in 1980 (International Union for Conservation of Nature and Natural Resources, 1980). This theme has become an accepted part of World Bank (1987) policy, which stated that environmental issues must be addressed as part of overall economic policy rather than project by project. The same philosophy was echoed in the Brundtland report (World Commission on Environment and Development, 1987) and at the United Nations summit on the environment held in Rio de Janeiro in 1992. The United Nations Economic Commission for Europe (1992) has recommended the extension of EIA principles to policies, plans and programmes. The European Commission has long espoused the desirability of extending EIA from projects to higher tiers of action and began consultations on a SEA directive in 1991 (Chapter 3).

In the United States, where the National Environmental Policy Act 1969 provided for SEA from the outset, there has been an upsurge in interest in programmatic environmental impact statements (Sigal and Webb, 1989; Webb and Sigal, 1992). California probably has the most developed SEA system in the world and several hundred SEAs have been undertaken to date. An advantage of assessing the impacts of higher tier actions in California has been that certain aspects of subsequent projects have not then needed to be assessed in detail (Bass, 1991).

In Canada, submissions to Cabinet must contain an assessment of any

267

significant environmental effects. While it is anticipated that many submissions will not require any comment, others will require detailed annexes setting down the environmental impacts. Announcements of new policies are to be accompanied by the release of a public statement on their impacts (Federal Environmental Assessment Review Office, 1993b). In New Zealand local authorities are now obliged to undertake a form of SEA on their plans and policies.

There are several reasons for the perceived need for SEA. SEA can increase the weight given to the environment in decision making, facilitate and increase consultation and participation on environmental matters and establish principles for the development of certain classes of project. Alternative approaches, cumulative impacts and synergistic impacts (which may be cross-sectoral in nature), ancillary impacts, regional or global impacts and non-project impacts (e.g. impacts resulting from management practices) may all be better assessed initially at policy, plan or programme level, rather than at the project level. The various arguments advanced in favour of the environmental assessment of policies, plans and programmes are summarised in Box 19.1 and the

Box 19.1 Potential benefits of strategic environmental assessment

Encourages the consideration of environmental objectives during policy, plan and programme-making activities within non-environmental organisations;

Facilitates consultations between authorities on, and enhances public involvement in, evaluation of environmental aspects of policy, plan and programme formulation;

May render some project EIAs redundant if impacts have been assessed adequately;

May leave examination of certain impacts to project EIA;

Allows formulation of standard or generic mitigation measures for later projects;

Encourages consideration of alternatives often ignored or not feasible in project EIA;

Can help determine appropriate sites for projects subsequently subject to EIA;

Allows more effective analysis of cumulative effects of both large and small projects;

Encourages and facilitates the consideration of synergistic effects;

Allows more effective consideration of ancillary or secondary effects and activities;

Facilitates consideration of long range and delayed impacts;

Allows analysis of the impacts of policies which may not be implemented through projects.

Source: Wood and Djeddour, 1992, p. 7.

direct and indirect impacts of higher order actions are shown in Fig. 19.2.

The main elements of the EIA process and its most tangible output – the EIA report – are, in principle, applicable to all levels of decision making, including policies, plans and programmes. Notwithstanding, in

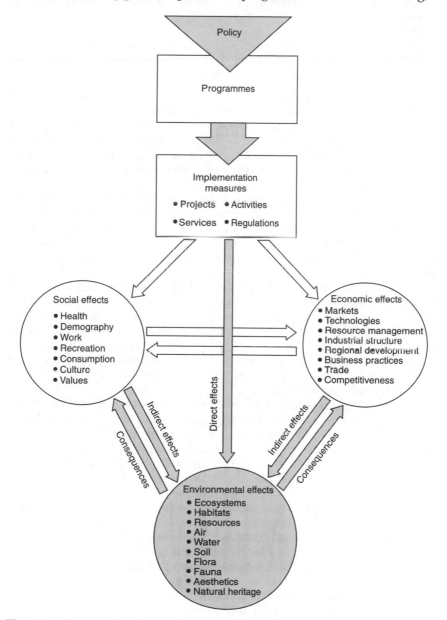

Fig. 19.2   Direct and indirect effects of policies and programmes.
Source: Federal Environmental Assessment Review Office (1994).

practice it is likely that the scope and purpose of the environmental assessment of policies, plans and programmes will be different from that of projects in five main ways:

1. The precision with which spatial implications can be defined is less.
2. The amount of detail relating to the nature of physical development is less.
3. The lead time is greater.
4. The decision-making procedures and the organisations involved may differ, requiring a greater degree of coordination.
5. The degree of confidentiality may well be greater.

These variations indicate that the nature of strategic environmental assessment will differ in detail from the nature of EIA projects. It is apparent that the likely significant environmental impacts of higher order actions can only be assessed at a more strategic level. Indeed, it is probable that, in many instances, the detailed nature of the environmental assessment of different types of policies, plans and programmes will also differ.

When certain alternatives and significant environmental impacts cannot be adequately assessed at the project level, it may well be possible to assess them at the programme, plan or policy level, utilising a form of SEA basically similar in nature to that employed for projects. Thus, SEA would involve screening, scoping, prediction, consultation, public participation, mitigation of impacts and monitoring (UN Economic Commission for Europe, 1992). SEA could deal with the alternatives and significant impacts not covered adequately at project level in more detail than other types of impacts. These last might be assessed only very generally, or not at all. The assessment of alternatives and significant impacts need not be confined to one level of action. Different types of alternatives or different types of impact can be assessed at different tiers in the planning process, provided that such assessment takes place as early as feasible in this process.

If alternatives are adequately assessed in project EIA, and if all the various significant impacts are adequately examined, then there is no need to carry out a SEA. Almost by definition, however, this is seldom the case. In many countries where there already is a project-level EIA system, the most sensible course of action might be to supplement these EIAs with higher tier SEAs largely confined to issues, such as cumulative impacts, which cannot be adequately assessed at the project level.

It is sometimes argued that there are severe methodological problems associated with SEA. These are supposedly related to difficulties of predicting impacts, lack of definition, monitoring of on-going environmental change, absence of specific SEA methods and consultation and participation. It is true that SEA methodologies are not well developed (Therivel *et al.*, 1992, p. 73). However, nearly all the tasks involved in SEA are similar to those in project-level EIA. It follows that many of the methods employed are directly transferable, though many will differ in detail and in level of specificity.

As with project EIA, the skill of the assessor comes to bear in selecting an appropriate mix from all the different approaches, tools and techniques available. Box 19.2 lists the key considerations which it has been suggested that assessors should bear in mind in designing their approach to SEA. They could equally apply to project EIA.

An EIA system which covers higher tier actions could apply to the significant environmental impacts of policies, plans and programmes relating to the various sectoral activities, especially (but not only): agriculture, forestry, fishing, extractive industry, energy industry, manufacturing industry, transport, non-transport infrastructure, housing, the environment, and recreation and tourism. It could also relate to policies, plans and programmes concerning research and development generally. These policies, plans and programmes must, for reasons of practicality, be restricted to those prepared by or for public or quasi-public bodies or those requiring approval by a public body.

Many alternative arrangements of juxtaposed land uses, and some significant synergistic and cumulative impacts, cannot be satisfactorily considered in sectoral or project environmental assessments, because of the effects of new activities in other sectors, or because of the cumulative effects of many activities not subject to project EIA. They can be considered in the SEA of land use and other spatial plans. Indeed, it could be argued that the environmental assessment of land use plans is the

---

Box 19.2 Key considerations in choosing SEA techniques

1. Will this technique or approach help achieve the objectives of this step of the process? What is the best technique at this stage for:
   - identifying linkages?
   - estimating and forecasting effects and consequences?
   - assessing significance?

2. Does the magnitude and potential significance of the impacts warrant the level of effort required by the technique?
   - cost?
   - timing?
   - involvement of key personnel?
   - involvement of peers, outside experts and public stakeholders?

3. Is it possible and practical to utilise techniques under consideration?
   - are peers, experts and stakeholders available and willing to participate?
   - do adequate and reliable data exist?

4. Are there any other factors that may influence selection of approaches and techniques?
   - strictures of confidentiality?
   - skill levels and capacity to design and implement given techniques?
   - personal preferences of parties involved?

Source: Federal Environmental Assessment Review Office (1994)

most easily implementable of all types of SEA. Certainly, in California most SEAs have been related to land use plans.

While many jurisdictions possess permissive powers for SEA to be undertaken, few outside the United States possess any clear triggering mechanism. Very few SEAs have related to policies, as opposed to plans or programmes (Therivel *et al.*, 1992, p. 71). There have been few attempts to enforce the SEA provisions in the United States in relation to policies and most other jurisdictions possess powers which are discretionary in application (if they have any SEA powers at all). To date, there has been little evaluation of the costs and benefits of SEA, very little SEA training has been undertaken, little research into SEA appears to have been commissioned and little SEA guidance has been issued. As a New Zealand/Australia/Canada meeting on the environmental assessment of policy concluded:

> The lack of action probably stems from the fact that too many people see the EA of policy as being 'too hard'. People not accustomed to thinking in environmental terms shy away when confronted with the idea.
>
> (Fookes, 1991, p. 5)

However, the methodological difficulties in undertaking SEA (above) appear to be secondary to political difficulties, namely the reluctance of politicians and senior bureaucrats in powerful departments voluntarily to cede any role in the making of decisions to external environmental authorities by permitting the SEA of their activities. Bregha *et al.* (1990) concluded that a number of barriers constrained the EA of policy:

> A lack of clear objectives, insufficient political will, the narrow definition of issues, the existing organisational structure, absence of accountability, bureaucratic politics, lack of information and absence of incentives.

To overcome these barriers, it appears that some type of action-forcing mechanism, or a formal framework to ensure that the SEA process works, is needed. However, for SEA to take place, it appears that there needs to be both a specific and unambiguous requirement to undertake it and either a legal system to enforce it (as in the United States or California) or an environmental authority with sufficient strength of purpose, and professional competence, to ensure that the requirement is discharged (as in Canada).

A number of other steps are also necessary to overcome these barriers:

- Increasing the general understanding of SEA: for example, the types of actions to which SEA could usefully be applied and its relation to existing EIA and sustainable development policies.
- Clarifying procedural issues: for example, at which decision points in a planning process should SEA be applied, how should SEA findings be integrated with other policy and planning considerations in decision making?
- Clarifying methodological issues by adapting existing methods (including EIA methods) for SEA use.

- Strengthening the capacity for the practical application of appropriate SEA methods: for example, undertaking 'trial runs', diffusing examples of good SEA practice, preparing SEA guidance and providing training in its use.
- Reviewing existing environmental data sources to assess their potential use in SEA and prioritising measures for correcting any deficiencies.

In many ways, the current situation regarding SEA resembles that before and during the early years of EIA implementation when similar fears about delays, duplication and difficulty were raised. Experience has shown that project level EIA is feasible, that EIA has altered decision making to give more weight to the environment and that EIA costs very little in relation to the costs of implementing the actions assessed and that it is widely believed to have delivered net benefits. There is every likelihood that SEA, if implemented judiciously and at the appropriate level in the various planning processes, would establish itself, in the same way as EIA, as a cost-effective tool of environmental management.

The current provisions for, and practice in, SEA in the seven jurisdictions are now reviewed.

United States of America

The US National Environmental Policy Act of 1969 (NEPA) requires EIA for 'major Federal actions significantly affecting the quality of the human environment' (Box 2.1). The term 'major Federal action' has subsequently been defined in the Council on Environmental Quality (CEQ) Regulations as including projects and programmes, rules, regulations, plans, policies, or procedures; and legislative proposals advanced by federal agencies (40 CFR 1508.18). SEA procedures are not distinguished from project EIA procedures in the United States. The EISs for policies, plans and programmes are called variously programmatic, regional, cumulative or generic EISs or sometimes simply EISs (Sigal and Webb, 1989).

CEQ indicated as early as 1972 that programmatic EISs should be prepared for federal programmes that might involve numerous actions to ensure that cumulative impacts were addressed. Such EISs would be broad in nature and cover 'basic policy issues that would not have to be repeated in subsequent impact statements for individual actions within a program' (Sigal and Webb, 1989). This was subsequently termed 'tiering'. Tiering involves the preparation of an EIS to cover general issues, and particularly alternatives, in a broader policy or programme-activated analyses. EISs at subsequent stages then incorporate by reference the general discussions from the broader EIS, while concentrating on the issues specific to the action being evaluated (Bass and Herson, 1993a, pp. 79–80).

The CEQ Regulations and published advice on EIA (CEQ, 1981a) deal specifically with tiering. Programmatic EISs may be required for broad federal actions which can be grouped geographically (e.g. covering a metropolitan area), generically (e.g. actions having similar methods of implementation) or by stage of technological development (e.g. federally assisted research on new energy technologies) (40 CFR 1502.4). Programmatic EISs are considered to be particularly relevant for actions that are complex or give rise to environmental effects of unknown extent where the comparative analysis of alternatives can highlight potential problems (Mandelker, 1993b, Chapter 9.02). An example of geographical tiering would be the preparation of a forest plan EIS, followed by a watershed programme EIS for part of the forest and then an individual timber harvest or road EIS or environmental assessment within that watershed (Bass and Herson, 1993a). The programmatic EIS should be prepared at a point in the agency planning process when it can highlight potential environmental problems and allow a wide range of alternatives to be evaluated.

Sigal and Webb (1989) estimated that about 325 programmatic EISs were issued between 1979 and 1987 (around 36 per annum), of which the majority related to policies and programmes but a significant minority were regulatory. The largest categories were resource management (40 EISs), pest control (32) and flood control (30). Other important types of programmatic EIS types related to wilderness, permits, water development, technology development and mineral/timber decisions.

Webb and Sigal (1992) gave brief outlines of the programmatic EISs prepared for the Mount Baker – Snoqualamie National Forest Land and Resource Management Plan, for the seven hydro-electric projects proposed in the Owens River Valley in California and for the destruction of deteriorating chemical agents and munitions stored at eight locations throughout the United States. While inevitably somewhat general, and containing less detail and quantification than project EISs, the quality of programmatic EISs is undoubtedly increasing. Webb and Sigal (1992) stressed the role of public participation and, especially, public meetings in the SEA process. They pointed out that although SEA practice is developing, it still has not been formally applied at the broadest level of policy.

Webb and Sigal (1992) pointed out that agencies have frequently resisted preparation of programmatic EISs because of perceptions of cost and lack of understanding of timing and scope. However, they believed that there was a growing realisation that a well-prepared, timely programmatic EIS can 'highlight and anticipate potential environmental problems, prevent future delays, assist in long-range planning and prevent or simplify litigation'.

California

Like the US National Environmental Policy Act, the California Environmental Quality Act (CEQA) of 1970 applies not only to individual development projects but also to the adoption of plans, programmes, policies and ordinances of public agencies. An environmental impact report (EIR) prepared for these broad government actions is, as under

NEPA, referred to as a 'program' EIR (Bass and Herson, 1993b). The CEQA Guidelines suggest that a program EIR, rather than a project EIR, should be prepared when an agency proposes a series of related actions, including one or more of the following:

- activities that are linked geographically
- activities that are logical parts of a chain of contemplated events
- rules, regulations or plans that govern the conduct of a continuing programme
- individual activities regulated by the same body and having similar environmental effects that can be mitigated in similar ways.

(Guidelines, Section 15168(a))

Bass (1990a) stated that some 340 program EIRs were received by the Californian State Clearinghouse in 1988–90, an average of about 11 per month. City and county land use plans make up the vast majority of the SEAs in California.

Guidance on the EIA of general land use plans has been issued (Office of Planning and Research, 1990). This stresses the need for environmental impact assessment to be an integral part of preparing or revising a general plan, not an after-the-fact exercise. A well-prepared general plan EIR covering broad geographic areas can increase the possibility that negative declarations can be issued at a later time for specific project proposals within the planning area (p. 153).

The guidance also sets out the required contents of a draft EIR (Box 19.3). It can be seen that these requirements are identical to those for a project EIR save for the addition of the two sustainability components

---

Box 19.3 Required contents of a draft Californian EIR for a general plan

- Description of the project
- Description of the environmental setting
- The significant environmental effects of the proposed project
- Any significant environmental effects which cannot be avoided if the proposal is implemented
- Mitigation measures proposed to minimise the significant effects
- Alternatives to the proposed action
- The relationship between local short-term uses of man's environment and the maintenance and enhancement of long-term productivity
- Any significant irreversible environmental changes which would be involved in the proposed action should it be implemented
- The growth-inducing impact of the proposed action
- Effects found not to be significant
- A list of organisations and persons consulted

Source: Office of Planning and Research (1990, p. 154).

relating to long-term productivity and irreversible impacts. The requirements relating to impact monitoring (Chapter 14) apply to land use plans also.

As Bass and Herson (1993b) stated, there are considerable differences in practice between program and project EIRs, despite the similarities in their content requirements. Program EIRs tend to contain a more general discussion of impacts, alternatives and mitigation measures. Once a reasonable range of assumptions about the future has been developed in the plan or programme, quantitative and qualitative methods similar to those used for project-specific EIRs are employed. Bass and Herson (1993b, p. 45) stated that some agencies have developed successful and innovative methods to evaluate programmatic impacts. Others, however, have used qualitative methods that are so general that the information has been of little use to decision-makers (Bass, 1990a).

EIRs may be presented either integrated within the general plans or as separate documents. In any event, they frequently run to several hundred pages in length and cost tens of thousands of dollars to produce. There has been no evaluation of the effectiveness of EIRs for land use plans but their very integration into the land use planning process should have resulted in more environmentally sensitive planning. Practice appears to be variable, from observance of the letter to the acceptance of the spirit of SEA. Bass (1990a) gave examples of the West Sacramento general plan in which wetland protection policies were incorporated early on as a result of the SEA which was commenced at the outset of the planning process.

The introduction in 1993 of master EIRs to evaluate cumulative, growth-inducing and irreversible significant environmental effects of subsequent projects to the greatest extent feasible and set out appropriate mitigation measures is interesting. CEQA now provides for the preparation of focused EIRs to address issues, mitigation measures or alternatives not included in the master EIR. The intention was that, by tiering, the CEQA process would be streamlined. However, while the concept is laudable, master EIR procedures are complex to operate and, since master EIRs are similar to (but do not replace) programme EIRs, somewhat confusing.

United Kingdom

As mentioned in Chapter 7, there is no formal requirement for the strategic environmental assessment of policies, plans and programmes in the United Kingdom. However, it is now accepted that the environmental impacts of government policies should be considered in their formulation (Her Majesty's Government, 1990) and a guide to incorporating environmental factors in policy appraisal has been published. This suggested an iterative process involving consideration of the environment from the outset, identification of policy options, seeking advice, identification of impacts, analysis of environmental effects, and monitoring and evaluation (Department of the Environment – DOE, 1991). The main steps in the process are shown in Box 19.4. While the guide was intended principally for central government, local government has also been referred to it (DOE, 1992, para. 6.25). There is no published SEA report on programmes, plans or policies in the United Kingdom, except for some environmental appraisals of land use plans.

It is now a requirement that local planning authorities (LPAs) 'have regard to environmental considerations' in preparing their land use plans. Policies in land use plans should reflect national environmental policies (DOE, 1992, para. 6.24). DOE also advises that the environmental implications of policies and proposals in land use plans should be appraised. DOE issued a limited amount of guidance on the appraisal of the environmental impacts of land use plans in 1992. *Inter alia*, it stated:

> Such an environmental appraisal is the process of identifying, quantifying, weighing up and reporting on the environmental costs and benefits of the measures which are proposed ... But the requirement to 'have regard' does not require a full environmental impact statement of the sort needed for projects likely to have serious environmental effects.
>
> (DOE, 1992, para. 5.52)

The results of the appraisal are to be set down in the plan documentation.

Box 19.4 Advisory steps in UK policy appraisal

- **Summarise the policy issue**: seek expert advice to augment your own knowledge as necessary.

- **List the objectives**: give them priorities, and identify any conflicts and trade-offs between them.

- **Identify the constraints**: indicate how binding these are, and whether they might be expected to change over time or be negotiable.

- **Specify the options**: seek a wide range of options, including the do-nothing or do-minimum options; continue to look at new options as the policy develops.

- **Identify the costs and benefits**, including the environmental impacts; do not disregard likely costs or benefits simply because they are not easily quantifiable.

- **Weigh up the costs and benefits**, concentrating on those impacts which are material to the decision.

- **Test the sensitivity of the options** to possible changes in conditions, or to the use of different assumptions.

- **Suggest the preferred option**, if any, identifying the main factors affecting the choice.

- **Set up any monitoring necessary** so that the effect of the policy may be observed, and identify any further analysis needed at project level.

- **Evaluate the policy at a later stage**, and use the evaluation to inform future decision making.

Source: Department of the Environment (1991, p. 2).

		Global sustainability					
	Criteria	1	2	3	4	5	6
Policies		Transport energy: Efficiency: trips	Transport energy: Efficiency: models	Built environment Energy: efficiency	Renewable energy potential	Rate of CO$_2$ 'fixing'	Wildlife habitats
1	To provide a network for open space corridors	●	✔	●	●	✔	✔
2	To concentrate residential development on an existing public transport corridor of the city	●	✔	✔	●	●	✔?
or							
3	To concentrate residential development on a new rural 'green' settlement (c. 8000 pop)	X	X	✔	✔?	✔?	✔?

Context: District-wide plan for a city of 150,000 and its hinterland
Illustrative policies: 1 For open space; 2 and 3 Represent options
for the location of housing

Suggested
impact
symbols:

  No relationship or insignificant impact

X Significant adverse impact

Fig. 19.3 Policy impact matrix for environmental appraisal of UK land use plans.
Source: Department of the Environment (1993, p. 24).

Natural resources				Local environmental quality				
7	8	9	10	11	12	13	14	15
Air quality	Water conservation and quality	Land and soil quality	Minerals conservation	Landscape and open land	Urban environmental 'liveability'	Cultural heritage	Public access open space	Building quality
✔	●	✔	●	✔?	✔	✔	✔	●
X	●	●	●	✔?	✔	✔?	X	✔
●	✔?	X	✔?	X	✔	✔?	✔	X

✔?	Likely, but unpredictable impact
✔	Significant beneficial impact
?	Uncertainty of prediction or knowledge

The 1992 DOE guidance was not found to be helpful by LPAs, several of which embarked upon their own appraisals of structure plans relying mainly upon matrices for the identification of impacts of individual policies (see, for example, Pinfield, 1992; Kent County Council, 1993). Partly based upon these appraisals, research commissioned in 1993 resulted in the publication of a guide to the environmental appraisal of development plans (DOE, 1993). DOE has strenuously publicised this guidance and, by mid-1994, other appraisals of structure plans had been published.

The 1993 Guide suggested that a policy impact analysis should form the core of the environmental appraisal. The use of a policy impact matrix was recommended so that an explicit identification of the impact of each policy option on each aspect of the environmental stock could be made. This matrix could then be used to record whether there was a positive (enhancing), negative (harmful), or neutral impact. This analysis could work in three ways:

- It can work in a iterative way with the initial evaluation of impacts providing an indication of ways in which the policy needs to be refined;
- It can provide a basis for comparing relative performance when choices need to be made;
- It can work as a tool of policy-checking or review, in the consideration of proposed modifications after plan examination or inquiry.

(DOE, 1993, p. 22)

Figure 19.3 shows a worked example of the use of the policy impact matrix, with the elements of the environmental stock expressed as criteria on one axis and illustrated policies on the other axis. In practice, scores of policies would have to be tested in this way. The early indications, in mid-1994, were that LPAs were finding the guide to be of some help.

The Netherlands

The Netherlands has some experience of SEA dating back to the early 1980s (Wood, 1988a). As a result the EIA system was designed to encompass SEA. Article 3 of the amended Environmental Impact Assessment Decree makes provision for certain types of plans and programmes to be subject to EIA. These relate to structure plans for electricity supply, industrial and drinking water supply, landscaping, nature conservation and outdoor recreation, to provincial waste management proposals, mineral extraction plans and certain types of land use plans (Verheem, 1993). The Decree does not apply to national policy plans. Experience of the EIA of over 30 plans and programmes has been gained to date (van Eck, 1993). However, decisions have been reached on only a few of these plans and programmes and even fewer have been implemented.

An EIA was carried out, on a voluntary basis, of the National Programme on Waste Management 1992–2002. The EIS compared the environmental impact of the various options for waste disposal, including incineration and various land disposal methods by using a number of environmental indicators such as energy use, land use, pollutant emissions, etc. Other EIAs have been undertaken for the national structure plan on electricity supply utilising different fuels and different sites for power stations. EIAs have also been carried out for several regional physical plans, especially where site selection for major housing areas was involved. The criteria utilised included pollution of land and water, nature and landscape protection, transport usage and economy of resource use. Several EIAs for local land use plans, which frequently also involve site selection for development, have also been undertaken (van Eck, 1993).

Several difficulties were encountered in undertaking these EIAs, especially in relation to the poor quality of information and uncertainty inherent in SEA. However, experience has shown that the EIA approach for projects generally worked satisfactorily for plans and programmes, largely as a result of formal scoping and the preparation of guidelines. Generally, it was felt that multi-criteria analysis involving environmental parameters was very useful in choosing sites. Finally, it was concluded that the experience of undertaking SEA was invaluable, that important lessons had been learned and that practice was bound to improve (van Eck, 1993).

The importance of discussing alternatives in the Dutch SEA process has been stressed by Verheem (1992), notwithstanding the difficulties in predicting their environmental effects because of their frequently ill-specified nature. The Evaluation Committee on EIA (1990) felt that EIA was less effective where it was used to assess the general principles of policy plans than it was for projects. However, it suggested that all policy plans which affect the environment should include an obligatory section on the environment and that the EIA Commission should be consulted about them.

Canada

To a certain degree, Canada has been assessing the environmental implications of policies, plans and programmes since 1984 when the federal Environmental Assessment and Review Process Guidelines Order 1984 defined 'proposal' as including 'any initiative, undertaking or activity for which the Government of Canada has a decision-making responsibility'. While the extension of this definition from projects to earlier stages in the planning process has not been manifest in the publication of environmental impact statements for policies, plans or programmes, there has been considerable interest in SEA in Canada (Federal Environmental Assessment Review Office – FEARO, 1988b). The arguments for and against the codification of SEA requirements in law culminated in the Government's commitment, in June 1990, that a non-legislated environmental assessment process was to apply to proposals for policy or programme initiatives submitted to the Cabinet for consideration. This EA process was seen as a demonstration of Canada's commitment to sustainable development, but it was felt by its many opponents that the theory and practice of SEA (for example in determining compliance) was insufficiently developed to legislate for it. External commentators did not always agree (Gibson, 1993).

The Government decided that:

a public statement outlining the anticipated environmental effects of a policy or programme initiative, which would be determined through an environmental assessment, would, as appropriate, accompany that announcement of the initiative.

(FEARO, 1993b)

This statement was to be a means of demonstrating that the assessment had been undertaken. In practice, very few public statements have actually been published, since there are several provisions for exemptions and it was assumed that about 75 per cent (probably an over-estimate) of Cabinet business could not be regarded as environmentally relevant. Those public statements which have been made have generally consisted of a sentence or paragraph stating that the initiative will have no significant adverse effects.

While practice in SEA in Canada is obviously rudimentary, as elsewhere, there has been a significant published environmental review of the North American Free Trade Agreement (Government of Canada, 1992). Although this has been criticised for its lack of detail and failure to influence the outcome of the agreement, it provides a useful indication of what could be achieved in such assessments. The environmental assessment of certain land use plans has been undertaken by the National Capital Commission.

Various research and guidance initiatives have been undertaken in Canada. The research undertaken by Bregha et al. (1990) was based upon case studies and workshops and raised issues about scope, responsibility, criteria, process, monitoring and accountability, public consultation and methodologies. It concluded that the current barriers to integrated policy assessment could only be overcome by measures such as:

the setting of sustainable development objectives in relevant policy sectors, a larger commitment to environmental science, a public consultation process, the development of indicators to gauge success, and government staff training and support.

More recently, FEARO commissioned work to provide draft guidance for SEA in Canada. This was intended not only to explain the nature of SEA but to provide details of methods for undertaking each step in the process. These steps are summarised in Fig. 19.4. The draft guide was found to be too prescriptive by many federal departments and agencies, leading to a decision not to publish it. There is to be a series of discussions between the Canadian Environmental Assessment Agency and other departments to review their guidance needs for implementing the policy EA process.

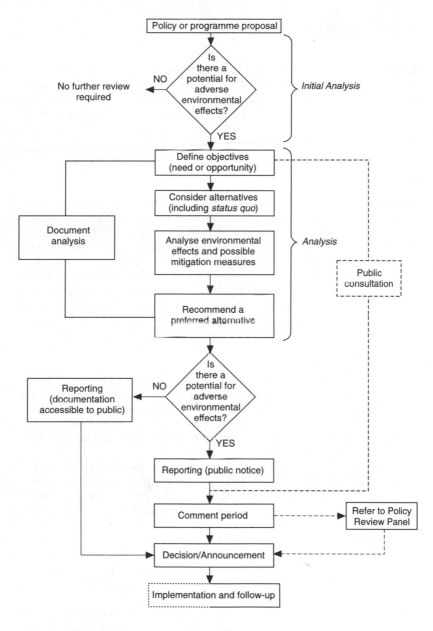

Fig. 19.4 Proposed Canadian approach to the EA of policy and programme proposals.
Source: Federal Environmental Assessment Review Office (1994).

Commonwealth of Australia

The object of the Environment Protection (Impact of Proposals) Act 1974 is to examine and take into account environmental matters in relation to, *inter alia*, 'the formulation of proposals', 'the making of ... decisions and recommendations' and 'the incurring of expenditure' (Section 5 (2)). It was clearly intended that the Act would apply to policies and other Commonwealth actions as well as to projects (Fowler, 1982, p. 17).[2] As the responsible Minister stated in Parliament in debating the Bill: 'The Bill will affect many facets of the Australian Government's activities' (Cass, 1974, p. 4082).

In practice, the degree of discretion provided by the Act and the reluctance of politicians and administrators to extend EIA to non-project actions has restricted its coverage almost entirely to projects (Anderson, 1990). Of the 132 Commonwealth directives that environmental impact statements be prepared between 1974 and June 1993 none related obviously to policy, plan or programme EISs (Wood, 1992, updated). There have, however, been a few examples of recommendations being made under the provisions of the Impact of Proposals Act in relation to programmes without requiring the preparation of formal EIA reports. These include the assessment of Commonwealth-funded highway construction programmes and of systematic applications to the Australian overseas aid programme.

There have been several recent notable recommendations about, and commitments to, the use of SEA. Thus, the Ecologically Sustainable Development Working Groups (1991, p. 258) recommended that EIA 'be applied to transport policies, programs and projects as an essential part of transport planning and decision making and to facilitate the application of ESD principles to transport'.

Similarly, the Australian and New Zealand Environment and Conservation Council (ANZECC, 1991, p. 9) has agreed that the principles and objectives of its national approach to EIA 'would be as effectively, and often more effectively, applied to [the assessment of] policies and major programmes' as to projects, with major benefits. The 1992 Australian Intergovernmental Agreement on the Environment, part of which drew upon the ANZECC national approach agreed between its Commonwealth and State governments, specifically mentioned impact assessment in relation to programmes and policies (Commonwealth Environment Protection Agency, 1992).

Finally, the Commonwealth Government committed the recently established Commonwealth Environment Protection Agency to 'ensuring environmental impact assessment processes are applied to policy and program proposals (often not done now) as well as to specific projects' (Department of the Arts, Sport, the Environment, Tourism and Territories, 1991, p. 16).

In view of the lack of SEA reports to date, it appears that the Commonwealth Government's phrasing was somewhat disingenuous here.

New Zealand

The 1974 Environmental Protection and Enhancement Procedures applied in principle, but not in practice, to the 'management policies of all departments which may affect the environment' and to certain other government higher tier actions (Ministry for the Environment, 1987, p. 1).[3] In the absence of any other forum, EIA often provided a focus for debate on the environmental impacts of policies in New Zealand (Morgan, 1988), principally as a result of the comments by the Parliamentary Commissioner for the Environment (PCE) in published 'audits' and of debate at hearings before the Planning Tribunal. The Ministry for the Environment reports to Cabinet on all national policies by other ministries.

Building upon this experience, the Resource Management Act 1991 has ensured that environmental assessment applies to policies and plans prepared under the Resource Management Act provisions (but not to other policies, plans and programmes). The Act provides that central government can prepare national policy statements, that regional councils must prepare regional policy statements and regional coastal plans and may prepare other types of regional plan and that territorial (city and district) authorities must prepare district plans. In each case the focus is upon effects rather than on uses and an environmental evaluation of the policy statement or plan must be carried out (Sections 62(1)(g) and 75(1)(g)). These policy statements and plans, of course, also provide a framework for the implementation of project EIA. Section 75(1)(f) of the Act requires district plans to specify the information that should be submitted with an application for a resource consent, including the circumstances in which the powers to ask for further environmental information may be used (MfE, 1991c, d). This is a practical example of the tiering of environmental assessment. Section 32(1)(b) of the Act also requires an 'evaluation which ... is appropriate in the circumstances, of the likely environmental costs and benefits of the principal alternative(s)'.

It should be noted that costs are defined to include environmental and social, as well as monetary, costs and that policies and plans must include a statement of the environmental results anticipated. Plans should also set out the content and consultation requirements for any EIA required for a consent. In this way the plan can be used to ensure the EIA requirements are better tailored to selected types of activities and focus only on the relevant environmental issues.

As mentioned in Chapter 5, several guides relating to the environmental assessment of policy statements and plans have been issued (MfE, 1991c, d, 1992a). A guide to good practice under Section 32 has also been issued which, inter alia, advocated defining the options, defining the evaluation framework by identifying relevant costs and benefits (including social and environmental impacts), analysis involving comparison of options, and decision making (MfE, 1993).

A review of a list of some 95 environmental impact audits produced by the PCE and its predecessor, the Commission for the Environment, between 1973 and 1990 revealed that all but one appeared to relate to location-specific projects. The exception concerned the proposed introduction of myxomatosis to New Zealand for rabbit control. When the New Zealand Agricultural Pests Destruction Council controversially recommended to the Minister of Agriculture that the virus myxomatosis be introduced as an additional means of controlling infestations of wild rabbit in marginal pastureland in 1985, the Minister directed that an environmental impact report (EIR) be undertaken. This was duly prepared by consultants and the revised version, which examined the environmental case for and against introduction, was published in 1987. The PCE audited the report, recommended against introduction (principally because of biological risks) and suggested that a number of alternative management actions be taken (Wood, 1992).

Dixon (1993a) has reported that, unsurprisingly, district councils are lagging behind regional councils in the production of plans. Separate EIA reports are not being prepared. Rather, EIA is being incorporated into the plan preparation and policy analysis process by land use planners. Regional councils are generally organising discussion under the headings:

- issue
- objective
- policy
- methods of implementation
- environmental results anticipated.

(Dixon, 1993a, p. 5)

However, practice among districts is very variable.

Generally, it appears that expertise is very weak in policy and plan EIA, and that there is a need for development of, and research into, appropriate EIA methods. Further, monitoring is being neglected, public scrutiny is lacking and there is a need for a wider range of professional skills in policy and plan EIA (Dixon, 1993a).

It is apparent that the generally small regional and local authority professional staffs are struggling to come to terms not only with the SEA provisions but with the EIA provisions of the Act and that there is an acute need for further guidance and for training. There is a need for the Ministry for the Environment to continue to work informally with practitioners, and use the Minister's 'call-in' powers appropriately. In all probability, the Parliamentary Commissioner for the Environment may also need to play an active role.

Summary

It would be possible to apply all the evaluation criteria utilised in the earlier chapters of this book to SEA. However, rather like using the Box 1.2 criteria to evaluate developing country EIA systems in general, this would be a fruitless exercise at present because of the generally undeveloped state of SEA systems. For the sake of symmetry, however, a single undemanding evaluation criterion has been utilised in Table 19.1 to summarise the SEA situation in the seven jurisdictions.

Table 19.1 The treatment of SEA in the EIA systems

Criterion 14: Does the EIA system apply to significant programmes, plans and policies, as well as to projects?

Jurisdiction	Criterion met?	Comment
United States	Yes	1969 Act provides clear legal provisions for SEA. SEA practice developing steadily: several hundred programmatic EISs prepared.
California	Yes	1971 Act interpreted to include provisions for SEA. SEA practice developing steadily (several hundred land use plan EIRs prepared).
United Kingdom	No	No formal requirement for SEA. Guidance on environmental appraisal of both central government policy and of local land use plans exists. Some practice.
The Netherlands	Yes	EIA Decree defines 'proposal' to include certain policies, plans and programmes. Growing SEA practice.
Canada	No	Non-legislated SEA process applies to Cabinet and other proposals. SEA research and guidance commissioned but little practice.
Commonwealth of Australia	No	1974 Act provides SEA powers but no SEA reports prepared. Commitment to undertake SEA of policies and programmes in future.
New Zealand	Yes	1991 Act requires SEA of certain regional and local policies and plans. Some guidance. Practice developing but varies.

This SEA criterion has not been applied harshly. Thus, if there is legal provision for the SEA of policies, plans *or* programmes and at least some evidence of practice at any of these stages in the planning process, a 'yes' has been awarded. The United States, California, The Netherlands and New Zealand all meet this rather generous criterion. However, the United Kingdom, Canada and the Commonwealth of Australia do not meet it, despite some SEA practice in the first two countries and legal provisions for SEA in Australia. Most developing countries would fail to meet the SEA criterion, though work on the World Bank's regional and sectoral environmental assessments is advancing in some of these countries (see Appendix).

It is clear that SEA practice is developing around the world and that almost all the points made in earlier chapters and in Chapter 20 in relation to the different stages of the EIA process could be applied to SEA. The increasing emphasis on cumulative and indirect impacts and on the sustainability of development make further progress on SEA, and the sharing of experience to date, imperative.

Notes

1. This section draws heavily upon Wood and Djeddour (1992).
2. This section is an edited version of the account presented by Wood (1992).
3. This section relies partly upon Wood (1992).

CHAPTER 20

Conclusions

Introduction

As mentioned in Chapter 1, several developments in EIA have taken place as experience with its use has grown. These may be summarised as:

- increasing emphasis on the relationship of EIA to its broader decision-making and environmental management context
- increasing codification of EIA requirements
- increasing adoption of additional EIA requirements
- increasing emphasis on maximising the benefits and minimising the costs of EIA
- increasing recognition that some form of strategic environmental assessment is necessary.

These developments are evident in each of the seven EIA systems reviewed.

This final chapter draws together the main threads of the earlier chapters by summarising the performance of the EIA systems in the United States, California, the United Kingdom, The Netherlands, Canada, the Commonwealth of Australia and New Zealand against the evaluation criteria listed in Box 1.2. Table 20.1 summarises the overall performance of the seven EIA systems against these criteria. Several suggestions for improving the various EIA systems by overcoming the shortcomings identified are advanced. These suggestions derive directly from the comparative review of the EIA systems and provide one of the principal justifications for such a study.

United States of America

The US EIA system meets 11 of the 14 evaluation criteria and partially meets another. The main shortcomings of the system relate to its coverage (which is confined to federal actions), to lack of centrality to decision making (notwithstanding the requirement to publish a record of decision) and to the mitigation and monitoring of impacts. Other weaknesses relate to the lack of oversight of environmental assessments, to lengthy descriptive and derivative environmental impact statements (which neglect the treatment of cumulative impacts) and to the court-driven procedural nature of the system. Because the system is operated by federal agencies, the general level of expertise is high but there is still a perceived need for training and guidance (Clark, 1993). The roles of the Council on Environmental Quality, of the Environmental Protection Agency (EPA) and of public interest groups in maintaining and refining the system and in ensuring that federal agencies perform are pivotal.

Broadening the coverage of NEPA to cover other actions would provoke a constitutional outcry. Increasing the centrality of EIA to the decision-making process could be achieved by amending NEPA to require that an action could only be taken if all feasible mitigation measures were included in the proposal (Blumm, 1990). Such a solution could be applied following both an environmental impact statement (EIS) and a finding of no significant impact (FONSI). This would go some way to achieving the initial intention of NEPA's authors (Caldwell, 1989b; Yost, 1990). A further improvement would be to ensure that EISs and environmental assessments (EAs) are made shorter and more readable and thus accessible to decision-makers (Bear, 1989; Blumm, 1990).

Other improvements include mandatory public review of FONSIs (Herson, 1986), the provision of guidance on EA generally and the role of public participation in particular (Blumm, 1990; Blaug, 1993; Reed and Cannon, 1993). EPA has prepared a sourcebook that provides guidance on the whole EIA process (EPA, 1993) and CEQ is coordinating the development of a handbook on cumulative impact assessment. Rendering post-decision monitoring compulsory is a necessary reform (Blumm, 1990; Reed and Cannon, 1993). Additional auditing studies are also needed to ensure that innovative mitigation measures are effective. The referral of more cases to CEQ and an extension of CEQ's role in EIA oversight might also be helpful in increasing the effectiveness of EIA (Blumm, 1990). There is also a need to improve the quality of predictions by developing methodologies (e.g. for cumulative impacts and effects on ecosystems), by preparing technical guidance and by peer review (Dickerson and Montgomery, 1993; Reed and Cannon, 1993).

California

The Californian EIA system meets most of the evaluation criteria (Table 20.1). The main weaknesses of the frequently modified California Environmental Quality Act (CEQA) are:

- lack of centrality to decision making on projects
- over-long and descriptive environmental impact reports (EIRs)
- weaknesses in project monitoring
- marked absence of EIA system monitoring and system evaluation

Evaluation criterion	Criterion met within jurisdiction						
	United States	California	United Kingdom	The Netherlands	Canada	Australia	New Zealand
1. Legal basis	■	■	■	■	■	■	■
2. Coverage	▢	■	▢	■	○	▢	■
3. Alternatives in design	■	■	○	■	■	■	■
4. Screening	■	■	■	■	■	○	■
5. Scoping	■	■	○	■	■	■	▢
6. Content of EIA report	■	■	▢	■	■	■	○
7. Review of EIA report	■	■	▢	■	■	■	■
8. Decision making	○	○	○	■	○	○	○
9. Impact monitoring	○	▢	○	▢	▢	○	○
10. Mitigation	■	■	■	■	■	■	■
11. Consultation and participation	■	■	▢	■	▢	▢	▢
12. System monitoring	■	○	○	■	■	○	○
13. Costs and benefits	■	■	■	■	■	■	■
14. Strategic EA	■	■	○	■	○	○	■

■ Yes   ▢ Partially   ○ No

Table 20.1 The overall performance of the EIA systems

- EIA system overly bureaucratic and slow, driven by legal defensibility.

The problems of lengthy and sometimes somewhat irrelevant EIRs have been overcome to some extent by modifications to the Guidelines and by the use of experienced consultants. While there is no universally employed methodology, it is usual for the issues to be identified, at least to a degree, not only by the proponent but by the consultees and the public and for the project manager for the EIR within the consulting firm to assemble a team possessing appropriate training and expertise. The principal system deficiencies probably lie in the lack of training in EIA, and in the lack of knowledge of detailed procedures, exhibited by many of the local government officials appointed to oversee the EIA process. There is still scope for speeding up the assessment process and for improving the relevance of EIRs. The role of citizens' groups in monitoring the process and ensuring that EIRs are of adequate quality is crucial.

It is generally accepted that the EIA process has resulted in improved projects in California, though the financial, time and paper costs have been substantial. However, improvements are possible, although it is difficult to see how the centrality of EIA to decision making can be further increased by legislation. Perhaps a reduction in length of EIA documents, an increase in readability and the production of illustrated, separate, non-technical summaries would lead to a greater political awareness of EIA, especially among elected representatives in local government. Improvements in the training of, and in guidance to, local government officials would also be helpful and could reduce the overly bureaucratic approach to EIA. Learning from previous experience would be much improved in California if better EIA system monitoring were carried out, perhaps following the example of the US Council on Environmental Quality. Certainly, more resources are needed to manage and monitor the Californian EIA system effectively.

United Kingdom

The UK EIA system fully meets four and partially meets another four of the 14 evaluation criteria employed in this comparative review (Table 20.1). It performs worst of all the eight EIA systems.

Britain's is probably a fairly typical first generation EIA system with screening, environmental statement (ES) publication and public participation provisions integrated into existing town and country planning decision-making processes but without scoping, early participation, unpenalised rights to further information, true centrality of EIA to the decision, third party appeal or monitoring provisions. Obviously, experience of EIA is being gathered by local planning authorities (LPAs), developers and consultants as time elapses and the diffusion of practice takes place, especially in regard to ESs (Wood et al., 1991; Lee and Brown, 1992).

However, while the range of experience within consultancies is growing, local authority experience of EIA is still very limited in many cases. This situation will obviously improve with time, with the publication and diffusion of the promised guidance on preparation and review of ESs and with the greater incorporation of environmental considerations into land use plans.

The shortcomings of the UK EIA system relate to impact coverage, to the consideration of alternatives in design, to scoping, to the proponent's response to public comments, to the use made of EA in decision making, to project monitoring, to consultation and participation, to formal system monitoring and to strategic environmental assessment (see also Glasson *et al.*, 1994; Sheate, 1994). They are a reflection of the UK's implementation almost to the letter of the somewhat rudimentary compromise requirements of the European Directive (Wathern, 1988b; Lambert and Wood, 1990). In these and other aspects of the environmental assessment (EA) process practice varies very considerably from the exemplary to the unprofessional.

These shortcomings mean that the aims of EA, better quality project planning and better quality decision making, are not being universally achieved. If weaknesses continue to be evident as practice evolves, then changes more radical than the provision of EA guidance for those preparing ESs (released in draft form in 1994 – Department of the Environment, 1994b) and for those receiving them will be necessary. As a first step, measures relating to:

- better diffusion of EA information and, in particular, of ESs
- further training provision
- clarification of screening criteria
- improvement of LPA procedures for coping with EA
- better provision of information to the public
- briefing of planning inspectors on the acceptability of ESs
- research into several aspects of the EA process.

could be taken (Wood and Jones, 1991).

If practice subsequently failed to improve sufficiently then the EA system itself would need to be strengthened, as has happened over the years in many mature EIA systems (e.g. the United States, California, The Netherlands, Canada, Commonwealth of Australia and New Zealand) which now satisfy far more criteria than does the UK's. Changes to the EA system (and in particular to the treatment of alternatives, to scoping, to formal ES review, to project monitoring and to SEA) should be designed to ensure that the evaluation criteria employed in this review are met more fully than at present.

The Netherlands

The Netherlands EIA system meets almost every one of the evaluation criteria utilised in this review (Table 20.1). The only criterion which is not met relates to monitoring and even here the legal provisions meet the criteria: it is practice which falls short. On the other hand, the mitigation of environmental impacts is not separately specified in the law but there appears to be no inherent weakness in the treatment of mitigation in the EIA system. Mitigation is subsumed under the very extensive coverage of alternatives in the Dutch EIA system and, in particular, in the environmentally preferable alternative. However, since Article 5(2) of the European EIA Directive requires the EIS to include 'a description of the measures envisaged in order to avoid, reduce and, if possible, remedy significant adverse effects' (Chapter 3) it is somewhat surprising that the omission of mention of mitigation measures in the Dutch EIA legislation persists.

Glaring omissions in the implementation of the monitoring and auditing provisions of the Environmental Management Act 1994 exist. It would assist the operation of the EIA system, and the knowledge base for EIA, if these provisions were strengthened, perhaps by time-limiting permissions and requiring the submission of satisfactory monitoring information. This would improve the feedback of knowledge into the EIA system, where weaknesses in utilising previous experience are apparent. This is, perhaps, symptomatic of the failure, in some instances, fully to integrate the results of the EIA into the proponent's own planning and project development at a sufficiently early stage to genuinely influence project design. It may also reflect the willingness of the competent authorities to leave too much of the operation of the EIA process to the increasingly influential EIA Commission and not to make EIA truly central to their decisions.

The Evaluation Committee on EIA (1990) made a number of recommendations relating to the EIA system including broadening the types and numbers of activities subject to EIA, improving safeguards where the proponent and the competent authority are one and the same, expanding the scope of the notification of intent, spelling out the least damaging alternative more fully and expanding the coverage 'of EIA to include energy use. All these recommendations have now been implemented.

Further improvements in the effectiveness of the Dutch EIA system can therefore be anticipated. It has sometimes been criticised as overly cumbersome, expensive, time consuming and limited in application: a Rolls Royce where a Ford would suffice. The Dutch would generally argue that the wealth of their concern about the environment of their small and vulnerable country justifies their extensive preparation for, and introduction of, the best EIA system their money could purchase. They would also argue that the EIA system is flexible enough to allow them only to use a car when absolutely necessary: that detailed EIA is needed to address only the most acute problems. As the Evaluation Committee on EIA (1990, p. 13) put it:

> ... the Evaluation Committee can only express its satisfaction about the way the regulation has operated in the first years since its introduction. Experience to date gives ground for optimism.

Canada

The Canadian EA system delineated by the Canadian Environmental Assessment Act (CEAA) meets many of the evaluation criteria used in this review. Table 20.1 summarises the strengths and weaknesses of this second generation federal EA system. The main weaknesses are:

- limitations of coverage to federal actions
- inter-jurisdictional limitations on the coverage of environmental effects
- marked distinction between likely effectiveness of screening and panel review procedures
- lack of means for ensuring that EA centrally influences decisions
- lack of effective mechanism for ensuring compliance with mitigation and monitoring requirements
- lack of formal SEA requirement.

Gibson (1992) also identified deficiencies relating to alternatives, coverage of socio-economic and cultural impacts and screening and felt that CEAA was too discretionary. Under the federal Environmental Assessment and Review Process (EARP) panel reviews have generally been seen as providing an excellent means of cooperatively identifying and mitigating the environmental effects of a handful of projects. Screening, on the other hand, has been seen as a largely clandestine and ineffective means of assessing the effects of the vast majority of projects. The panel review has been an elaborate ice castle constructed on an assessment iceberg nine-tenths of which has been submerged in cold, impenetrable waters.

There are varying views on the extent to which CEAA will lead to improvements in EA practice in Canada. On the one hand, the problems of lack of environmental political clout at federal Cabinet level and of disagreement with the provinces about the extent of federal jurisdiction in environment matters remain. On the other hand, CEAA contains provisions for bilateral agreements and for determining when comprehensive studies will be undertaken. These allow more intervention by the Minister of the Environment and expose much more of the EA system to public scrutiny than EARP. These provisions provide a much firmer foundation for the construction of an effective EA system than EARP, notwithstanding their weaknesses.

How well CEAA works in practice will, like the environmental assessment of policies, plans and programmes, depend on the efforts and resourcefulness of the CEA Agency and on the growing influence and expertise of environmental professionals within the responsible authorities and proponent organisations. Above all, however, the strength of the CEAA system depends upon the bedrock of public environmental concern and vigilance.

Commonwealth of Australia

The Commonwealth EIA system fully meets half the evaluation criteria and partially meets another two (Table 20.1). The main weaknesses relate to coverage, to screening, to decision making, to monitoring, to public participation, to system monitoring and to SEA. Apart from these, several criticisms have been made of the legal basis of the system, of scoping, of review, of uncertainty (principally caused by the use of discretion, duplication of rules between the Commonwealth and the states, changes of rules, etc.) and of delay (Bureau of Industry Economics, 1990; Australian and New Zealand Environment and Conservation Council – ANZECC, 1991; Kinhill and Phillips Fox, 1991). It appears that Porter (1987) was exaggerating the merits of the Australian EIA system, at least in regard to the United States, when he claimed that 'We seem to be far ahead of the larger nations such as the United States, Britain and the Soviet Union, and to have embraced the most worthwhile aspects of EIA'.

Although there are no proposals to strengthen the legal basis of the EIA system, a large number of suggestions for overcoming the identified weaknesses have been advanced. The Ecologically Sustainable Development Working Group (ESDWG) Chairs (1992) argued that such reforms would also help to buttress EIA as a major component in Australia's ecologically sustainable development strategy. They recommended that:

a high priority be accorded to work on the evaluation and development of techniques for environmental impact assessment, particularly on aspects of the technique associated with the framing of the terms of reference, social impact assessment, economic analysis, cumulative effects and post-approval assessment (p. xxii).

Other recommendations for EIA system and organisation improvement include strengthening screening, scoping and system monitoring (Chapters 9, 10 and 17), the accreditation of EIS preparers (ESDWG, 1991) and increased training and communication (Department of the Arts, Sport, the Environment, Tourism and Territories, 1991). The enforcement of time limits (Bureau of Industry Economics, 1990), the use of government project coordination units (Kinhill and Phillips Fox, 1991) and the use of a national approach to EIA (see, for example, ANZECC, 1991) have all been proposed.

In response to these latter suggestions the 1992 Intergovernmental Agreement on the Environment (Commonwealth Environment Protection Agency, 1992) stated that time schedules for all stages of Commonwealth and state EIA processes would be set for proposals and that the Commonwealth would accredit state EIA systems, thus removing much of the remaining need for joint working on EIA. The comprehensive public review of the Commonwealth EIA system (Chapter 18) may well result in measures to resolve many of these shortcomings. However, the crucial measure, increasing the centrality of EIA to decision making, would require both public and political will and a strengthening of the Environment Protection (Impact of Proposals) Act.

New Zealand

The highly complicated Resource Management Act is one of the first attempts in the world to achieve sustainable management. It is, perhaps, all the more surprising that New Zealand should have relied quite so completely on local discretion in its Mark II EIA system. While the temptation to devolve responsibilities has proved irresistible to a reformist, monetarist, central government (Memon, 1993), the implementation gap is potentially enormous.

New Zealand is famous for 'do it yourself' activities but to leave local authorities with tiny professional staffs and little or no experience of EIA to evolve screening, scoping, review and decision-making procedures individually seems courageous. Quite apart from the huge task faced by planners in coping with EIA (and other aspects of the Resource Management Act requiring significant professional reorientation (Montz and Dixon, 1993; Morgan, 1993)) 'there is a danger of turning the clock back 20 years to when central government first embarked on environmental assessment procedures' (Hughes, 1992). As each local authority strives to put appropriate EIA provisions in place Morgan (1988) felt that significant variations would be inevitable, resulting in confusion and uncertainty for proponents and public alike, and so it has proved.

It can be seen from Table 20.1 that the New Zealand EIA system fails to meet several of the evaluation criteria. This may, to some extent, be an inevitable consequence of making the EIA system applicable to such an extraordinary range of activities from the trivial to the highly significant. The EIA system necessarily provides considerable flexibility to ensure that the scale of the EIA is appropriate to the likely severity of the impacts. In some instances (e.g. scoping) a failure is recorded because the requirement is not mandatory. The EIA activity may nevertheless be very strongly encouraged and, for notified projects, in practice almost universally undertaken.

Overall, however, there are some weaknesses in scoping, in EIA report preparation, in the centrality of EIA to the decision, in monitoring, in public participation and in EIA system monitoring. A number of suggestions have been made to overcome these shortcomings. In particular, Morgan (1993) has suggested greater consultation in scoping and in EIA report preparation. Dixon (1993b) has stressed the need to develop EIA prediction methods and to enhance skills in, for example, the assessment of cumulative impacts. Improvements in local authority monitoring and enforcement procedures and methods are also needed (Chapter 14). Equally, there would be many benefits in diffusing information and practice through better EIA system monitoring by, for example, maintaining a central record of all EIA documents.

Because of the lack of experience of EIA among most applicants, their consultants and local authority officers and elected members there is a pressing need for training and for more specific guidance on the different stages in the EIA process for the various participants. Training and encouragement to adapt are particularly necessary in overcoming the resistance of traditional land use planners to the new EIA procedures which require reorientation of practice from control of use to control of effects (Morgan, 1993). Training and growing experience, further guidance, Planning Tribunal decisions, Parliamentary Commissioner interventions, and the implementation of regional policy statements and plans and district plans and, perhaps, the results of Ministry for the Environment 'call-ins' will all undoubtedly help to realise the potential of New Zealand's innovative EIA system.

Improving EIA

It can be seen from Table 20.1 that the EIA systems do not perform equally well, and that certain shortcomings become more evident when they are seen overall, rather than partially as in the earlier chapters. Not surprisingly, all the EIS systems perform much better against the criteria than do the developing country EIA systems reviewed in the Appendix.

Certain general shortcomings in the current state of EIA practice can be observed. These may be summarised as:

- weaknesses in coverage
- weaknesses in integrating EIA into decision making
- weaknesses in impact monitoring and enforcement
- weaknesses in public participation
- weaknesses in system monitoring
- weaknesses in SEA.

In addition, there are widely acknowledged weaknesses in the quality of many EIA reports. These weaknesses also apply to most developing country EIA systems. A number of specific measures can be used to strengthen the different EIA systems by introducing or bolstering appropriate procedural requirements (above). There continues to be a need, in each EIA system, for three other elements to strengthen EIA practice: guidance, training and research.

The existence of published guidance on the EIA systems as a whole is clearly useful to those responsible for preparing EIA reports, to those reviewing them and making decisions, to those consulted and to the public. Such guidance provides a valuable aid in undertaking any stage of the EIA process. While the practitioners most closely involved do not tend to need general guidance unless changes to EIA procedures are introduced, there are always those new to EIA or some aspect of the EIA process who do require such assistance. Guidance materials can include manuals, leaflets, computer programs and video tapes. The provision of guidance of this type tends to vary from EIA system to EIA system, just as the provision of more detailed guidance on the different stages of the EIA process varies, but could be strengthened to assist in overcoming the weaknesses identified above.

The provision of EIA training for EIA project managers, for technical specialists and for others involved in the EIA process is an effective method of increasing the standard of practice even in mature EIA systems. While EIA training need not be provided only by the agency responsible for EIA, encouragement of, and participation in, such training by the agency is clearly desirable (Lee and Wood, 1985). A variety of different training methods is appropriate in most EIA systems: courses, manuals, guides (e.g. Lee, 1989), case studies (e.g. Wood, 1989a), video tapes, computer programs, etc. The involvement of responsible authorities in EIA training tends to vary from jurisdiction to jurisdiction but the need for further training remains. This is especially true in developing countries.

There is a continuing need for research on various aspects of EIA, both general and specific. Generally, research on the treatment of alternatives, on scoping, on forecasting, on review methods, on the integration of EIA in decision making, on monitoring, on public participation, on system monitoring and on strategic environmental assessment is clearly needed. Such research needs to be concerned with both substantive (methodo-

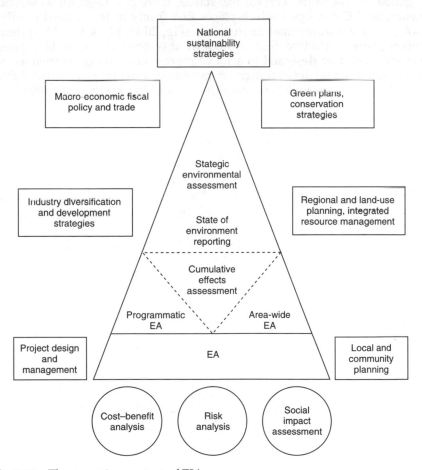

Fig. 20.1 The emerging context of EIA.
Source: Sadler (1994a, p. 9).

logical) and procedural issues. Although research on EIA methods is likely to be of most application across EIA systems, the results of procedural research tend also to be widely disseminated since there is a need to share knowledge and insights. Clearly, each EIA system is likely to have its own specific research needs in addition to those identified here. The EIA research needs of developing countries should not be neglected.

Several of the EIA systems reviewed in this book have progressed from Mark I to Mark II versions (for example, the Canadian and New Zealand systems). The steps outlined above should enable progress to be made towards Mark III versions. However, it is necessary not only to improve EIA systems but to ensure that they receive real public and political endorsement.

There are undoubtedly pressing demands from some quarters for the achievement of sustainability goals and to meet the requirements of Agenda 21. As Sadler (1994a) has stated, there is a need for a 'second-generation' EIA process which places EIA firmly in the context of other policy and environmental instruments (Fig. 20.1). Mark III EIA systems incorporating effective SEA would fit this context admirably. These systems could be designed to achieve more ambitious goals such as 'no net environmental deterioration' or 'net environmental gain'. If the public and the politicians will the ends, EIA can help to provide the means.

EIA in developing countries

Just as there are huge differences in EIA systems in the developed world (where some countries have still not adopted EIA) so there are between EIA systems in developing countries.[1] Thus, there are enormous variations between the situations in Latin America and in SE Asia (where many countries have developed EIA systems) and in Africa (where many countries have not). Just as in Europe, however, the situation in different countries within continents varies considerably. Within Africa, for example, while the South African EIA system is sophisticated and EIA is becoming important in Zimbabwe, as yet EIA is unimportant in Somalia.

Despite these variations, it remains true that, on the whole, environmental impact assessment in developing countries tends to be very different from EIA in the developed world. The most conspicuous difference relates to the fact that the first EIAs to be carried out were usually demanded by development assistance agencies on a project-by-project basis, not as a response to a widespread indigenous demand for better environmental conditions.

There are, however, exceptions and the EIA requirements in, for example, Columbia (1974) and the Philippines (1977) pre-date those in many developed countries. Over the past decade several developing countries have established their own formal legislative bases for EIA. Often, however, the necessary organisation to enforce it is absent. EIA is therefore commonly a 'top-down' requirement imposed by external agencies (Rayner, 1993). While this may have been the case with the implementation of the European Directive on EIA in certain countries (Chapter 3), the 'bottom-up' demand for environmental controls and the organisational capacity to implement them, which existed in many of the European Union countries, are often absent in developing countries.

This lack of demand is a consequence of the lack of political priority accorded to the environment in general, and EIA in particular, in many developing countries. While many officials in environmental ministries, and others, may appreciate the relationship between rational management of the environment and long-term economic development (and thus be enthusiastic about EIA), most politicians do not. This lack of political will is allied both to existing systems, in which pressing environmental concerns (frequently fuelled by severe environmental degradation) often cannot be effectively represented politically, and to widespread corruption.

Bisset (1992) and the EIA Centre (1993) have distinguished about a dozen differences between EIA systems in developing countries and in developed countries. Bisset (1992, p. 217) has surmised that these differences lead to two main consequences:

1. Fewer EIAs are undertaken than legal and other requirements would seem to indicate.
2. Most EIAs seem to have been a function of justifying a decision (usually to develop) which has been made and are concerned only with remedial measures. Rarely do they consider alternative courses of action at an early stage of the project planning cycle, in order to choose the most environmentally favourable.

Rayner (1993, p. 678), rather depressingly, has concluded that:

For the Third World, EIA remains, at best, a Band-Aid to mitigate the worst consequences of rapid industrial development because it is wealth, not legislation, that leads to indigenous demands for clean energy, stable populations, and stewardship of the land and water.

While the importance of wealth in determining environmental awareness can hardly be exaggerated, there are examples of EIA being undertaken successfully in many developing countries. These include EIAs in Brazil, China, Egypt, India, Indonesia, Malaysia, Pakistan, the Philippines, South Africa and Thailand (Biswas and Geping, 1987; Biswas and Agarwala, 1992; Turnbull, 1992; Hildebrand and Cannon, 1993). However, it is true that, with some exceptions, the EIA systems in many developing countries would fail to meet virtually every one of the criteria in Box 1.2. This appendix discusses the problems and improvements which need to be made in relation to each stage of aspect of the EIA process.

The *legal basis* of EIA systems in many developing countries may be weak, non-mandatory or non-existent. In addition, the organisations responsible for EIA are frequently new, lacking in status and political clout and working in a culture where an absence of information sharing considerably reduces their influence. Environment ministries are often 'bypassed' by other, more powerful, ministries. As in the developed world, it is clearly necessary to put in place, first, the appropriate institutional framework and only then the regulatory requirements for EIA (Biswas, 1992). It is also necessary for government environmental ministries in developing countries to be granted the means to acquire new or existing environmental information to a greater extent than in the past (McCormick, 1993).

The *coverage* of EIA systems in developing countries is patchy both in relation to the projects covered and, especially, in relation to the impacts assessed. The top-down nature of EIA requirements in many developing countries means that it frequently applies only to grant-aided projects or that systems are put in place purely to satisfy the development assistance agency. The mandatory EIA system which is needed in each developing country must clearly be designed to cover all the types of actions which have the potential to cause environmental damage in the local circumstances. It appears generally to be accepted that social as well as environmental impacts should be included in EIA and that positive as well as negative impacts should be emphasised (World Bank, 1991; Biswas, 1992; Organisation for Economic Cooperation and Development – OECD, 1992).

The consideration of *alternatives* in developing country EIAs is frequently weak. The no-action alternative is often not a viable choice in circumstances where the alleviation of poverty and starvation may be the predominant goal and, in practice, the environmentally preferable alternative may not be considered either. However, the choice of an alternative which minimises damage to the environment and/or the use of mitigation measures should, in principle, usually be possible. The importance of utilising EIA in developing countries is that it should result in development which is more sustainable than would otherwise be the case.

The *screening* of actions for the applicability of EIA is not undertaken satisfactorily in many developing countries. Frequently, the main criterion is whether or not the development assistance agency requires an EIA, perhaps after application of its own screening procedure (World Bank, 1991; Commission of the European Communities, 1993a). Kennedy (1988a) has indicated the need to have a simple and effective screening system in place in developing countries. In many cases, this should extend to the use of simplified EIA for appropriate projects.

It is generally considered that, as in EIA systems in developed countries, *scoping* is a very important step (Ahmad and Sammy, 1985; Kennedy, 1988a; Bisset, 1992; OECD, 1992). It is, however, frequently missing in developing countries, at least in so far as public consultation is concerned. Few EIAs in developing countries appear to be produced with the assistance of project-specific guidelines. However, the World Bank, which has a major influence on EIA practice, now demands scoping.

There are several prominent differences in practice between developing and developed countries in relation to *EIA report preparation*:

1. There is a lack of trained human resources and of financial resources which often leads to the preparation of inadequate and irrelevant EIA reports in developing countries.
2. Environmental conditions in tropical or near-tropical areas render many of the environmental assumptions, models and standards derived in temperate zones inappropriate.
3. Baseline socio-economic and environmental data may be inaccurate, difficult to obtain or non-existent in developing countries.
4. The significance attached to particular environmental impacts may be either much less or much greater (especially where cultural effects are involved) in developing countries than in developed countries.

Various means of addressing these difficulties have been suggested.

First, there is a need to focus both on capacity building and on training. The development assistance agencies are turning their attention to the need to create not only the institutional framework to administer EIA (above) but local centres of EIA expertise. Some of the agencies are now trying to encourage the development of the capacity to deal with EIA without great expense or complexity. In the past, many agencies have failed to recognise the importance of ensuring that skilled EIA personnel exist in environmental consultancies and research institutes in developing countries.

Numerous commentators have stressed the importance of training to increase the human resource capacity to undertake and review EIAs (Ahmad and Sammy, 1985; Biswas, 1992; Wilbanks et al., 1993). There is considerable agreement that training should be provided within the developing country. As McCormick (1993, p. 726) put it:

> Training in the home country makes it nearly impossible for trainers to ignore cultural differences that influence the effectiveness of training ... To provide training for those from developing countries, go there.

Such training needs to relate not just to government officials but to personnel in environmental consultancies and research institutes. Both longer-term and specialised short courses are necessary. Courses, as in the developed world, need to be multidisciplinary and focused on the practical and operational aspects of EIA and not on the theory of EIA (Biswas, 1992). In practice, courses have often failed to be practically orientated. The importance of assembling an appropriate interdisciplinary team to prepare EIA reports in developing countries has been stressed (Kennedy, 1988a; United Nations Environment Programme – UNEP, 1988).

Second, there is a need to employ EIA methodologies appropriate to conditions in developing countries. Biswas (1992, p. 240) has blamed the use of inappropriate, imported, methodologies for EIA reports which are 'too academic, bureaucratic, mechanistic and voluminous'. The World Bank (1991), the Overseas Development Administration (1992), OECD (1992) and the Commission of the European Communities (1993a) have all produced valuable guides. In many ways, however, UNEP (1988) enunciates the simplest and most important principles:

1. Focus on the main issues.
2. Involve the appropriate persons and groups.
3. Link information to decisions about the project.
4. Present clear options for the mitigation of impacts and for sound environmental management.
5. Provide information in a form useful to the decision-makers.

Methodologies, techniques and standards must be selected with the observance of these principles as the primary objective.

Third, while the absence of reliable baseline data in developing countries is a hindrance (Wilbanks et al., 1993), there are often more data available than people believe. However, problems of poor or non-existent data retrieval and management systems, inter-ministerial and/or inter-institutional rivalry, unnecessary classification of data as secret or confidential and inaccuracy of data need to be overcome (Biswas, 1992). It is often easier for an external expert to gain access to environmental data than an internal one. Training, personal motivation and public pressure are the main keys to unlocking data sources in developing countries.

Fourth, this lack of relevant baseline data, together with the different significance attached to impacts in different countries, are two of the strongest reasons for ensuring that indigenous experts undertake EIA and that local people participate in the EIA process. Local people can assist not only by helping to determine significance but by providing baseline environmental data.

The *review* stage appears to be missing in the EIAs in many developing countries. EIA reports are often confidential (Bisset, 1992). Some EIA reports in developing countries are bound like PhD theses, are about as indigestible, are produced with similarly limited numbers of copies and are not even available through inter-library loan. This is hardly an appropriate climate for peer and public review and needs to be addressed by development assistance agencies and governments as a matter of urgency. The World Bank's disclosure arrangements now address this issue.

Decision making on projects may be made both by development assistance agencies and by governments, is frequently closed to external scrutiny, and may be influenced not only by economic and social factors but by corruption. Too many examples exist in developing countries of mechanistic EIA reports being produced which have little or no effect on decisions. There appear to be two principal problems: the lack of willingness to integrate EIA either into project planning or into decision making; and the secretive nature of EIA and of decision making. Two possible solutions present themselves.

First, the problem of top-down EIA can be partially overcome by a real (rather than a 'lip-service') commitment by leaders of developing countries to use EIAs in decision making (Ahmad and Sammy, 1985). However, countries are much more likely to use EIAs in decision making if the EIA system responds to their needs and is designed and implemented by their own nationals (McCormick, 1993). In other words, the EIA process needs to be rooted in the indigenous culture for decision making, even if this is indirectly top-down (Wilbanks *et al.*, 1993), and to generate simple, easy to use, focused EIA reports. UNEP has also suggested that EIA must be integrated into the process of designing and implementing projects in the country concerned, or in what it terms the 'project cycle'. It has stated that:

> The key seems to lie in the management of the EIA: by designing the process so that it provides useful information to decision-makers at just the right time in the project cycle, EIA can have a real effect on projects.
>
> (UNEP, 1988, p. 5)

Figure A.1 represents this process diagrammatically.

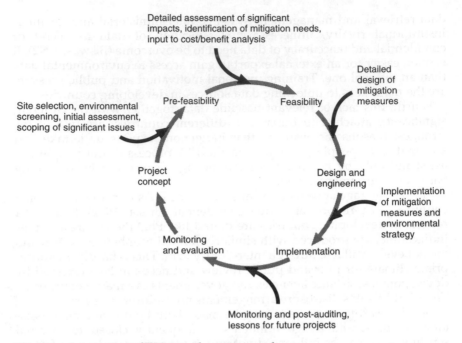

Fig. A.1   Integration of EIA into the project cycle.
Source: United Nations Environment Programme (1988, p. 5).

Second, piercing the shroud of secrecy could begin if development assistance agencies took the lead in publicising the way in which EIA influenced their own decisions. This would involve considerable, but necessary, change in many cases. An agency could then reasonably demand, perhaps as a condition of aid for a project, that the government department published the EIA and the reasons for its decision. As Wilbanks *et al.* (1993, p. 740) have stated 'rewards should be offered for doing the environmental impact assessment right'. The best reward would be to relate further aid to EIA performance. This would require close coordination and the sharing of EIA information between the various offices of the development assistance organisation. In some instances, this would cut across development agencies' political agendas and alter the way their staff operated. However, the World Bank has begun to make reforms in this area which have had a considerable effect.

As in the developed world, *monitoring* has been a missing step in EIA in developing countries. Projects may change substantially between author-isation and implementation and environmental controls may not be observed or monitored. Not only does appropriate compliance monitor-ing need to be made a condition of assistance but case studies of post-auditing the impacts of completed developments are urgently needed (Biswas, 1992). There is a gradual move towards insistence on management plans by development assistance agencies (OECD, 1992;

Overseas Development Administration, 1992) reflecting trends in the developed world (Chapter 15).

Mitigation of the impacts of some projects in developing countries is considered during the EIA process and is implemented in fewer cases. Too often, there is little opportunity for changes to be made to previously designed projects: mitigation is frequently an after-thought. Like the treatment of alternatives, mitigation is given less emphasis than in the developed world and, in many instances, mitigation measures remain on the unread pages of the EIA report. Since mitigation is, in many mature EIA systems in developed countries, the name of the game, this situation needs to be addressed urgently (Overseas Development Administration, 1992). In the first instance, this must involve the development assistance agencies and government departments demanding the inclusion of adequate mitigation measures in EIA reports and then the enforcement of their implementation.

In many developing countries there may be no tradition of *consultation and participation*. This lack of experience is often exacerbated by lack of knowledge about EIA, the confidentiality of EIA reports, lack of a culture of participation and low levels of literacy. While the notion of public participation in decision making may be revolutionary in developing countries (Wilbanks *et al.*, 1993) without some form of real participation EIA is meaningless (Chapter 1). Useful first steps to consultation might include encouragement to consult certain designated authorities, the establishment of one or more public interest environmental groups and the active involvement of local universities and research establishments in EIA. Targeted overseas funding might well be necessary to facilitate the taking of these steps. There is also a need to develop public participation methods appropriate to local circumstances.

There is very little *EIA system monitoring* in developing countries, though there are exceptions. Not only is there little information about EIA but EIA reports are not widely available and there is little interest in reviewing the operation of the system. This will change as EIA organisational capacity and regulatory requirements are expanded but motivation and external pressure will also be necessary (Ahmad and Sammy, 1992).

There is possibly a majority view in developing countries that EIA system *costs* exceed the *benefits*, though opinions vary from country to country. It is noticeable that the same arguments about delays, financial resources, lack of expertise, lack of data and confidentiality which were rehearsed in developed countries when project EIA was being introduced are surfacing again in developing countries. It is likely, given the firm encouragement and support of the international community for cost-effective EIA, that the outcome may be similar: that EIA will come to be accepted as an essential part of the development process, as it already has been in some developing countries.

There is very little experience of *strategic environmental assessment* (SEA) in developing countries but there is considerable interest in its application, especially in relation to regional development plans and land use plans for developing areas. Work on the World Bank's national environmental action plans, regional and sectoral environmental assessments, however, means that this type of activity is probably more advanced in some developing countries than in many developed countries. While the various problems relating to the use of SEA in developed countries are more acute in developing countries, the potential advantages of using SEA are all the greater precisely because development is taking place so quickly in many areas. The need for external support in this endeavour is apparent.

There would be considerable merit in testing individual developing country EIA systems against evaluation criteria similar to those used in this book as a basis for determining whether the current situation is still one of 'token compliance' and for establishing the most important steps towards improvement. The lessons learned in one developing country (and by one development assistance agency) could then be passed on to others.

There is a need for a database on EIA regulations, organisations and experience in developing countries to prevent countries from 'reinventing the wheel' in preparing their EIA legislation when examples which could easily be adapted already exist. Such a database would aid communication, which is often very poor, between environmental agencies in developing countries.

Ortolano (1993, p. 361) has stressed how important the role of development assistance agencies is in improving EIA in developing countries:

> Development aid agency control has great potential for bringing about effective EIA in developing countries, particularly those without national EIA requirements. ... However, this potential has not yet been fully realised because aid agencies have been slow to impose EIA requirements on recipients and even slower to enforce consistently nontrivial compliance with their own requirements. Under these circumstances, project proponents receiving development aid have often been able to get by with token compliance with the EIA requirements of donors.

However, the biggest constraint to effective EIA in developing countries is lack of political will. As in the developed world, only widespread popular demand for environmental improvement will ensure that effective EIA systems are introduced in developing countries. The development assistance agencies have a major role to play here. Since the pace of change is so much greater in developing countries, it would be appropriate for a greater proportion of the world's EIA expertise (appropriately adapted) and resources to be devoted to them if real progress towards sustainable development is to be made.

Note

1. I am grateful to Ron Bisset, Tim Clarke, Bill Kennedy, Mark King and Chris West for their helpful comments on a draft of this appendix.

References

Ahmad Y J and Sammy G K (1985) *Guidelines to Environmental Impact Assessment in Developing Countries.* Hodder and Stoughton, London

Anderson E (compiler) (1994) Australian Country Paper, 7th Tripartite Workshop on Environmental Impact Assessment. Commonwealth Environment Protection Agency, Canberra

Anderson E M (1990) Environmental impact assessment – the dream that failed? Paper to Conference on Government, Engineering and the Nation. Institution of Engineers, Canberra

Andreen W L (1992) The evolving law of environmental protection in the United States: 1970–1991. *Environmental and Planning Law Journal* 9: 96–110

Arnstein S R (1969) A ladder of citizen participation in the USA. *Journal of the American Institute of Planners* 35: 216–24

Australian and New Zealand Environment and Conservation Council (1991) *A National Approach to Environmental Impact Assessment in Australia.* ANZECC, Canberra, Australia

Bailey J M and English V (1991) Western Australian environmental impact assessment: an evolving approach to environmentally sound development. *Environmental and Planning Law Journal* 8: 190–9

Bailey J M and Hobbs V (1990) A proposed framework and database for EIA auditing. *Journal of Environmental Management* 31: 163–72

Bailey J M, Hobbs V and Saunders A (1992) Environmental auditing: artificial waterway developments in Western Australia. *Journal of Environmental Management* 34: 1–13

Baldwin J H (1985) *Environmental Planning and Management.* Westview Press, Boulder, CO

Bartlett R V (1989) Impact assessment as a policy strategy. In Bartlett R V (ed.) *Policy through Impact Assessment.* Greenwood Press, New York, NY

Bartlett R V and Baber W F (1989) Bureaucracy or analysis: implications of impact assessment for public administration. In Bartlett R V (ed.) *Policy through Impact Assessment*. Greenwood Press, New York, NY

Bass R E (1990a) California's experience with environmental impact reports. *Project Appraisal* **5**: 220–4

Bass R E (1990b) Mitigation monitoring: California closes a loophole. *Proceedings International Association for Impact Assessment Conference, Lausanne.* IAIA, Ecole Polytechnique Federale de Lausanne

Bass R E (1991) Policy, plan and programme EIA in California. *EIA Newsletter* **5**: 4,5 (EIA Centre, Department of Planning and Landscape, University of Manchester)

Bass R E and Herson A I (1989) The California Environmental Quality Act, Chapters 20–3. In Manaster K A and Selmi P P (eds) *California Environmental Law and Practice*. Matthew Bender, New York, NY

Bass R E and Herson A I (1993a) *Mastering NEPA: a Step-by-Step Approach*. Solano Press, Point Arena, CA

Bass R E and Herson A I (1993b) *Successful CEQA Compliance: a Step-by-Step Approach*. Solano Press, Point Arena, CA

Bass R E and Herson A I (1993c) What CEQA practitioners must know about NEPA. *Land Use Forum* **2**: 132–9

Beanlands G (1988) Scoping methods and baseline studies in EIA. In Wathern P (ed.) *Environmental Impact Assessment: Theory and Practice*. Unwin Hyman, London

Beanlands G E and Duinker P N (1983) *An Ecological Framework of Environmental Impact Assessment in Canada*. Federal Environmental Assessment Review Office, Hull, Quebec

Beanlands G E and Duinker P N (1984) An ecological framework for environmental impact assessment. *Journal of Environmental Management* **18**: 267–77

Bear D (1988) Does NEPA make a difference? *EPA Journal* **14**(1): 34–5

Bear D (1989) NEPA at 19: a primer on an 'old' law with solutions to new problems. *Environmental Law Reporter* **19**: 10060–9

Bear D and Blaug E (1991) Recent EIA developments in the United States of America. *EIA Newsletter* **6**: 18–19 (EIA Centre, Department of Planning and Landscape, University of Manchester)

Bendix S (1979) A short introduction to the California Environmental Quality Act. *Santa Clara Law Review* **19**: 521–39

Bingham G (1986) *Resolving Environmental Disputes: a Decade of Experience*. Conservation Foundation, Washington, DC

Bisset R (1978) Quantification, decision-making and environmental impact assessment in the United Kingdom. *Journal of Environmental Management* **7**: 43–58

Bisset R (1980) Methods of environmental impact assessment: recent trends and future prospects. *Journal of Environmental Management* **11**: 27–43

Bisset R (1981) Problems and issues in the implementation of EIA audits. *Environmental Impact Assessment Review* **1**: 379–96

Bisset R (1984) Post development audits to investigate the accuracy of environmental impact predictions. *Zeitschrift für Umweltpolitik* **7**: 463–84

Bisset R (1988) Developments in EIA methods. In Wathern P (ed.) *Environmental Impact Assessment: Theory and Practice.* Unwin Hyman, London

Bisset R (1992) Devising an effective environmental assessment system for a developing country: the case of the Turks and Caicos Islands. In Biswas A K and Agarwala S B C (eds) *Environmental Impact Assessment for Developing Countries.* Butterworth-Heinemann, Oxford

Bisset R and Tomlinson P (1988) Monitoring and auditing of impacts. In Wathern P (ed.) *Environmental Impact Assessment: Theory and Practice.* Unwin Hyman, London

Biswas A K (1992) Summary and recommendations. In Biswas A K and Agarwala S B C (eds) *Environmental Impact Assessment for Developing Countries.* Butterworth-Heinemann, Oxford

Biswas A K and Agarwala S B C (eds) (1992) *Environmental Impact Assessment for Developing Countries.* Butterworth-Heinemann, Oxford

Biswas A K and Geping Q (eds) (1987) *Environmental Impact Assessment for Developing Countries.* Tycooly International, London

Blaug E A (1993) Use of the environmental assessment by federal agencies in NEPA implementation. *The Environmental Professional* **15**: 57–65

Blumm M C (1988) The origin, evolution and direction of the United States National Environmental Policy Act. *Environmental and Planning Law Journal* **5**: 179–93

Blumm M C (1990) The National Environmental Policy Act at twenty: a preface. *Environmental Law* **20**: 447–83

Bregha F, Bendickson J, Gamble D, Shillington T and Weick E (1990) *The Integration of Environmental Considerations into Government Policy.* Canadian Environmental Assessment Research Council, Hull, Quebec

Broili R T (1993) Identification, tracking and closure of NEPA commitments: the Tennessee Valley Authority Process. In Hildebrand S G and Cannon J B (eds) *Environmental Analysis: the NEPA Experience.* Lewis, Boca Raton, FL

Brown A L (1992) Beyond EIA – incorporating environment into the engineering design process. Paper to National Conference on Environmental Engineering, Gold Coast, Queensland, Australia

311

Buckley R (1989) *Precision in Environmental Impact Prediction.* Resource and Environmental Studies Paper 2. Centre for Resource and Environmental Studies, Australian National University, Canberra

Buckley R (1990) Adequacy of current legislative and institutional frameworks for environmental impact audit in Australia. *Environmental and Planning Law Journal* 7: 142–6

Buckley R (1991a) Auditing the precision and accuracy of environmental impact predictions in Australia. *Environmental Monitoring and Assessment* 18: 1–23

Buckley R (1991b) Environmental planning legislation: court back-up better than Ministerial discretion. *Environmental and Planning Law Journal* 8: 250–7

Bureau of Industry Economics (1990) *Environmental Assessment – Impact on Major Projects.* Research Report 35, AGPS, Canberra

Bush M (1990) *Public Participation in Resource Development after Project Approval.* Canadian Environmental Assessment Research Council, Hull, Quebec

Caldwell L K (1989a) A constitutional law for the environment: 20 years with NEPA indicates the need. *Environment* 31(10): 6–11, 25–8

Caldwell L K (1989b) NEPA revisited: a call for a constitutional amendment. *The Environmental Forum* 6(6): 18–22

Caldwell L K (1989c) Understanding impact analysis: technical process, administrative reform, policy principle. In Bartlett R V (ed.) *Policy through Impact Assessment.* Greenwood Press, New York, NY

Caldwell L K, Bartlett R V, Parker D E and Keys D L (1983) *A Study of Ways to Improve the Scientific Content and Methodology of Environmental Impact Analysis.* Document No PB 83-222851, National Technical Information Service, Springfield, VA

Callies P L (1984) *Regulating Paradise: Land Use Controls in Hawaii.* University of Hawaii Press, Honolulu, HI

Canadian Environmental Assessment Research Council (1988a) *Evaluating Environmental Impact Assessment: an Action Prospectus.* CEARC, Hull, Quebec

Canadian Environmental Assessment Research Council (1988b) *The Assessment of Cumulative Effects: a Research Prospectus.* CEARC, Hull, Quebec

Canter L W (1977) *Environmental Impact Assessment.* McGraw-Hill, New York

Canter L W, Robertson J M and Westcott R M (1991) Identification and evaluation of biological impact mitigation measures. *Journal of Environmental Management* 33: 35–50

Cass W H (1974) Speech by the Minister for the Environment and

Conservation. *Parliamentary Debates: House of Representatives* **92**: 4081–3, 26 November 1974, AGPS, Canberra

Catlow J and Thirwall C G (1976) *Environmental Impact Analysis*. Research Report 11, Department of the Environment, London

Cheshire County Council (1989) *The Cheshire Environmental Assessment Handbook*. Planning Practice Notes No. 2, Environmental Planning Series, Environmental Planning Department, CCC, Chester

Clark B D, Chapman K, Bisset R and Wathern P (1976) *Assessment of Major Industrial Applications: a Manual*. Research Report 13, Department of the Environment, London

Clark B D, Bisset R and Wathern P (1980) *Environmental Impact Assessment: a Bibliography with Abstracts*. Mansell, London

Clark B D, Chapman K, Bisset R, Wathern P and Barrett M (1981) *A Manual for the Assessment of Major Development Proposals*. Department of the Environment, HMSO, London

Clark R (1993) The National Environmental Policy Act and the role of the President's Council on Environmental Quality. *The Environmental Professional* **15**: 4–6

Cocklin C (1989) The restructuring of environmental administration in New Zealand. *Journal of Environmental Management* **28**: 309–26

Cocklin C, Parker S and Hay J (1992) Notes on cumulative environmental change: concepts and issues. *Journal of Environmental Management* **35**: 31–49

Coenen R (1993) NEPA's impact on environmental impact assessment in European Community member countries. In Hildebrand S G and Cannon J B (eds) *Environmental Analysis: the NEPA Experience*. Lewis, Boca Raton, FL

Coles T (1991) Experience to date. *Proceedings of Advances in Environmental Assessment Conference*. Legal Studies and Services, London, and Institute of Environmental Assessment, Grantham

Commission of the European Communities (1977) European Community Policy and Action Programme on the Environment for 1977–1981. *Official Journal of the European Communities* **C139**: 1–46, 13 June 1977

Commission of the European Communities (1979) *State of the Environment: Second Report*. CEC, Brussels

Commission of the European Communities (1980) Proposal for a Council Directive concerning the assessment of the environmental effects of certain public and private projects. *Official Journal of the European Communities* **C169**: 14–22, 9 July 1980

Commission of the European Communities (1982) Proposal to amend the proposal for a Council Directive concerning the assessment of the environmental effects of certain public and private projects. *Official Journal of the European Communities* **C110**: 5–11, 1 May 1982

Commission of the European Communities (1985) Council Directive of 27 June 1985 on the assessment of the effects of certain public and private projects on the environment. *Official Journal of the European Communities* **C175**: 40–9, 5 July 1985

Commission of the European Communities (1993a) *Environmental Manual: Environmental Procedures and Methodology Governing Lomé IV Development Cooperation Projects.* DGVII, CEC, Brussels

Commission of the European Communities (with the assistance of Lee N and Jones C E) (1993b) *Report from the Commission of the Implementation of Directive 85/337/EEC and Annex for the United Kingdom.* COM (93) 28, Volume 12. CEC, Brussels

Commission of the European Communities (with the assistance of Lee N and Jones C E) (1993c) *Report from the Commission of the Implementation of Directive 85/337/EEC and Annexes for all Member States.* COM (93) 28, Volume 13. CEC, Brussels

Commonwealth of Australia (1987) *Environment Protection (Impact of Proposals) Act 1974 Administration Procedures.* Order under Section 6 of the Act, 29 May 1987. AGPS, Canberra

Commonwealth Environment Protection Agency (1992) *Inter-governmental Agreement on the Environment.* EPA, Canberra

Commonwealth Environment Protection Agency (1993) *Setting the Direction.* Initial Discussion Paper, Review of Commonwealth EIA Process, EPA, Canberra

Couch, W J (ed.) (1988) *Environmental Assessment in Canada: 1988 Summary of Current Practice.* Federal Environmental Assessment Review Office, Hull, Quebec

Council on Environmental Quality (1973) *Environmental Quality 1973: Fourth Annual Report.* USGPO, Washington, DC

Council on Environmental Quality (1974) *Environmental Quality 1974: Fifth Annual Report.* USGPO, Washington, DC

Council on Environmental Quality (1975) *Environmental Quality 1975: Sixth Annual Report.* USGPO, Washington, DC

Council on Environmental Quality (1976) *Environmental Quality 1976: Seventh Annual Report.* USGPO, Washington, DC

Council on Environmental Quality (1977) *Environmental Quality 1977: Eighth Annual Report.* USGPO, Washington, DC

Council on Environmental Quality (1978) Regulations for implementing the procedural provisions of the National Environmental Quality Act *40 Code of Federal Regulations 1500–1508* (reproduced in Council on Environmental Quality, 1992 and in many other texts)

Council on Environmental Quality (1980) *Environmental Quality 1980: Eleventh Annual Report.* USGPO, Washington, DC

Council on Environmental Quality (1981a) Memorandum: forty most asked questions concerning CEQ's National Environmental Policy Act Regulations (40 questions) *46 Federal Register* 18026 (23 March 1981), as amended by *51 Federal Register* 15618 (25 April 1986) (reproduced in Bass and Herson, 1993a, and Mandelker, 1993b)

Council on Environmental Quality (1981b) *Memorandum: Scoping Guidance*. CEQ, Washington, DC (reproduced in Bass and Herson, 1993a)

Council on Environmental Quality (1990) *Environmental Quality 1989: Twentieth Annual Report*. USGPO, Washington, DC

Council on Environmental Quality (1992) *Environmental Quality 1991: Twenty-second Annual Report*. USGPO, Washington, DC

Council on Environmental Quality (1993) *Environmental Quality 1992: Twenty-third Annual Report*. USGPO, Washington, DC

Culhane P J (1987) The precision and accuracy of US environmental impact statements. *Environmental Monitoring and Assessment* **8**: 218–38

Culhane P J, Friesema H P and Beecher J A (1987) *Forecasts and Environmental Decision Making: the Content and Predictive Accuracy of Environmental Impact Statements*. Westview Press, Boulder, CO

de Boer J J (1991) EIA training in The Netherlands. In Wood C M and Lee N (eds) *Environmental Impact Assessment Training and Research in the European Communities*. Occasional Paper 27, Department of Planning and Landscape, University of Manchester

Dee N, Baker J K, Drobny N L, Duke K M, Whitman I and Fahringer D C (1973) Environmental evaluation system for water resource planning. *Water Resources Research* **9**: 523–35

Department of the Arts, Sport and the Environment, Tourism and Territories (1991) *Proposed Commonwealth Environment Protection Agency*. Position paper for public comment, DASETT, Canberra

Department of the Environment (1975) *Review of the Development Control System: Final Report by Mr George Dobry, QC*. Circular 113/75, HMSO, London

Department of the Environment (1977) *Press Notice 68*. DOE, London

Department of the Environment (1978) *Press Notice 488*. DOE, London

Department of the Environment (1986a) *Implementation of the European Directive on Environmental Assessment*. Consultation Paper, DOE, London

Department of the Environment (1986b) *The Planning System: the Environmental Assessment of Major Projects in England and Wales*. Draft Advisory Booklet, DOE, London

Department of the Environment (1988a) *Environmental Assessment: Implementation of EC Directive*. Consultation Paper, DOE, London

Department of the Environment (1988b) *Environmental Assessment*. Circular 15/88, HMSO, London

Department of the Environment (1988c) *Environmental Assessment of Projects in Simplified Planning Zones and Enterprise Zones.* Circular 23/88, HMSO, London

Department of the Environment (1989) *Environmental Assessment: a Guide to the Procedures.* HMSO, London

Department of the Environment (1991) *Policy Appraisal and the Environment: a Guide for Government Departments.* HMSO, London

Department of the Environment (1992) *Development Plans and Regional Planning Guidance.* Planning Policy Guidance (PPG)12, HMSO, London

Department of the Environment (1993) *The Environmental Appraisal of Development Plans: a Good Practice Guide.* HMSO, London

Department of the Environment (1994a) *Environmental Assessment: Amendment of Regulations.* Circular 7/94, HMSO, London

Department of the Environment (1994b) *Guide on Preparing Environmental Statements for Planning Projects.* Consultation Draft, DOE, London

Department of Transport (1983) *Manual for Environmental Appraisal.* DOT, London

Department of Transport (1993) *Environmental Assessment.* Design manual for roads and bridges, Volume 11. HMSO, London

Dickerson W and Montgomery J (1993) Substantive scientific and technical guidance for NEPA analysis: pitfalls in the real world. *The Environmental Professional* **15**: 7–11

Dixon J (1993a) EIA in policy and plans: new practice in New Zealand, unpublished paper. Centre for Environmental and Resource Studies, University of Waikato

Dixon J (1993b) The integration of EIA and planning in New Zealand: changing process and practice. *Journal of Environmental Planning and Management* **36**: 239–51

Dobry G (1975) *Review of the Development Control System: Final Report.* DOE, HMSO, London

Dorais M (1993) Environmental assessment in the 1990s: facing the test of relevance. *Proceedings of International Association for Impact Assessment Conference, Shanghai.* IAIA, Belhaven, NC

Ecologically Sustainable Development Working Groups (1991) *Final Report-Executive Summaries.* AGPS, Canberra

Ecologically Sustainable Development Working Group Chairs (1992) *Intersectoral Issues Report.* AGPS, Canberra

EIA Centre (1993) *EIA in Developing Countries.* Leaflet 15, EIA Centre, Department of Planning and Landscape, University of Manchester

Elkin T J and Smith P G R (1988) What is a good environmental impact statement? Reviewing screening reports from Canada's national parks. *Journal of Environmental Management* **26**: 71–89

Environmental Impact Assessment Commission (1994) *Annual Report for 1993*. Ministry of Housing, Physical Planning and Environment (VROM) and Ministry of Agriculture, Nature Management and Fisheries, The Hague (in Dutch, English summary)

Environmental Law Institute (1981) *NEPA in Action: Environmental Offices in 19 Federal Agencies*. Council on Environmental Quality, Washington, DC

Environmental Law Institute (1989) *NEPA Deskbook*. ELI, Washington, DC

Environmental Protection Agency (1984) *Policy and Procedures for the Review of Federal Actions Impacting the Environment*. Office of Federal Activities, EPA, Washington, DC

Environmental Protection Agency (1993) *Sourcebook for the Environmental Assessment (EA) Process*. EPA, Washington, DC

Essex County Council (1994) *The Essex Guide to Environmental Assessment*. Essex Planning Officers' Association, Essex County Council, Chelmsford (revised edition)

Evaluation Committee on EIA (1990) *Towards a Better Procedure to Protect the Environment: Report on the Working of the Regulation on Environmental Impact Assessment Contained in the Environmental Protection (General Provisions) Act: Summary*. ECW Report 3, Ministry of Housing, Physical Planning and Environment (VROM), The Hague

Fairfax S K (1978) A disaster in the environmental movement. *Science* **199**: 743–48, 17 February 1978

Fairfax S K and Ingram H M (1981) The United States experience. In O'Riordan T and Sewell W R D (eds) *Project Appraisal and Policy Review*. Wiley, Chichester

Federal Environmental Assessment Review Office (1988a) *Public Review: Neither Judicial, nor Political, but an Essential Forum for the Future of the Environment*. FEARO, Hull, Quebec

Federal Environmental Assessment Review Office (1988b) *The National Consultation Workshop on Federal Environmental Assessment Reform: Report of Proceedings*. FEARO, Hull, Quebec

Federal Environmental Assessment Review Office (1990) *Guidelines for the Preparation of an Environmental Impact Statement on Air Transportation Proposals for the Toronto Area*. FEARO, Hull, Quebec

Federal Environmental Assessment Review Office (1991) *Impact: FEARO Newsletter* **1**(1), FEARO, Hull, Quebec

Federal Environmental Assessment Review Office (1992a) *Bulletin of Initial Assessment Decisions 13*, FEARO, Hull, Quebec

Federal Environmental Assessment Review Office (1992b) *Oldman River Dam: Report of the Environmental Assessment Panel.* Report 42, FEARO, Hull, Quebec

Federal Environmental Assessment Review Office (1993a) *A Guide to the Canadian Environmental Assessment Act.* FEARO, Hull, Quebec

Federal Environmental Assessment Review Office (1993b) *The Environmental Assessment Process for Policy and Program Proposals.* FEARO, Hull, Quebec

Federal Environmental Assessment Review Office (1994) Unpublished tables and figures. FEARO, Hull, Quebec

Fenge T and Rees W E (1987) *Hinterland or Homeland: Land-use Planning in Northern Canada.* Canadian Arctic Resources Committee, Ottawa, Ontario

Fenge T and Smith L G (1986) Reforming the Federal Environmental Assessment and Review Process. *Canadian Public Policy* **12**: 596–605

Fogleman V M (1990) *Guide to the National Environment Policy Act: Interpretations, Applications and Compliance.* Quorum, New York, NY

Fookes T W (1987a) A comparison of environmental impact assessment in South Australia and proposed United Nations Environment Programme goals and principles. *Environmental and Planning Law Journal* **4**: 204–15

Fookes T W (1987b) New environmental administration. *Environmental Policy and Law [1987]*: 129–34

Fookes T W (1991) Findings of the Third Group, on 'how we make the environmental assessment of policy happen'. New Zealand/Australia/Canada Workshop on Environmental Impact Assessment, Wellington, New Zealand

Formby J (1987) Australian Government's experience with environmental impact assessment. *Environmental Impact Assessment Review* **3**: 207–26

Fortlage C A (1990) *Environmental Assessment: a Practical Guide.* Gower, Aldershot

Fowler R J (1982) *Environmental Impact Assessment, Planning and Pollution Measures in Australia.* Department of Home Affairs and Environment, AGPS, Canberra

Fowler R J (1985) Legislative bases of environmental impact assessment. *Environmental and Planning Law Journal* **2**: 200–5

Fowler R J (1991) *Proposal for a Federal Environment Protection Agency.* Faculty of Law, University of Adelaide

Frieden B J (1979) *The Environmental Protection Hustle.* MIT Press, Cambridge, MA

Frost R, Therivel R, Metton J and McKenzie C (1994) *Directory of*

Environmental Impact Statements July 1988–September 1993. Impacts Assessment Unit, School of Planning, Oxford Brookes University

Gibson R B (1992) The New Canadian Environmental Assessment Act: possible responses to its main deficiencies. *Journal of Environmental Law and Practice* **2**: 223–55

Gibson R B (1993) Environmental assessment design: lessons from the Canadian experience. *The Environmental Professional* **15**: 12–24

Glasson J, Therivel R and Chadwick A (1994) *Introduction to Environmental Impact Assessment*. UCL Press, London

Government of Canada (1984) Environmental Assessment and Review Process Guidelines Order. *Canada Gazette* Part II **118**(14): 2794–802

Government of Canada (1992) *North America Free Trade Agreement: Canadian Environment Review*. Environment Canada, Hull, Quebec

Haigh N (1991) *Manual of European Policy: the EC and Britain*. Longman, Harlow

Hart S L (1984) The costs of environmental review: assessment methods and trends. In Hart S L, Enk G A and Hornick W F (eds) *Improving Impact Assessment: Increasing the Relevance and Utilisation of Scientific and Technical Information*. Westview Press, Boulder, CO

Hart S L and Enk G A (1980) *Green Goals and Greenbacks: State-Level Environmental Programs and their Associated Costs*. Westview Press, Boulder, CO

Hart S L, Enk G A and Hornick W F (1984) *Improving Impact Assessment: Increasing the Relevance and Utilisation of Scientific and Technical Information*. Westview Press, Boulder, CO

Her Majesty's Government (1990) *This Common Inheritance: Britain's Environmental Strategy*. Cm 1200, HMSO, London

Her Majesty's Government (1994) *Sustainable Development: the UK Strategy*. Cm 2426, HMSO, London

Hernandez J, Ziebron W S, O'Hare T, Dakin N and Ness S (1993) *A Practical Guide to Implementing the California Environmental Quality Act*. California Environmental Publications, San Rafael, CA

Herson A I (1986) Project mitigation revisited: most courts approve findings of no significant impact justified by mitigation. *Ecology Law Quarterly* **13**(1): 51–72

Hildebrand S G and Cannon J B (eds) (1993) *Environmental Analysis: the NEPA Experience*. Lewis, Boca Raton, FL

Hollick M (1981a) Enforcement of mitigation measures resulting from environmental impact assessment. *Environmental Management* **5**: 507–13

Hollick M (1981b) The role of quantitative decision-making methods in environmental impact assessment. *Journal of Environmental Management* **12**: 65–78

Hollick M (1986) Environmental impact assessment: an international evaluation. *Environmental Management* **10**: 157–78

Holling C S (ed.) (1978) *Adaptive Environmental Assessment and Management*. Wiley, Chichester

House of Lords (1981a) *Environmental Assessment of Projects*. Select Committee on the European Communities, 11th Report, Session 1980–81, HMSO, London

House of Lords (1981b) EEC 11th Report: Environment. *Parliamentary Debates (Hansard) Official Report, Session 1980–81*. 30 April 1981, 1311–47, HMSO, London

Hughes H R (1992) The Resource Management Act: new responsibilities for the young and not-so-young planners. Paper to New Zealand Planning Institute Seminar, Napier, April

Hyman E L and Stiftel B (1988) *Combining Facts and Values in Environmental Impact Assessment: Theories and Techniques*. Westview Press, Boulder, CO

Institute of Environmental Assessment (1993) *Digest of Environmental Statements*. Sweet and Maxwell, London, 2 volumes

International Union for Conservation of Nature and Natural Resources (1980) *World Conservation Strategy – Living Resource Conservation for Sustainable Development*. IUCN, Gland, Switzerland

Jacobs P and Sadler B (eds) (1990) *Sustainable Development and Environmental Assessment: Perspectives and Planning for a Common Future*. Canadian Environmental Assessment Research Council, Hull, Quebec

Jeffery M I (1987) Accommodating negotiation in environmental impact assessment and project approval processes. *Environmental and Planning Law Journal* **4**: 244–52

Johnston R A and McCartney W S (1991) Local government implementation of mitigation requirements under the California Environmental Quality Act. *Environmental Impact Assessment Review* **11**: 53–67

Jones C E (1993) *UK Environmental Statements 1991: a Comparative Analysis*. Occasional Paper 36, Department of Planning and Landscape, University of Manchester

Jones C E, Lee N and Wood C M (1991) *UK Environmental Statements 1988–1990: an Analysis*. Occasional Paper 29, Department of Planning and Landscape, University of Manchester

Jones M G (1984) The evolving EIA procedure in The Netherlands. In Clark B D, Gilad A, Bisset R and Tomlinson P (eds) *Perspectives on Environmental Impact Assessment*. Reidel, Dordrecht

Kaplan-Wildmann J and McBride J (1992) The California Environmental

Quality Act: current practice and prospects for reform. *California Land Use* **1**: 190–8

Keating P J (1992) *One Nation: Statement by the Prime Minister The Honourable P J Keating, MP.* 26 February 1992, AGPS, Canberra

Kennedy W V (1988a) Environmental impact assessment and bilateral development aid: an overview. In Wathern P (ed.) *Environmental Impact Assessment: Theory and Practice.* Unwin Hyman, London

Kennedy W V (1988b) Environmental impact assessment in North America, Western Europe: what has worked where, how and why? *International Environment Reporter* **11**(4): 257–62

Kent County Council (1991) *Kent Environmental Assessment Handbook.* County Planning Department, Maidstone

Kent County Council (1993) *Strategic Environmental Appraisal of Policies.* Kent Structure Plan Third Review: Technical Working Paper 1/93, County Planning Department, Maidstone

Kinhill Engineers Pty Ltd and Phillips Fox Solicitors (1991) *Approval Systems for Major Projects in Australia – a Comparative Review.* Prepared for a Joint Steering Committee chaired by Federal Department of Industry, Technology and Commerce, Canberra

Krawetz N M, MacDonald W R and Nichols P (1987) *A Framework for Effective Monitoring.* Canadian Environmental Assessment Research Council, Hull, Quebec

Lambert A and Wood C M (1990) Environmental assessment implementation: spirit or letter? *Town Planning Review* **61**: 247–61

Land Use Consultants (1992) Unpublished table. LUC, London

Lawrence D P (1994) Designing and adapting the EIA planning process. *The Environmental Professional* **16**: 2–21

Lee N (1989) *Environmental Impact Assessment: a Training Guide.* Occasional Paper 18, Department of Planning and Landscape, University of Manchester (2nd edition)

Lee N and Brown D (1992) Quality control in environmental assessment. *Project Appraisal* **7**(1): 41–5

Lee N and Colley R (1991) Reviewing the quality of environmental statements: review methods and findings. *Town Planning Review* **62**: 239–48

Lee N and Colley R (1992) *Reviewing the Quality of Environmental Statements.* Occasional Paper 24, Department of Planning and Landscape, University of Manchester (2nd edition)

Lee N and Walsh F (1992) Strategic environmental assessment: an overview. *Project Appraisal* **7**: 126–36

Lee N and Wood C M (1976) *The Introduction of Environmental Impact*

Statements in the European Community. ENV/197/76. Commission of the European Communities, Brussels

Lee N and Wood C M (1978a) EIA – a European perspective. *Built Environment* **4**: 101–10

Lee N and Wood C M (1978b) Environmental impact assessment of projects in EEC countries. *Journal of Environmental Management* **6**: 57–71

Lee N and Wood C M (1985) Training for environmental impact assessment within the European Economic Community. *Journal of Environmental Management* **21**: 271–86

Leopold L B, Clark F E, Hanshaw B B and Balsley J R (1971) *A Procedure for Evaluating Environmental Impact.* US Geological Survey Circular 645, Department of the Interior, Washington, DC

Lundquist L J (1978) The comparative study of environmental politics: from garbage to gold? *International Journal of Environmental Studies* **12**: 89–97

Lynn S and Wathern P (1991) Intervenor funding in the environmental assessment process in Canada. *Project Appraisal* **5**: 169–73

MacLaren V W and Whitney J B (eds) (1985) *New Directions in Environmental Impact Assessment in Canada.* Methuen, Toronto

Macrory R (1989) UK pollution control – legal perspectives. In Jain R and Clark A (eds) *Environmental Technology, Assessment and Policy.* Ellis Horwood, Chichester

Mandelker D R (1993a) Environmental policy: the next generation. *Town Planning Review* **64**: 107–17

Mandelker D R (1993b) *NEPA Law and Litigation.* Clark Boardman Callaghan, Deerfield, IL (2nd edition)

Martyn A, Morris M L and Downing F (1990) *Environmental Impact Assessment Process in Australia.* Paper for Environmental Institute of Australia National Workshop on EIA, Adelaide

McCallum P R (1987) Follow-up to environmental impact assessment: learning from the Canadian Government experience. *Environmental Monitoring and Assessment* **8**: 199–215

McCormick J F (1993) Implementation of NEPA and environmental impact assessment in developing countries. In Hildebrand S G and Cannon J B (eds) *Environmental Analysis: The NEPA Experience.* Lewis, Boca Raton, FL

Memon P A (1993) *Keeping New Zealand Green: Recent Environmental Reforms.* University of Otago Press, Dunedin

Miller C E and Wood C M (1983) *Planning and Pollution.* Oxford University Press, Oxford

Ministry for the Environment (1987) *Environmental Protection and Enhancement Procedures.* MfE, Wellington

Ministry for the Environment (1991a) *Assessment of Environmental Effects.* Information Sheet 12, MfE, Wellington

Ministry for the Environment (1991b) *Guide to the Act.* MfE, Wellington

Ministry for the Environment (1991c) *Guideline for District Plans.* MfE, Wellington

Ministry for the Environment (1991d) *Regional Policy Statements and Plans.* MfE, Wellington

Ministry for the Environment (1992a) *Hinengaro Bay: a Fictional Case Study of Policy and Plan Making.* MfE, Wellington

Ministry for the Environment (1992b) *Scoping of Environmental Effects: a Guide to Scoping and Public Review Methods in Environmental Assessment.* MfE, Wellington

Ministry for the Environment (1993) *Section 32 – a Guide to Good Practice.* MfE, Wellington

Ministry of Housing, Physical Planning and Environment (1981) *Scoping and Guidelines.* MER Series 3, MHPPE (VROM), The Hague

Ministry of Housing, Physical Planning and Environment (1984) *Prediction in Environmental Impact Assessment.* MER Series 17, MHPPE (VROM), The Hague

Ministry of Housing, Physical Planning and Environment (1985) *Handling Uncertainty in Environmental Impact Assessment.* MER Series 18, MHPPE (VROM), The Hague

Ministry of Housing, Physical Planning and Environment and Ministry of Agriculture, Nature Management and Fisheries (1991) *Environmental Impact Assessment: The Netherlands – Fit for Future Life.* MHPPE (VROM), The Hague

Montz B E and Dixon J E (1993) From law to practice: EIA in New Zealand. *Environmental Impact Assessment Review* **13**: 89–108

Morgan R K (1983) The evolution of environmental impact assessment in New Zealand. *Journal of Environmental Management* **16**: 139–52

Morgan R K (1988) Reshaping environmental impact assessment in New Zealand. *Environmental Impact Assessment Review* **8**: 293–306

Morgan R K (1993) An evaluation of progress with implementing the environmental assessment requirements of the Resource Management Act. *International Proceedings of Association for Impact Assessment Conference, Shanghai.* IAIA, Belhaven, NC

Morgan R K and Memon P A (1993) *Assessing the Environmental Effects of Major Projects: a Practical Guide.* Publication 4, Environmental Policy and Management Research Centre, University of Otago

Morgan R K, Memon P A and Miller M A (eds) (1991) *Implementing the Resource Management Act.* Publication 1, Environmental Policy and Management Research Centre, University of Otago

Munn R E (ed.) (1979) *Environmental Impact Assessment: Principles and Procedures.* SCOPE 5, Wiley, Chichester

Munro D A, Bryant T J and Matte-Barker A (1986) *Learning from Experience: a State-of-the-Art Review and Evaluation of Environmental Impact Assessment Audits.* Canadian Environmental Assessment Research Council, Hull, Quebec

Murthy K S (1988) *National Environmental Policy Act (NEPA) Process.* CRC Press, Boca Raton, FL

Nay Htun (1988) The EIA process in Asia and the Pacific region. In Wathern P (ed.) *Environmental Impact Assessment: Theory and Practice.* Allen and Unwin, London

North Atlantic Treaty Organization (1993) *Methodology, Evaluation and Scope of Environmental Impact Assessment.* Report 197, Committee on the Challenges of Modern Society, NATO, Brussels

Office of Planning and Research (1983) *Preparing an Environmental Impact Report for a General Plan.* OPR, Sacramento, CA

Office of Planning and Research (1989) *Tracking CEQA Mitigation Measures under AB 3180.* OPR, Sacramento, CA

Office of Planning and Research (1990) *General Plan Guidelines.* OPR, Sacramento, CA

Office of Planning and Research (1992) *CEQA: the California Environmental Quality Act: Statutes and Guidelines.* OPR, Sacramento, CA

Office of Planning and Research (1993) Project Review Summary. OPR, Sacramento, CA (database printout)

Olshansky R B (1991) *CEQA and Planning Practice: Tabulated Results.* Department of Urban and Regional Planning, University of Illinois, Urbana, IL

Olshansky R B (1992) The California Environmental Quality Act: implications for local land use planning. *Environmental Assessor* 3(1): 1–4

Olshansky R B (1993) Department of Urban and Regional Planning, University of Illinois at Urbana – Champaign, personal communication, 11 October 1993

Organisation for Economic Cooperation and Development (1992) *Good Practices for Environmental Impact Assessment of Development Projects.* Development Assistance Committee, OECD, Paris

O'Riordan T and Sewell W R D (eds) (1981) *Project Appraisal and Policy Review.* Wiley, Chichester

Ortolano L (1984) *Environmental Planning and Decision Making.* Wiley, New York, NY

Ortolano L (1993) Controls on project proponents and environmental impact assessment effectiveness. *The Environmental Professional* **15**: 352–63

Ortolano L, Jenkins B and Abracosa R P (1987) Speculations on when and why EIA is effective. *Environmental Impact Assessment Review* 7: 287–92

Overseas Development Administration (1992) *Manual of Environmental Appraisal*. ODA, London (2nd edition)

Passenger Transport Executive Group (1991) *Environmental Assessment Guide for Passenger Transport Schemes*. PTEG, Manchester (2 volumes)

Peterson E B, Chan Y H, Peterson N M, Constable G S, Caton R B, Davis C S, Wallace R R and Yarranton G A (1987) *Cumulative Effects Assessment in Canada: an Agenda for Action and Research*. Canadian Environmental Assessment Research Council, Hull, Quebec

Petts J and Hills P (1982) *Environmental Assessment in the UK: a Preliminary Guide*. Institute of Planning Studies, University of Nottingham

Pinfield G (1992) Strategic environmental assessment and land-use planning. *Project Appraisal* 7: 157–64

Planning and Conservation League (and others) (1993) *In Defense of CEQA*. Leaflet Folder, PCL, Sacramento, CA

Porter C F (1985) *Environmental Impact Assessment: a Practical Guide*. University of Queensland Press, St Lucia

Porter C F (1987) Environmental impact assessment. In Hundloe T and Neumann R (eds) *Environmental Practice in Australia*. Environmental Institute of Australia, Canberra

Rand S D and Tawater M S (1986) *Environmental Referrals and the Council on Environmental Quality*. Environmental Law Institute, Washington, DC

Rayner S (1993) Introduction: the international influence of NEPA. In Hildebrand S G and Cannon J B (eds) *Environmental Analysis: the NEPA Experience*. Lewis, Boca Raton, FL

Reed R M and Cannon J B (1993) Introduction to the NEPA process. In Hildebrand S G and Cannon J B (eds) *Environmental Analysis: the NEPA Experience*. Lewis, Boca Raton, FL

Rees J A (1990) *Natural Resources: Allocation, Economics and Policy*. Methuen, London (2nd edition)

Rees W E (1980) EARP at the crossroads: environmental assessment in Canada. *Environmental Impact Assessment Review* 1: 355–77

Rees W E (1987) Introduction: a rationale for northern land-use planning. In Fenge T and Rees W E (1987) *Hinterland or Homeland: Land-use Planning in Northern Canada*. Canadian Arctic Resources Committee, Ottawa, Ontario

Reilly W (1974) New directions in federal land use legislation. In Listokin D (ed.) *Land Use Controls: Present Problems and Future Reform*. Centre for Urban Policy Research, Rutgers University, New Brunswick, NJ

Remy M H, Thomas T A, Moose J G and Yeates J W (1993) *Guide to the California Environmental Quality Act (CEQA).* Solano Press, Point Arena, CA (7th edition)

Renwick W H (1988) The eclipse of NEPA as environmental policy. *Environmental Management* **12**: 267–72

Roberts J A (1991) *Just What is EIR?* Global Environmental Management Services, Sacramento, CA

Roques D (1993) Mitigation monitoring programs (Association of Environmental Professionals). *Environmental Monitor*, Fall: 8–13

Ross W A (1987) Evaluating environmental impact statements. *Journal of Environmental Management* **25**: 137–47

Royal Commission on Environmental Pollution (1976) *Fifth Report: Air Pollution Control: an Integrated Approach.* Cm 6371, HMSO, London

Royal Commission on Environmental Pollution (1988) *Twelfth Report: Best Practicable Environmental Option.* Cm 310, HMSO, London

Royal Town Planning Institute (1981) *Environmental Assessment of Projects.* Report E53, 27 April 1981, RTPI, London

Sadler B (1988) The evaluation of assessment: post-EIS research and process development. In Wathern P (ed.) *Environmental Impact Assessment: Theory and Practice.* Unwin Hyman, London

Sadler B (1989) *An Evaluation of the Beaufort Sea Environmental Assessment Review Panel.* Federal Environmental Assessment Review Office, Hull, Quebec

Sadler B (1994a) *International Study of the Effectiveness of Environmental Assessment: Proposed Framework.* Federal Environmental Assessment Review Office, Hull, Quebec

Sadler B (1994b) Mediation provisions and options in Canadian environmental assessment. *Environmental Impact Assessment Review* **13**: 375–90

Salford City Council (1994) *Greater Manchester Environmental Assessment Handbook.* SCC, Salford

Salter J R (1992a) Environmental assessment: the challenge from Brussels. *Journal of Planning and Environment Law [1992]:* 14–20

Salter J R (1992b) Environmental assessment – the need for transparency. *Journal of Planning and Environment Law [1992]:* 214–21

Scholten J J and van Eck M (1994) Reviewing EISs in Environmental Impact Assessment Commission. *EIA Methodology in the Netherlands: Views of the Commission for EIA.* EIAC, Utrecht

Sewell G H and Korrick S (1984) The fate of EIS projects: a retrospective study. In Hart S L, Enk G A and Hornick W F (eds) *Improving Impact Assessment: Increasing the Relevance and Utilisation of Scientific and Technical Information.* Westview Press, Boulder, CO

Sewell W R D (1981) How Canada responded: the Berger Inquiry. In O'Riordan T and Sewell W R D (eds) *Project Appraisal and Policy Review*. Wiley, Chichester

Sewell W R D and Coppock J T (eds) (1976) *Public Participation in Planning*. Wiley, Chichester

Sheate W R (1991) Public participation: the key to effective environmental assessment. *Environmental Policy and Law* **21**: 156–60

Sheate W R (1994) *Making an Impact: a Guide to EIA Law and Policy*. Cameron May, London

Sheate W R and Macrory R B (1989) Agriculture and the EC environmental assessment directive: lessons for community policy making. *Journal of Common Market Studies* **28**: 68–81

Shopley J B and Fuggle R F (1984) A comprehensive review of current environmental impact assessment methods and techniques. *Journal of Environmental Management* **18**: 25–34

Sigal L L and Webb J W (1989) The programmatic environmental impact statement: its purpose and use. *The Environmental Professional* **11**: 14–24

Smith L G (1990) Canada's changing impact assessment provisions. *Environmental Impact Assessment Review* **11**: 5–9

Smith L G (1993) *Impact Assessment and Sustainable Resource Management*. Longman, Harlow

Social Impact Unit (1991) *Working with Communities: a Guide for Proponents*. SIU, Perth

Sonntag N C, Everitt R R, Rattie L P, Colnett D L, Wolf C P, Truett J C, Dorcey A H J and Holling C S (1987) *Cumulative Effects Assessment: a Context for Further Research and Development*. Canadian Environmental Assessment Research Council, Hull, Quebec

Stiles R, Wood C M and Groome D M (1991) *Environmental Assessment: the Treatment of Landscape and Countryside Recreation Issues*. CCP326, Countryside Commission, Cheltenham

Talbot A R (1983) *Settling Things: Six Case Studies in Environmental Mediation*. Conservation Foundation, Washington, DC

Taylor S (1984) *Making Bureaucracies Think: the Environmental Impact Statement Strategy of Administrative Reform*. Stanford University Press, Stanford, CA

Therivel R, Wilson E, Thompson S, Heaney D and Pritchard D (1992) *Strategic Environmental Assessment*. Earthscan, London

Thompson M A (1990) Determining impact significance in EIA: a review of 24 methodologies. *Journal of Environmental Management* **30**: 235–56

Tomlinson P and Atkinson S F (1987a) Environmental audits: proposed terminology. *Environmental Monitoring and Assessment* **8**: 187–98

Tomlinson P and Atkinson S F (1987b) Environmental audits: a literature review. *Environmental Monitoring and Assessment* 8: 239–61

Turnbull R G H (ed.) (1992) *Environmental and Health Impact Assessment of Development Projects.* Elsevier, London

United Nations Economic Commission for Europe (1992) *Application of Environmental Impact Assessment Principles to Policies, Plans and Programmes.* Environmental Series 5, UNECE, Geneva

United Nations Environment Programme (1988) *Environmental Impact Assessment: Basic Procedures for Developing Countries.* UNEP, Regional Office for Asia and the Pacific, Bangkok

United States Congressional Record (1974) **120**(26): 35515–16

van Eck M (1993) Environmental impact assessment for policy plans and programmes in The Netherlands. *Proceedings of the International Association for Impact Assessment Conference, Shanghai.* IAIA, Belhaven, NC

Verheem R (1991) EIA and EIA training in The Netherlands. In Wood C M and Lee N (eds) *Environmental Impact Assessment Training and Research in the European Communities.* Occasional Paper 27, Department of Planning and Landscape, University of Manchester

Verheem R (1992) Environmental assessment at the strategic level in The Netherlands. *Project Appraisal* 7: 150–6

Verocai Moreira I (1988) EIA in Latin America. In Wathern P (ed.) *Environmental Impact Assessment: Theory and Practice.* Unwin Hyman, London

Vig N and Kraft M (eds) (1984) *Environmental Policy in the 1980s: Reagan's New Agenda.* CQ Press, Washington, DC

von Moltke K (1984) Impact assessment in the United States and Europe. In Clark B D, Gilad A, Bisset R and Tomlinson P (eds) *Perspectives on Environmental Impact Assessment.* Reidel, Dordrecht

Wandesforde-Smith G (1979) Environmental impact assessment in the European Community. *Zeitschrift für Umwelt Politik* 1: 35–76

Wandesforde-Smith G (1981) The evolution of environmental impact assessment in California. In O'Riordan T and Sewell W R D (eds) *Project Appraisal and Policy Review.* Wiley, Chichester

Wandesforde-Smith G (1989) Environmental impact assessment, entrepreneurship, and policy change. In Bartlett R V (ed.) *Policy through Impact Assessment.* Greenwood Press, New York, NY

Wandesforde-Smith G and Kerbavaz J (1988) The co-evaluation of politics and policy: elections, entrepreneurship and EIA in the United States. In Wathern P (ed.) *Environmental Impact Assessment: Theory and Practice.* Unwin Hyman, London

Wathern P (ed.) (1988a) *Environmental Impact Assessment: Theory and Practice.* Unwin Hyman, London

Wathern P (1988b) The EIA directive of the European Community. In Wathern P (ed.) *Environmental Impact Assessment: Theory and Practice.* Unwin Hyman, London

Wathern P (1988c) Introduction. In Wathern P (ed.) *Environmental Impact Assessment: Theory and Practice.* Unwin Hyman, London

Wathern P (1989) Implementing supranational policy: environmental impact assessment in the United Kingdom. In Bartlett R V (ed.) *Policy through Impact Assessment.* Greenwood Press, New York, NY

Webb J W and Sigal L L (1992) Strategic environmental assessment in the United States. *Project Appraisal* **7**: 137–42

Wells C and Fookes T (1988) *Impact Assessment in Resource Management.* Resource Management Law Reform Working Paper 20, Ministry for the Environment, Wellington

Weston S M C (1991) *The Canadian Federal Environmental Assessment and Review Process: an Analysis of the Initial Assessment Phase.* Canadian Environmental Assessment Research Council, Hull, Quebec

Wilbanks T J, Hunsaker D B Jr, Petrich C H and Wright S B (1993) Potential to transfer the US NEPA experience in developing countries. In Hildebrand S G and Cannon J B (eds) *Environmental Analysis: The NEPA Experience.* Lewis, Boca Raton, FL

Wood C M (1982) The impact of European Commission's directive on environmental planning in the United Kingdom. *Planning Outlook* **24**: 92–8

Wood C M (1985) United States of America. In Lee N, Wood C M and Gazidellis V (eds) *Implications in the European Communities and North America: Country Studies.* Occasional Paper 13, Department of Planning and Landscape, University of Manchester

Wood C M (1988a) EIA in plan-making. In Wathern P (ed.) *Environmental Impact Assessment: Theory and Practice.* Unwin Hyman, London

Wood C M (1988b) The genesis and implementation of environmental impact assessment in Europe. In Clark M and Herington J (eds) *The Role of Environmental Assessment in the Planning Process.* Mansell, London

Wood C M (1989a) *Environmental Impact Assessment: Five Training Case Studies.* Occasional Paper 19, Department of Planning and Landscape, University of Manchester (2nd edition)

Wood C M (1989b) *Planning Pollution Prevention.* Heinemann Newnes, Oxford

Wood C M (1990) European influences on the environment. *Cheshire Environment Conference '89 Proceedings.* Cheshire County Council, Chester

Wood C M (1992) Strategic environmental assessment in Australia and New Zealand. *Project Appraisal* **7**: 143–9

Wood C M (1993a) Antipodean environmental assessment: a New Zealand/United Kingdom comparison. *Town Planning Review* **64**: 119–38

Wood C M (1993b) Environmental impact assessment in Australia: can the Old World learn from the New? *International Environmental Affairs* **5**: 256–74

Wood C M (1993c) Environmental impact assessment in Victoria: Australian discretion rules EA! *Journal of Environmental Management* **39**: 281–95

Wood C M and Bailey J (1994) Predominance and independence in environmental impact assessment: the Western Australia model. *Environmental Impact Assessment Review* **14**: 37–59

Wood C M and Djeddour M (1992) Strategic environmental assessment: EA of policies, plans and programmes. *Impact Assessment Bulletin* **10**: 3–22

Wood C M and Jones C E (1991) *Monitoring Environmental Assessment and Planning.* Department of the Environment, HMSO, London

Wood C M and Jones C E (1992) The impact of environmental assessment on local planning authorities. *Journal of Environmental Planning and Management* **35**: 115–27

Wood C M and Lee N (eds) (1991) *Environmental Impact Assessment Training and Research in the European Communities.* Occasional Paper 27, Department of Planning and Landscape, University of Manchester

Wood C M and McDonic G (1989) Environmental assessment: challenge and opportunity. *The Planner* **75**(11): 12–18

Wood C M, Lee N and Jones C E (1991) Environmental statements in the UK: the initial experience. *Project Appraisal* **6**: 187–94

World Bank (1987) *Environment, Growth and Development.* Development Committee Paper 14, World Bank, Washington, DC

World Bank (1991) *Environmental Assessment Sourcebook.* World Bank, Washington, DC (3 volumes)

World Commission on Environment and Development (1987) *Our Common Future.* Oxford University Press, Oxford

Yost N C (1981) Streamlining NEPA – an environmental success story. *Boston College Environmental Affairs Law Review* **9**: 507–12

Yost N C (1984) The 1978 CEQ NEPA Regulations. In Hart S L, Enk G A and Hornick W F (eds) *Improving Impact Assessment: Increasing the Relevance and Utilization of Scientific and Technical Information.* Westview Press, Boulder, CO

Yost N C (1990) NEPA's promise partially fulfilled. *Environmental Law* **20**: 533–49

Yost N C and Rubin J W (1989) The National Environmental Policy Act. In Environmental Law Institute *NEPA Deskbook.* ELI, Washington, DC

Index